T0296239

Op Amps for Everyone

Op Amps for Everyone

Fifth Edition

Bruce Carter
Analog Signal Chain and Power Supply Specialist,
Texas, USA

Ron Mancini
Consultant, USA

Newnes is an imprint of Elsevier
The Boulevard, Langford Lane, Kidlington, Oxford OX5 1GB, United Kingdom
50 Hampshire Street, 5th Floor, Cambridge, MA 02139, United States

Library of Congress Cataloging-in-Publication Data
A catalog record for this book is available from the Library of Congress

British Library Cataloguing-in-Publication Data
A catalogue record for this book is available from the British Library

ISBN: 978-0-12-811648-7

For information on all Newnes publications visit our
website at https://www.elsevier.com/books-and-journals

Working together
to grow libraries in
developing countries

www.elsevier.com • www.bookaid.org

Publisher: Mara E. Conner
Acquisition Editor: Tim Pitts
Editorial Project Manager: Charlotte Kent
Production Project Manager: Mohanapriyan Rajendran
Designer: Mark Rogers

Typeset by TNQ Books and Journals

This book is dedicated to
Dr. Tracy Fanara—a brilliant scientist and great role model for young women
(and young men) seeking a career in the sciences.

Contents

List of Figures

List of Tables

Foreword

The Changing World

In the 16 years since the first edition of this book, the world has changed in many ways. Geopolitical discussions aside, the world of analog design has changed.

In the year 2000, high-speed op amps were just beginning to be introduced. Now they are available in hundreds of part numbers from a number of manufacturers. Op amps have made inroads into new high-speed applications that would have been unthinkable in the year 2000, such as RF design.

Low power/low supply voltages are also a new area of application. From single-supply operation down to a volt and below, to op amps that operate from microamps of supply current, to new packaging that makes an entire op amp the size of small passive components, today's designer can fit op amp applications into those that are smaller, operate from lower supply currents, and off of smaller batteries than ever before. Op amps have also been integrated by the dozens or even hundreds onto one die, allowing multichannel applications such as medical ultrasound receivers to be integrated onto a single IC.

Op amps have also made inroads into harsh environment applications, operating to 200°C in downhole drilling applications, operating in spacecraft, and operating under extreme conditions of vibration and stress such as jet engines.

With all these changes, many of the fundamentals of op amp design have not changed. The fundamental feedback equations, gain techniques, and filter techniques presented here are as applicable for an op amp operating in a battery-operated toy as they are in a military weapon system. The engineer only needs to be aware of the special needs dictated by their application—be it low cost for a toy or the ultimate in reliability for a medical device implanted in a patient.

This book, as well, has undergone changes over its four editions:

- Edition 1 was a Texas Instruments (TI) design guide, written primarily by Ron Mancini with a few chapters contributed by myself and other authors.

- Edition 2 followed very quickly after when it was discovered that Edition 1 had many typos. Sadly, neither Edition 1 nor Edition 2 are still available from TI.
- Edition 3 was the first edition available from Elsevier and contained a lot of additional material contributed by me. It contained the best of the application material discovered during many hundreds of customer inquiries. I expanded Ron Mancini's original four cases of gain and offset to include all possible combinations of gain, attenuation, offset, and power—discovering a continuum that included not only Ron's four cases but also the simple cases of no offset, as well as realizing that offset-only cases are actually voltage regulators. I also included the new categories of op amps such as fully differential op amps.
- Edition 4 was the first edition written after my departure from TI. No longer limited to just the product offerings and application material available from TI, I was able to expand the scope to other manufacturers. It was also a significant downsizing from the first three editions, emphasizing simple methodologies for design of gain/offset circuits and filters.

Now, nearly 5 years later, I can see that I did a disservice to the original intent of the book. By eliminating some of Ron's excellent background material, I left some inexperienced engineers in the dark as to why certain things are true—such as the least stable operating point of an op amp and its lowest specified gain—and why they never make an inverting attenuator by making $R_G > R_F$. I work with engineers who have less experience and how I have wished I had a supply of *Op Amps for Everyone* editions 1—3 to give them! It has long been my concern that engineering education emphasizes areas other than analog design, and I have seen colleagues with doctoral degrees from prestigious universities who cannot tell the difference between instability and external noise pickup! Hence the need for a fifth edition—which will finally include the long-awaited troubleshooting chapter. Just as I did for the third edition, when I drew heavily on application problems and mistakes made by customers, I will draw heavily on misconceptions and mistakes made by colleagues over the years—in an attempt to spare you similar problems with your designs.

- Edition 5 will contain that troubleshooting chapter, as well as some other welcome additions. I spend a lot of time encouraging engineers to design circuits that operate on split supplies for better performance; then I had voltage regulator chapters that talked about only positive regulators. I have added exciting information about how to make negative regulators! Of course, all the material about how to make single-supply op amp circuits will still be included.

CHAPTER 1

The Op Amp's Place in the World

1.1 The Problem

In 1934, Harry Black [1] commuted from his home in New York City to work at Bell Labs in New Jersey by way of a railroad/ferry. The ferry ride relaxed Harry enabling him to do some conceptual thinking. Harry had a tough to solve; when phone lines were extended long distances, they needed amplifiers, and undependable amplifiers limited phone service. First, initial tolerances on the gain were poor, but that problem was quickly solved with an adjustment. Second, even when an amplifier was adjusted correctly at the factory, the gain drifted so much during field operation that the volume was too low or the incoming speech was distorted.

Many attempts had been made to make a stable amplifier, but temperature changes and power supply voltage extremes experienced on phone lines caused uncontrollable gain drift. Passive components had much better drift characteristics than active components had, thus if an amplifier's gain could be made dependent on passive components, the problem would be solved. During one of his ferry trips, Harry's fertile brain conceived a novel solution for the amplifier problem, and he documented the solution while riding on the ferry.

1.2 The Solution

The solution was to first build an amplifier that had more gain than the application required. Then some of the amplifier output signal was fed back to the input in a manner that makes the circuit gain (circuit is the amplifier and feedback components) dependent on the feedback circuit rather than the amplifier gain. Now the circuit gain is dependent on the passive feedback components rather than the active amplifier. This is called negative feedback, and it is the underlying operating principle for all modern day op amps. Harry had documented the first intentional feedback circuit during a ferry ride. I am sure unintentional feedback circuits had been built prior to that time, but the designers ignored the effect!

I can hear the squeals of anguish coming from the managers and amplifier designers of the era. I imagine that they said something like this, "it is hard enough to achieve 30 kHz gain—bandwidth (GBW), and now this fool wants me to design an amplifier with 3 MHz GBW. But, he is still going to get a circuit gain GBW of 30 kHz". Well, time has proven

Op Amps for Everyone. http://dx.doi.org/10.1016/B978-0-12-811648-7.00001-7

Harry right, but there is a minor problem that Harry did not discuss in detail, and that is the oscillation problem. It seems that circuits designed with large open loop gains sometimes oscillate when the loop is closed. A lot of people investigated the instability effect, and it was pretty well understood in the 1940s, but solving stability problems involved long, tedious, and intricate calculations. Years passed without anybody making the problem solution simpler or more understandable.

In 1945, H.W. Bode presented a system for analyzing the stability of feedback systems by using graphical methods. Until this time, feedback analysis was done by multiplication and division, so calculation of transfer functions was a time-consuming and laborious task. Remember, engineers did not have calculators or computers until the 1970s. Bode presented a logarithmic technique that transformed the intensely mathematical process of calculating a feedback system's stability into graphical analysis that was simple and perceptive. Feedback system design was still complicated, but it no longer was an art dominated by a few electrical engineers kept in a small dark room. Any electrical engineer could use Bode's methods to find the stability of a feedback circuit, so the application of feedback to machines began to grow. There really was not much call for electronic feedback design until computers and transducers become of age, however.

1.3 The Birth of the Op Amp

The first real-time computer was the analog computer! This computer used preprogrammed equations and input data to calculate control actions. The programming was hard wired with a series of circuits that performed math operations on the data, and the hard wiring limitation eventually caused the declining popularity of the analog computer. The heart of the analog computer was a device called an operational amplifier because it could be configured to perform many mathematical operations such as multiplication, addition, subtraction, division, integration, and differentiation on the input signals. The name was shortened to the familiar *op amp*, as we have come to know and love them. The op amp used an amplifier with a large open loop gain, and when the loop was closed, the amplifier performed the mathematical operations dictated by the external passive components. This amplifier was very large because it was built with vacuum tubes and required a high-voltage power supply, but it was the heart of the analog computer, thus its large size and huge power requirements were accepted as the price of doing business. Early op amps were designed for analog computers, and it was soon found out that op amps had other uses and were very handy to have around the physics lab.

At this time, general-purpose analog computers were found in universities and large company laboratories because they were critical to the research work done there. There was a parallel requirement for transducer signal conditioning in lab experiments, and op amps found their way into signal conditioning applications. As the signal conditioning applications expanded, the demand for op amps grew beyond the analog computer requirements, and even when the analog computers lost favor to digital computers, the op amp survived because of its importance in universal analog applications. Eventually digital computers replaced the analog computers (a sad day for real-time measurements), but the demand for op amps increased as measurement applications increased.

1.3.1 The Vacuum Tube Era

The first signal conditioning op amps were constructed with vacuum tubes prior to the introduction of transistors, so they were large and bulky. During the 1950s, miniature vacuum tubes that worked from lower voltage power supplies enabled the manufacture of op amps that shrunk to the size of a brick used in house construction, so the op amp modules were nicknamed *bricks*. Vacuum tube size and component size decreased until an op amp was shrunk to the size of a single octal vacuum tube.

One of the first commercially available op amps was the model K2-W, sold by George A. Philbrick Research. It consisted of two vacuum tubes, and operated from a $\pm$ 300 V power supplies! If that is not enough to make a modern analog designer cringe—then its fully differential nature would be sure to. A fully differential op amp, as opposed to the more familiar single-ended op amp, has two outputs—a noninverting output and an inverting output. It requires the designer to close two feedback paths, not just one. Before panic sets in—the two feedback pathways only require duplication of components, not an entirely new design methodology. Fully differential op amps are currently enjoying resurgence—because they are ideal components for driving the inputs of fully differential analog to digital converters. They also find use in driving differential signal pairs such as DSL and balanced 600 Ohm audio. Suffice it to say, op amps have come full circle since their original days.

1.3.2 The Transistor Era

Transistors were commercially developed in the 1960s, and they further reduced op amp size to several cubic inches, but the nickname brick still held on. Now the nickname brick is attached to any electronic module that uses potting compound or nonintegrated circuit (IC) packaging methods. Most of these early op amps were made for specific applications, so they were not necessarily general purpose. The early op amps served a specific purpose, but each manufacturer had different specifications and packages; hence, there was little second sourcing among the early op amps.

1.3.3 The IC Era

ICs were developed during the late 1950s and early 1960s, but it was not till the middle 1960s that Fairchild released the μA709. This was the first commercially successful IC op amp, and Robert J. Widler designed it. The μA709 had its share of problems, but any competent analog engineer could use it, and it served in many different analog applications. The major drawback of the μA709 was stability; it required external compensation and a competent analog engineer to apply it. Also, the μA709 was quite sensitive because it had a habit of self-destructing under any adverse condition. The self-destruction habit was so prevalent that one major military equipment manufacturer published a paper titled something like, *The 12 Pearl Harbor Conditions of the μA709*.

The legacy of the μA709 continues today, but it is a negative legacy. The μA709 would not work if applied incorrectly, primarily due to its external compensation. The engineers of today may not even know the part, but memory of its instability remains—few uncompensated amplifiers are sold today due to the problem of misapplication. Stability remains one of the least understood aspects of op amp design, and one of the easiest ways to misapply an op amp. Even engineers with years of analog design experience have differing opinions on the topic. The wise engineer, however, will look carefully at the op amp data sheet and not attempt a gain less than its specification. It may be counter intuitive, but the op amp is least stable at its lowest specified gain. Future chapters will delve deeply into this phenomenon.

The μA741 followed the μA709, and it is an internally compensated op amp that does not require external compensation if operated under data sheet conditions. Also, it is much more forgiving than the μA709.

The legacy of the μA741 is much more positive than that of its predecessor. In fact, the part number "741" is etched into the memory of practically every engineer in the world, much like the "2N2222" transistor and the "1N4148" diode. It is usually the first part number that comes to mind whenever an engineer thinks of an op amp. Unlike the μA709, the μA741 will work unless grossly misapplied—a fact that has endeared it generations of engineers. Its power supply requirements of ± 15 V have given rise to hundreds of power supply components that generate these levels, much as +5 V has been driven by TTL logic and ± 12 V has been driven by RS232 serial interfaces. For many years, every op amp introduced used the same ± 15 V power supplies as the μA741. Even today, the μA741 is an excellent choice where wide dynamic range and ruggedness are required.

There has been a never-ending series of new op amps released each year since the introduction of the μA741, and their performance and reliability has improved to the point where present day op amps can be used for analog applications by anybody.

The IC op amp is here to stay; the latest generation op amps cover the frequency spectrum from 5 kHz GBW for extremely low power devices to beyond 3 GHz GBW. The supply voltage ranges from guaranteed operation at 0.9 V to absolute maximum voltage ratings of 1000 V. The input current and input offset voltage has fallen so low that customers have problems verifying the specifications during incoming inspection. The op amp has truly become the universal analog IC because it performs all analog tasks. It can function as a line driver, amplifier, level shifter, oscillator, filter, signal conditioner, actuator driver, current source, voltage source, and many other applications. The designer's problem is how to rapidly select the correct circuit/op amp combination and then, how to calculate the passive component values that yield the desired transfer function in the circuit.

It should be noted that there is no op amp that is universally applicable. An op amp that is ideal for transducer interfaces will not work at all for RF applications. An op amp with good RF performance may have miserable DC specifications. The hundreds of op amp models offered by manufacturers are all optimized in slightly different ways, so the designer's task is to weed through those hundreds of devices and find the handful that are appropriate for their application. This edition includes a design methodology for doing so—at least in the case of signal chains.

This book deals with op amp circuits—not with the innards of op amps. It treats the calculations from the circuit level, and it does not get bogged down in a myriad of detailed calculations. Rather, the reader can start at the level appropriate for them, and quickly move on to the advanced topics. If you are looking for material about the innards of op amps, you are looking in the wrong place. The op amp is treated as a completed component in this book.

The op amp will continue to be a vital component of analog design because it is such a fundamental component. Each generation of electronics equipment integrates more functions on silicon and takes more of the analog circuitry inside the IC. Do not fear; as digital applications increase, analog applications also increase because the predominant supply of data and interface applications are in the real world, and the real world is an analog world. Thus, each new generation of electronics equipment creates requirements for new analog circuits; hence, new generations of op amps are required to fulfill these requirements. Analog design, and op amp design, is a fundamental skill that will be required far into the future.

Reference

[1] H.S. Black, Stabilized feedback amplifiers, BSTJ 13 (January 1934).

CHAPTER 2

Development of the Ideal Op Amp Equations*

2.1 Introduction

This chapter will delve into the first practical applications for the op amp. Before we delve into these applications, we will deal with the op amp in its simplest possible configuration—as a perfect or ideal component.

So what is an op amp, anyway? You have read the story of the genesis of the part, but for those of you who skipped the first chapter, I will present a shortened version in a bit more practical way.

An op amp is, first and foremost, an amplifier; hence the "amp" suffix. The "op" or "operational" part is a nod to its origins as a component in analog computers, now the "operational" part of the name probably means to most designers that it can do many things from gain to filtering to buffering. The closest many engineers will ever get to the analog computer origins are the "adder" and "subtractor" presented in this chapter.

Every inexperienced hobbyist begins with the op amp used as an amplifier as shown in Fig. 2.1.

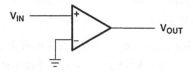

Figure 2.1
A first (and not very useful) circuit.

Even if you have never done this, I encourage you to try it once, to illustrate the problems Harry Black encountered in Chapter 1. Ignore the "−" (negative feedback) input by tying it off to ground. What you are left with is an amplifier with fixed gain in the thousands,

---

* This and subsequent chapters make frequent use of simple electronics rules that should be familiar to all engineers. Consult Appendix A if you wish to review the rules used in this book.

Op Amps for Everyone. http://dx.doi.org/10.1016/B978-0-12-811648-7.00002-9

7

perhaps as much as a million. Connected to a microphone in an ultraquiet environment, it might detect speech or other sounds from a great distance. Of course, any noise you generate or any ambient noise will swamp the amplifier and make it saturate and clip. But much more likely, it will oscillate or howl like an army of tomcats fighting in an alley, which is precisely the type of problems Harry Black encountered with his telephone line amplifiers in Chapter 1—at least the clipping part! Give Chapter 1 a read if you have not, it is fascinating!

Harry Black's solution was to essentially take advantage of that negative feedback ("−" input) to cancel most of the signal amplitude. You have all that gain, and promptly throw away the vast majority of it by feeding the amplified output back 180 degrees out of phase, canceling most of the signal. You use the voltage divider described in Appendix A to determine how much of that signal is canceled, and—presto—you have just started using the noninverting op amp configuration of 2.3! But before we delve into that, or other useful op amp circuits, we need to take a baby step. We need to introduce the parameters that define an op amp. And the perfect beginning point is the perfect—or ideal—op amp. Real-world op amp parameters, and their effect on circuit configurations, will be covered in subsequent chapters and Appendix B when the elementary analysis is complete.

2.2 Ideal Op Amp Assumptions

The name *ideal op amp* is applied to this and similar analysis because the salient parameters of the op amp are assumed to be perfect. This assumption simplifies the analysis, thus it clears the path for insight. It is so much easier to see the forest when the underbrush is cleared away. Although the ideal op amp analysis makes use of perfect parameters, the analysis is often valid because some op amps approach perfection. In addition, when working at low frequencies, several kilohertz, the ideal op amp analysis produces accurate answers. Voltage feedback op amps are covered in this chapter, and current feedback op amps are covered in a later chapter.

An engineer may wish that an ideal op amp existed at times, but if such a component actually did exist, it would destroy the known universe! See the end of this chapter for an explanation. Thankfully, there is no such thing as an ideal op amp, but present day op amps come so close to ideal that *ideal op amp* analysis approaches actual analysis. Op amps depart from the ideal in the following ways:

- The ideal assumes that input offset voltage is zero. DC parameters, such as input offset voltage, cause departure from the ideal.
- AC parameters such as gain are a function of frequency, so they go from large values at DC to small values at high frequencies.

- Assume that the current flow into the input leads of the op amp is zero. This assumption is almost true in FET op amps where input currents can be less than a picoampere, but this is not always true in bipolar high-speed op amps where tens of microamperes input currents are found.
- The op amp gain is assumed to be infinite, hence it drives the output voltage to any value to satisfy the input conditions. This assumes that the op amp output voltage can achieve any value. In reality, saturation occurs when the output voltage comes close to a power supply rail, but reality does not negate the assumption, it only bounds it.
- The gain drives the output voltage, until the voltage between the input leads (the error voltage) is zero, therefore the voltage between the input leads is zero. The implication of zero voltage between the input leads means that if one input is tied to a hard voltage source such as ground, then the other input is at the same potential.
- The current flow into the input leads is zero, so the input impedance of the op amp is infinite.
- The output impedance of the ideal op amp is zero. The ideal op amp can drive any load without an output impedance dropping voltage across it. The output impedance of most op amps is a fraction of an ohm for low current flows, so this assumption is valid in most cases.
- The frequency response of the ideal op amp is flat; this means that the gain does not vary as frequency increases. By constraining the use of the op amp to the low frequencies, we make the frequency response assumption true.

Table 2.1 lists these basic ideal op amp assumptions, and Fig. 2.2 shows the ideal op amp.

2.3 The Noninverting Op Amp

The noninverting op amp has the input signal connected to its noninverting input (Fig. 2.3), thus its input source sees infinite impedance. There is no input offset voltage because $V_{OS} = V_E = 0$, hence the negative input must be at the same voltage as the positive input. The op amp output drives current into R_F until the negative input is at the voltage, V_{IN}. This action causes V_{IN} to appear across R_G.

Table 2.1: Basic Ideal Op Amp Assumptions

Parameter Name	Parameters Symbol	Value
Input current	I_{IN}	0
Input offset voltage	V_{OS}	0
Input impedance	Z_{IN}	∞
Output impedance	Z_{OUT}	0
Gain	a	∞

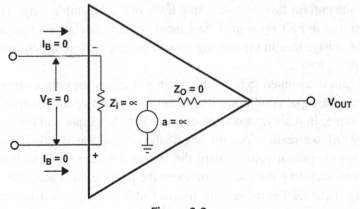

Figure 2.2
The ideal op amp.

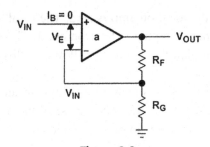

Figure 2.3
The noninverting op amp.

The voltage divider rule is used to calculate V_{IN}; V_{OUT} is the input to the voltage divider, and V_{IN} is the output of the voltage divider. Since no current can flow into either op amp lead, use of the voltage divider rule is allowed. Eq. (2.1) is written with the aid of the voltage divider rule, and algebraic manipulation yields Eq. (2.2) in the form of a gain parameter.

$$V_{IN} = V_{OUT} \frac{R_G}{R_G + R_F} \tag{2.1}$$

$$\frac{V_{OUT}}{V_{IN}} = \frac{R_G + R_F}{R_G} = 1 + \frac{R_F}{R_G} \tag{2.2}$$

When R_G becomes very large with respect to R_F, $(R_F/R_G) \Rightarrow 0$ and Eq. (2.2) reduces to Eq. (2.3).

$$V_{OUT} = 1 \tag{2.3}$$

Under these conditions $V_{OUT} = 1$ and the circuit becomes a unity gain buffer. R_G is usually deleted to achieve the same results, and when R_G is deleted, R_F can be made into a short (op amp output is connected to its inverting input with a wire). Some op amps are self-destructive when R_F is a short (particularly current feedback amplifiers), so R_F is used in many buffer designs. When R_F is included in a buffer circuit, its function is to protect the inverting input from an overvoltage to limit the current through the input electrostatic discharge structure (typically < 1 mA), and it can have almost any value.

2.4 The Inverting Op Amp

The noninverting input of the op amp circuit is grounded. The assumption is made that the input error voltage is zero, so the feedback keeps inverting the input of the op amp at a virtual ground (not actual ground but acting like ground). The current flow in the input leads is assumed to be zero, hence the current flowing through R_G equals the current flowing through R_F (Fig. 2.4). Using Kirchhoff's law, we write Eq. (2.4); and the minus sign is inserted because this is the inverting input. Algebraic manipulation gives Eq. (2.5).

$$I_1 = \frac{V_{IN}}{R_G} = -I_2 = -\frac{V_{OUT}}{R_F} \qquad (2.4)$$

$$\frac{V_{OUT}}{V_{IN}} = -\frac{R_F}{R_G} \qquad (2.5)$$

Notice that the gain is only a function of the feedback and gain resistors, so the feedback has accomplished its function of making the gain independent of the op amp parameters. The actual resistor values are determined by the impedance levels that the designer wants to establish. If $R_F = 10$ k and $R_G = 10$ k the gain is -1 as shown in Eq. (2.5), and if $R_F = 100$ k and $R_G = 100$ k the gain is still -1. The impedance levels of 10 or 100 k determine the current drain, the effect of stray capacitance, and a few other points. The impedance level does not set the gain; the ratio of R_F/R_G does.

One final note; the output signal is the input signal amplified and inverted. The circuit input impedance is set by R_G because the inverting input is held at a virtual ground.

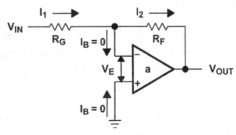

Figure 2.4
The inverting op amp.

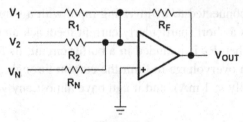

Figure 2.5
The adder circuit.

2.5 The Adder

An adder circuit can be made by connecting more inputs to the inverting op amp (Fig. 2.5). The opposite end of the resistor connected to the inverting input is held at virtual ground by the feedback; therefore, adding new inputs does not affect the response of the existing inputs.

Superposition is used to calculate the output voltages resulting from each input, and the output voltages are added algebraically to obtain the total output voltage. Eq. (2.6) is the output equation when V_1 and V_2 are grounded. Eqs. (2.7) and (2.8) are the other superposition equations, and the final result is given in Eq. (2.9).

$$V_{OUTN} = -\frac{R_F}{R_N}V_N \tag{2.6}$$

$$V_{OUT1} = -\frac{R_F}{R_1}V_1 \tag{2.7}$$

$$V_{OUT2} = -\frac{R_F}{R_2}V_2 \tag{2.8}$$

$$V_{OUT} = -\left(\frac{R_F}{R_1}V_1 + \frac{R_F}{R_2}V_2 + \frac{R_F}{R_N}V_N\right) \tag{2.9}$$

2.6 The Differential Amplifier

The differential amplifier circuit amplifies the difference between signals applied to the inputs (Fig. 2.6). Superposition is used to calculate the output voltage resulting from each input voltage, and then the two output voltages are added to arrive at the final output voltage.

The op amp input voltage resulting from the input source, V_1, is calculated in Eqs. (2.10) and (2.11). The voltage divider rule is used to calculate the voltage, V_+, and the

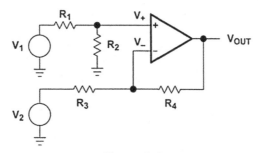

Figure 2.6
The differential amplifier.

noninverting gain equation (Eq. 2.2) is used to calculate the noninverting output voltage, V_{OUT1}.

$$V_+ = V_1 \frac{R_2}{R_1 + R_2} \tag{2.10}$$

$$V_{OUT1} = V_+(G_+) = V_1 \frac{R_2}{R_1 + R_2} \left(\frac{R_3 + R_4}{R_3} \right) \tag{2.11}$$

The inverting gain equation (Eq. 2.5) is used to calculate the stage gain for V_{OUT2} in Eq. (2.12). These inverting and noninverting gains are added in Eq. (2.13).

$$V_{OUT2} = V_2 \left(\frac{-R_4}{R_3} \right) \tag{2.12}$$

$$V_{OUT} = V_1 \frac{R_2}{R_1 + R_2} \left(\frac{R_3 + R_4}{R_3} \right) - V_2 \frac{R_4}{R_3} \tag{2.13}$$

Next, to simplify the equation, R_1 is made equal to R_3, and R_2 made equal to R_4:

$$V_{OUT} = (V_1 - V_2) \frac{R_4}{R_3} \tag{2.14}$$

It is now obvious that the differential signal $(V_1 - V_2)$ is multiplied by the stage gain, so the name differential amplifier suits the circuit. Because it only amplifies the differential portion of the input signal, it rejects the common-mode portion of the input signal. A common-mode signal is illustrated in Fig. 2.7. Because the differential amplifier strips off or rejects the common-mode signal, this circuit configuration is often employed to strip DC or injected common-mode noise off a signal.

The disadvantage of this circuit is that the two input impedances cannot be matched when it functions as a differential amplifier, thus there are two and three op amp versions of this circuit specially designed for high-performance applications requiring matched input impedances.

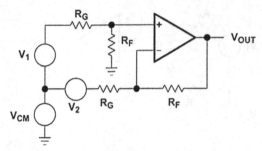

Figure 2.7
Differential amplifier with common-mode input signal.

2.7 Complex Feedback Networks

When complex networks are put into the feedback loop, the circuits get harder to analyze because the simple gain equations cannot be used. The usual technique is to write and solve node or loop equations. There is only one input voltage, so superposition is not of any use, but Thevenin's theorem can be used as is shown in the example problem given below.

Sometimes it is desirable to have a low-resistance path to ground in the feedback loop. Standard inverting op amps cannot do this when the driving circuit sets the input resistor value, and the gain specification sets the feedback resistor value. Inserting a *T* network in the feedback loop (Fig. 2.8) yields a degree of freedom that enables both specifications to be met with a low DC resistance path in the feedback loop.

Break the circuit at point X−Y, stand on the terminals looking into R_4, and calculate the Thevenin equivalent voltage as shown in Eq. (2.15). The Thevenin equivalent impedance is calculated in Eq. (2.16).

$$V_{TH} = V_{OUT} \frac{R_4}{R_3 + R_4} \tag{2.15}$$

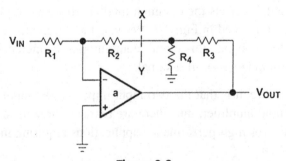

Figure 2.8
T network in feedback loop.

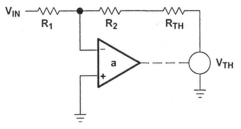

Figure 2.9
Thevenin's theorem applied to *T* network.

$$R_{TH} = R_3 || R_4 \qquad (2.16)$$

Replace the output circuit with the Thevenin equivalent circuit as shown in Fig. 2.9, and calculate the gain with the aid of the inverting gain equation as shown in Eq. (2.17).

Substituting the Thevenin equivalents into Eq. (2.17) yields Eq. (2.18).

$$-\frac{V_{TH}}{V_{IN}} = \frac{R_2 + R_{TH}}{R_1} \qquad (2.17)$$

$$-\frac{V_{OUT}}{V_{IN}} = \frac{R_2 + R_{TH}}{R_1}\left(\frac{R_3 + R_4}{R_4}\right) = \frac{R_2 + (R_3 || R_4)}{R_1}\left(\frac{R_3 + R_4}{R_4}\right) \qquad (2.18)$$

Algebraic manipulation yields Eq. (2.19).

$$-\frac{V_{OUT}}{V_{IN}} = \frac{R_2 + R_3 + \dfrac{R_2 R_3}{R_4}}{R_1} \qquad (2.19)$$

Specifications for the circuit you are required to build are an inverting amplifier with an input resistance of 10 k ($R_G = 10$ k), a gain of 100, and a feedback resistance of 20 K or less. The inverting op amp circuit cannot meet these specifications because R_F must equal 1000 k. Inserting a *T* network with $R_2 = R_4 = 10$ k and $R_3 = 485$ k approximately meets the specifications.

2.8 Impedance Matching Amplifiers

High-frequency op amp circuits may use coaxial cable to transmit and receive signals. The cable connecting these circuits has a characteristic impedance of 50 Ω. To prevent reflections, which cause distortion and ghosting, the input and output circuit impedances must match the 50 Ω cable.

Matching the input impedance is simple for a noninverting amplifier because its input impedance is very high; just make $R_{IN} = 50$ Ω. R_F and R_G can be selected as high values,

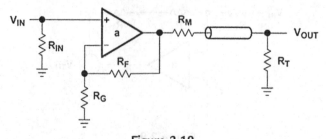

Figure 2.10
Impedance matching amplifier.

in the hundreds of ohms range, so that they have minimal affect on the impedance of the input or output circuit. A matching resistor, R_M, is placed in series with the op amp output to raise its output impedance to 50 Ω; a terminating resistor, R_T, is placed at the input of the next stage to match the cable (Fig. 2.10).

The matching and terminating resistors are equal in value, and they form a voltage divider of 1/2 because R_T is not loaded. Very often R_F is selected equal to R_G so that the op amp gain equals 2. Then the system gain, which is the op amp gain multiplied by the divider gain, is equal to 1 ($2 \times 1/2 = 1$).

2.9 Capacitors

Capacitors are a key component in a circuit designer's tool kit, thus a short discussion on evaluating their effect on circuit performance is in order. Capacitors have an impedance of $X_C = 1/2\pi fC$. Note that when the frequency is zero, the capacitive impedance (also known as reactance) is infinite, and that when the frequency is infinite the capacitive impedance is zero. These end points are derived from the final value theorem, and they are used to get a rough idea of the effect of a capacitor. When a capacitor is used with a resistor, they form what is called a break point. Without going into complicated math, just accept that the break frequency occurs at $f = 1/(2\pi RC)$, and the gain is -3 dB at the break frequency.

The low-pass filter circuit shown in Fig. 2.11 has a capacitor in parallel with the feedback resistor. The gain for the low-pass filter is given in Eq. (2.20).

$$\frac{V_{OUT}}{V_{IN}} = -\frac{X_C \| R_F}{R_G} \tag{2.20}$$

At very low frequencies $X_C \Rightarrow \infty$, so R_F dominates the parallel combination in Eq. (2.20), and the capacitor has no effect. The gain at low frequencies is $-R_F/R_G$. At very high frequencies $X_C \Rightarrow 0$, so the feedback resistor is shorted out, thus reducing the circuit gain to zero. At the frequency where $X_C = R_F$ the gain is reduced by $\sqrt{2}$ because complex impedances in parallel equal half the vector sum of both impedances.

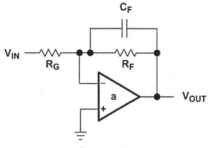

Figure 2.11
Low-pass filter.

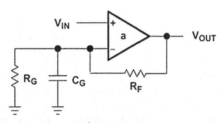

Figure 2.12
High-pass filter.

Connecting the capacitor in parallel with R_G where it has the opposite effect makes a high-pass filter (Fig. 2.12). Eq. (2.21) gives the equation for the high-pass filter.

$$\frac{V_{OUT}}{V_{IN}} = 1 + \frac{R_F}{X_C \| R_G} \tag{2.21}$$

At very low frequencies $X_C \Rightarrow \infty$, so R_G dominates the parallel combination in Eq. (2.21), and the capacitor has no effect. The gain at low frequencies is $1 + R_F/R_G$. At very high frequencies $X_C \Rightarrow 0$, so the gain setting resistor is shorted out thus increasing the circuit gain to maximum.

This simple technique is used to predict the form of a circuit transfer function rapidly. Better analysis techniques are presented in later chapters for those applications requiring more precision.

2.10 Why an Ideal Op Amp Would Destroy the Known Universe

I moved this somewhat humorous explanation to later in the chapter so as not to distract from the important discussions of circuit configuration above. This section provides an excellent review of ideal op amp parameters, in a way you can remember them in chapters to come. An ideal op amp has the following specifications:

• It draws no supply current, therefore has no power supplies. Therefore, it does not even have to be turned on to be dangerous!
• It has no V_{OH} and V_{OL} limitations because it has no power supplies. Therefore, its output voltage swings from $\pm \infty$ Volts.
• It has zero output resistance, and therefore it is capable of supplying infinite current at each voltage extreme.
• It has infinite open-loop gain; therefore the slightest input signal would allow it to swing to $\pm$ infinite voltage (that is, without feedback components).
• It has infinite slew rate, and therefore would swing to either rail—both equally destructive—instantly.

Therefore an ideal op amp, just lying on the table with no power applied, would instantly take a quantum difference between its $+$ and $-$ terminals and amplify it to an infinite voltage output at infinite current. The resulting surge of power would be a sphere of destruction radiating out from the op amp at the speed of light!

This somewhat humorous analysis is included to drive home a few points:

1. If the ideal op amp model is employed, the engineer must also know how real-world op amp parameters degrade and alter the ideal op amp model. An ideal op amp model is a useful tool for initial phases of simulation and analysis, but does not adequately explain real-world op amp behavior.
2. All of the mathematical analysis above can be boiled down to a simple concept: *The op amp will do whatever it has to do at its output to equalize the voltages at its inputs.* This is the entire content of this book, distilled down to its simplest form. This fundamental concept can be used to derive all of the behavior of all op amp circuits in all applications. Of course, it must be filtered through statement (1) for real-world op amps.
3. A very astute analog design engineer might see one fallacy in this "death star" scenario: "Where is the return path?" When a single-ended op amp is employed, the return path is to ground. But when no power supplies are utilized, there is no ground! So to be technically correct, the only type of ideal op amp that would destroy the known universe is a fully differential type, where the return for one output is the other output. This book will explain in detail the subject of proper return to ground for single-ended op amps and for single-ended op amps configured in single-supply operation.

2.11 Summary

When the proper assumptions are made, the analysis of op amp circuits is straightforward. These assumptions, which include zero input current, zero input offset voltage, and infinite gain, are realistic assumptions because the new op amps make them essentially true in most real-world applications.

When the signal is comprised of low frequencies, the gain assumption is valid because op amps have very high gain at low frequencies. When CMOS op amps are used, the input current is in the femtoamp range; close enough to zero for most applications. Laser-trimmed input circuits reduce the input offset voltage to a few microvolts; close enough to zero for most applications. The ideal op amp is becoming real; especially for undemanding applications.

CHAPTER 3

Single-Supply Op Amp Design Techniques

3.1 Single Supply Versus Dual Supply

The previous chapter assumed that all op amps were powered from dual or split supplies. This is not always the case in today's world of portable battery-powered equipment. When op amps are powered from dual supplies (see Fig. 3.1), the supplies are normally equal in magnitude and opposite in polarity. The center tap of the supplies is connected to ground. Any input sources connected to ground are automatically referenced to the center of the supply voltage, so the output voltage is automatically referenced to ground.

At this point, I would like to introduce the concept of virtual ground and DC operating point. Although virtual grounding was referred to in the discussion of the ideal op amp, in this case it means the actual voltage around which the signal swings in the op amp stage. Excursions of the input and output voltage swing equally positive and negative from this virtual ground point, instead of system ground. Your goal is to create a little localized "ecosystem" for your stage, in which the local ground as seen by the stage is different from the actual ground of the system. But within the stage, the virtual ground point is as real a reference to the signal as the system ground is for the split supply circuits previously discussed. This DC operating point is local to the stage. Even if subsequent stages are operated at the same DC operating point, direct connection is inadvisable, because DC offsets—real world parameters present in op amps discussed in

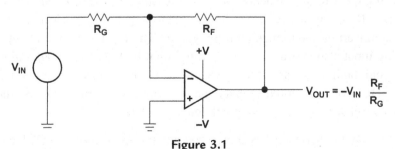

Figure 3.1
Split-supply inverting gain op amp circuit.

Op Amps for Everyone. http://dx.doi.org/10.1016/B978-0-12-811648-7.00003-0

21

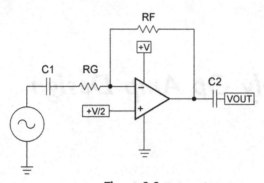

Figure 3.2
Single-supply op amp inverting gain circuit.

later chapters—tend to multiply quickly in gain stages. Therefore, the DC operating point should be isolated with capacitors to a single stage as shown in Fig. 3.2:

In this circuit, C_1 and C_2 isolate the rest of the circuit from the DC operating point of the stage. This point is labeled "+V/2" and is connected to the noninverting input of the op amp. This implies exactly what you think—the potential +V/2 is half of the supply voltage of the op amp.

In lieu of complex mathematical analysis, let us do some thought experiments on the circuit of Fig. 3.2. I believe this approach will give you a much clearer idea of what is going on.

As far as the op amp is concerned, it is operating at a positive supply of +V/2 and a negative supply of −V/2, with a virtual ground point in between. Because of C_1 and C_2, it is unaware of the fact that its "ground" is actually +V/2.

What is happening AC-wise on the inverting input? The transfer equation for an ideal op amp still applies, gain is still $-R_F/R_G$. This ignores the effect of the input and output capacitors C_1 and C_2, but I will get back to those later.

What is happening DC-wise on the noninverting input? Your first impulse would be to say that the DC operating point +V/2 should be amplified by $1 + R_F/R_G$, and the stage will not work (it will clip). But the capacitor C_1, which is isolating R_G from any DC potential, effectively makes R_G an open circuit. So the amplifier is operating like a noninverting buffer on the noninverting input, making it unity gain. The DC potential that appears at the noninverting input also appears at the inverting input by the ideal op amp model. So with no AC signal, both inputs are at the same DC level, as well as the output, which will go to whatever DC level is required to make the two inputs the same value. So the amplifier is balanced at a DC operating point equal to +V/2.

Hopefully this thought experiment has shown you how to apply your knowledge of the ideal op amp assumption to the problem of single-supply stage design.

There is no rule that the DC operating point has to be exactly half of the supply voltage. However, voltage swing—particularly in battery operated applications—is limited. The op amp output, before C_2, will swing (ideally) between $+V$ and ground, giving equal positive and negative excursions. If you place the DC operating point at a different level, you are limiting the output voltage swing for one direction or the other.

There are circumstances where the DC operating point might be something other than $+V/2$. The most common of these is the case where an op amp is the last stage in the signal chain and is directly driving an analog to digital converter (ADC). The ADC will have its own voltage reference that may be different from $+V/2$, in which case you would want to set the stage DC operating point to the same value—using the reference of the ADC, which is probably provided as an output.

There are many techniques for generating a potential exactly half that of the supply powering the op amp. The simplest is to employ a voltage divider composed of two equal value resistors. You must be careful, however, this technique places the resistors across the power supply and increases current. You can make the resistors larger, but then they tend to be more susceptible to noise pickup. A common method of combating this problem is to bypass the resistors with a decoupling capacitor as shown in Fig. 3.3.

R_1 and R_2 are made the same value, such as 10 k, C_3 is a decoupling capacitor—a common value is 0.1 μF. You might even be tempted to try to use this simple circuit to drive the noninverting inputs of *ALL* the op amps in your signal chain, but this is not advisable. Small amounts of leakage will occur and you can get cross talk between the stages. C_3 does act to decouple, but loses its effectiveness at low frequencies—and because a capacitor become inductive at very high frequencies, it will also lose its effectiveness for very high frequencies.

What you really need is a very low impedance voltage reference, one that is capable of driving many op amp stages. They make such a component, it is called a "voltage

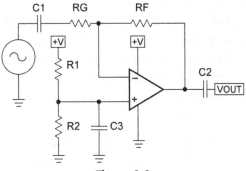

Figure 3.3
Single-supply op amp inverting gain circuit with internal DC operating point.

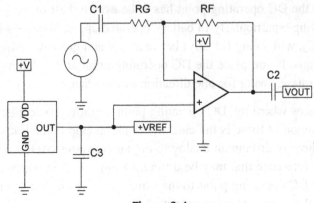

Figure 3.4
Single-supply op amp inverting gain circuit with external DC operating point.

reference"—imagine that! A voltage reference in a 5 V system, for example, should be selected to be 2.5 V, which it just so happens is a very common value for fixed voltage references. It is almost like somebody was anticipating your circuit's needs!

Fig. 3.4 shows the modification of the circuit to use an external voltage reference. C_3 decouples the output of the reference, and the circuit net "$+V_{REF}$" becomes available to all op amp stages in the signal chain—like a second power supply. Only instead of being equal and opposite $+V$, it is halfway between $+V$ and ground.

The use of an external voltage reference has some implications to the circuit. The first is that you are no longer guaranteed that the noninverting input is driven with exactly $+V/2$. However, with real world resistor values not being perfectly matched, you never had that guarantee with the circuit in Fig. 3.3. The advantage you have gained is not having to duplicate the voltage divider consisting of R_1, R_2, and C_3 in every stage. That saves board space, component cost, and increases reliability.

The second implication of using a voltage source is one of consolidation. If the ADC you are ultimately driving has a voltage reference output, you may be able to use it, and it alone, to drive all of the op amp stages in your signal chain. But be careful! It may have limited output current capability and not be able to drive more than one or two stages. The converse of that may be true—that is you may be able to supply the ADC with an external voltage reference—allowing you to use the same voltage reference that drove all of the signal chain to drive the ADC as well.

One very common design mistake that rookie engineers make involves using an op amp to buffer the voltage reference output of an ADC, so it can drive one or more stages in the signal chain.

At first glance, there is a lot that is correct about this circuit. The input is isolated from the DC operating point of the circuit by C_1. There is no output capacitor, because the intention is to operate the stage at the V_{REF} point of the ADC. Sometimes, the internal V_{REF} of an ADC is weak, and needs buffering, so that has been done. Where this circuit has a problem is in placing C_3 directly on the output of the buffer op amp. Most op amps cannot handle a capacitive load directly and will become unstable. A very few op amps are designed to handle capacitive loads, and if the data sheet says it can, then you can use this circuit. However, if you are not using one of those op amps specially designed for a capacitive load, you can fix the problem with one resistor (see Fig. 3.5).

In Fig. 3.6, the addition of resistor R_1 isolates the op amp output from the capacitance. R_1 does not even have to be very large, 10 Ω suffices in all but a handful of cases.

So far, I have been talking about the inverting op amp configuration for AC coupled circuits. There is one aspect of this circuit I have not talked about. Capacitors C_1 and C_2

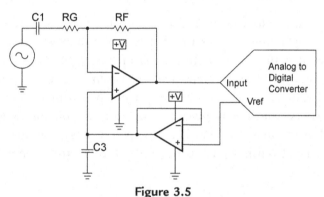

Figure 3.5
Incorrect voltage reference buffering.

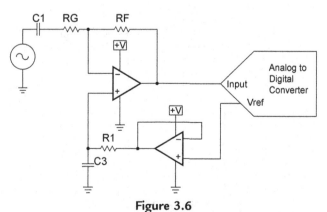

Figure 3.6
Correct voltage reference buffering.

are the DC blocking capacitors, but they also act as high pass filters! It is left to you to select a value that does not affect the signal—a good starting place is to consider the input network of C_1 and R_G as a single-pole high pass filter. Similarly, the output capacitor C_2 will form a high pass filter with the input impedance of the subsequent stage. A good rule of thumb is to place this high pass filter breakpoint 100 times lower in frequency as the lowest frequency of interest in the system—so it does not affect the signals you are interested in.

I will now turn my attention to the noninverting, single-supply gain stages. This is one of the most common misapplication of op amps. It has resulted in countless designs that do not work and endless hours of troubleshooting time. It probably also has a record of engineers being fired when their designs do not work as well! So pay careful attention to this section, it may save your career!

Fig. 3.7 is the circuit in question.

Make a mental note to yourself: *This is very bad, don't do it!*

So, what is wrong, you say to yourself? You can try the thought experiment method again: The stage DC operating point is isolated from the other stages by C_1 and C_2. But ask yourself—what is the DC operating point of this stage? If you have been following the discussion of ideal op amps and single-supply circuits, you will be saying that it is determined by the voltage divider comprised of R_F and R_G. If they are equal value resistors, which only gives a unity gain, the output swings to +V, the inverting input will swing to +V/2, and if the output swings to ground, the inverting input will be at ground. Even if R_F is 10 times R_G, the inverting input will be

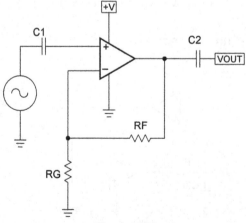

Figure 3.7
Incorrect noninverting single-supply stage.

between +V and ground, right? After all, the noninverting input will be at the same virtual voltage as the inverting input, right?

The answer is more complex than that and forces me to get a bit ahead of myself and talk about real world limitations of op amps.

- Limitation 1—the input range probably does *not* include ground of the op amp you are using. A few op amps use complex tricks to do it, and those amplifiers have other limitations that probably make them the wrong op amp for the job.
- Limitation 2—the noninverting input connects through a capacitor to the base of a transistor. It is a very good transistor. But what happens to any transistor if you leave the base floating? Not very much. Even if you hang a capacitor on the input and drive it with an AC signal, you probably will not get what you want. The input transistors require a small input bias current to "turn on" and work in the linear region. It does not have to be much, but it has to be there. The exact DC level does not matter to the transistor, it will turn on with a potential anywhere between the negative supply input of the op amp and the positive supply input of the op amps (consult absolute maximum table in an op amp data sheet).

So the main problem is input bias current—the noninverting input has none! Leakage through the capacitor may be enough. You may have constructed this circuit in the past and it (barely) worked. You got lucky! The first step in the effort to "fix" the circuit of Fig. 3.7 is adding two resistors as shown in Fig. 3.8.

Make a mental note to yourself: This will not work either!

Let us apply the thought experiment to this. DC-wise, the noninverting input is held at +V/2 by R_1 and R_2, giving it the required input bias current and setting the DC

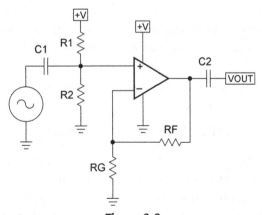

Figure 3.8
Another incorrect noninverting single-supply stage.

operating point. Unfortunately, the DC gain of the circuit is $1 + R_F/R_G$. So unless $R_F = R_G$, making a unity gain stage, this stage will not have the correct virtual ground point on the output, because the voltage divider composed of R_F and R_G will place a lower value than $+V/2$ on the inverting input, unbalancing the $+$ and $-$ inputs.

Obviously, the fix is not complete. The addition of another capacitor, C_3, will solve the problem:

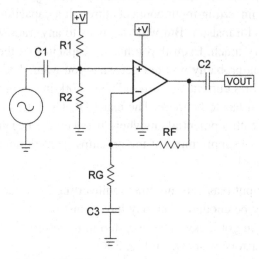

Applying our thought experiment to this:

- DC-wise, the noninverting input is held at $+V/2$ by R_1 and R_2, giving it the required input bias current and setting the DC operating point—just as it was before.
- The output of the op amp will also be at $+V/2$, which is correct.
- The inverting input will receive input bias current from the output, which is at $+V/2$, through R_F.
- The DC gain of the stage is unity, because C_3 isolates R_G, making it an open circuit, and therefore the DC gain of the stage is unity gain, because R_F forms a unity gain buffer circuit for DC.
- The AC gain of the stage is $1 + R_F/R_G$, assuming C_3 is large (low impedance) at the frequencies of interest to the circuit.

The circuit is fixed. But, it is a "fix" that is full of compromises:

- You cannot put a bypass capacitor on the voltage divider consisting of R_1 and R_2. Any capacitor will be an AC short on the input signal. The full force of any noise present on the $+V$ supply will be applied, divided by 2, to the input of the stage. BAD design, very bad! You amplify power supply noise along with signal. Your power supply voltage better be very clean. If you must do this, consider using a voltage reference the same potential as your power supply, but you need a higher system voltage to power the reference.

- You are, again loading your power supply through R_1 and R_2.
- There is no way to directly use an ADC reference voltage, unless it happens to match your stage's supply voltage.

Suffice it to say, go inverting if you can! There are a limited number of situations in an AC coupled systems that will require you to use a noninverting stage. One of these situations is with high speed amplifiers, where the resistor values tend to be smaller. For high gain stages, the value of R_G may be small enough to load down the input source.

I will not delver further into AC coupled op amp circuits, because at this point you have the basic skills required to set correct the DC operating point in an op amp stage. General rules for designing these stages:

- Analyze the DC operating point of the inverting and noninverting inputs separately— they should always be at $+V/2$.
- Resistors to ground should probably have a DC blocking capacitor, so that the elements they connect to will not form voltage dividers forcing the DC operating point lower. I will come back to this point when discussing fully differential amplifiers in a later chapter.

The next chapter will delve into an even more complex topic—what you have to do when you cannot AC couple a circuit!

CHAPTER 4

DC-Coupled Single-Supply Op Amp Design Techniques

4.1 An Introduction to DC-Coupled, Single-Supply Circuits

The previous chapter assumed that the circuit does not require DC gain as well as AC gain. This is not always the case with applications such as transducer amplifiers, etc. These are measurement type of circuits, where the DC coming from the transducer IS the measurement, and probably does not change very rapidly. DC must be preserved in any gain circuit, and DC accuracy is of paramount importance. The job is much more complex when the circuit must be operated from a single supply, such as a battery. This is often the case when the amplifier circuit is located remotely from the rest of the system—which is often times done to minimize noise pickup that would happen in a long cable.

The requirement to account for inputs connected to ground or different reference voltages makes it difficult to design single-supply op amp circuits. Unless otherwise specified, all op amp circuits discussed in this chapter are single-supply circuits.

Use of a single-supply limits the polarity of the output voltage. When the supply voltage $V_{CC} = 10$ V, the output voltage is limited to the range $0 \leq V_{OUT} \leq 10$. This limitation precludes negative output voltages when the circuit has a positive supply voltage, but it does not preclude negative input voltages when the circuit has a positive supply voltage. As long as the voltage on the op amp input leads does not become negative, the circuit can handle negative input voltages.

Beware of working with negative input voltages when the op amp is powered from a positive supply because op amp inputs are highly susceptible to reverse voltage breakdown. Also, insure that all possible start-up conditions do not reverse bias the op amp inputs when the input and supply voltage are opposite polarity.

4.2 Simple Application to Get You Started

Consider the circuit of Fig. 4.1. At first glance, it looks like an insurmountable challenge to add a reference voltage V_{REF} in series with V_{IN}—the same V_{REF} that is applied to the noninverting input of the op amp as well. How can this be practical? However, this is actually one of the most common circuits in DC-coupled applications. It is the circuit that

Op Amps for Everyone. http://dx.doi.org/10.1016/B978-0-12-811648-7.00004-2

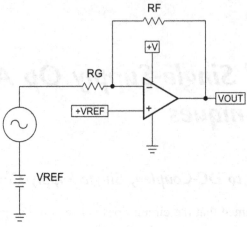

Figure 4.1
A simple transducer interface example.

buffers the output of one type of transducer, where the DC offset is included in the signal output by the design of the transducer. You can breathe a sigh of relief if this is your DC-coupled application, because all you need to do in this case is insure that the $+V_{REF}$ applied to the noninverting op amp input is the same potential coming from the transducer. Any difference will lead to significant DC offset on the output and therefore an error in the measurement.

As wonderful as this simple circuit is, it is very prone to the effects of drift and temperature changes. It is time to move to more practical applications.

The challenge above is to balance the DC potentials on the inverting and noninverting inputs of the op amp. An input bias voltage is used to eliminate the reference voltage when it must not appear in the output voltage (see Fig. 4.2).

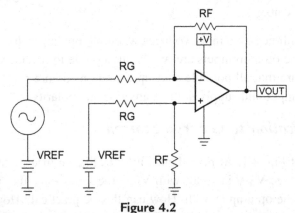

Figure 4.2
Split-supply op amp circuit with common-mode voltage.

The circuit of Fig. 4.2 is not very practical, but I will use it to illustrate several points:

- You should recognize this as the differential amplifier circuit from an earlier chapter, and in this case the two voltages connected to the inputs are the fixed V_{REF} and a fixed V_{REF} with a varying signal added to it.
- The voltage, V_{REF}, is in both input circuits; hence it is named a common-mode voltage. Voltage feedback op amps reject common-mode voltages because their input circuit is constructed with a differential amplifier (chosen because it has natural common-mode voltage rejection capabilities).
- The DC operating point of this circuit is $V_{REF}/2$, and not $+V/2$, so you would be wise to select V_{REF} such that $V_{REF}/2 = +V/2$, in other words V_{REF} becomes the circuit supply voltage $+V$.

4.3 Circuit Analysis

The complexities of single-supply op amp design are illustrated with the following example. Notice that the biasing requirement complicates the analysis by presenting several conditions that are not realizable. It is best to wade through this material to gain an understanding of the problem, especially since a cookbook solution is given later in this chapter. The previous chapters assumed that the op amps were ideal, but this chapter starts to deal with op amp deficiencies. The input and output voltage swings of many op amps are limited, but if one designs with rail-to-rail op amps, the input/output swing problems are minimized.

Before proceeding, I need to mention the following points:

- All real world op amp application circuits should include decoupling capacitors on the op amp power leads. For single-supply op amp circuits, there is only one power supply pin actively used; the negative supply pin can be connected to ground without a decoupling capacitor—since it is already directly tied to ground.
- All real world op amp application circuits connect to a load of some sort. In the subsequent discussion, I will assume that this load is high impedance compared to the component values in the circuit being analyzed. If this is not the case, the component values need to be scaled, or a buffer stage added after the circuits shown.
- All of the circuits shown in this chapter are some variation of the differential amplifier circuit. It is helpful to separate out what is happening to the signal applied to each one, as it affects and is affected by the other.

The inverting circuit shown in Fig. 4.3 is analyzed first.

Before delving directly into the math, I will take a step back and do thought experiments so the equations below make a bit more sense. The reference input above is a voltage

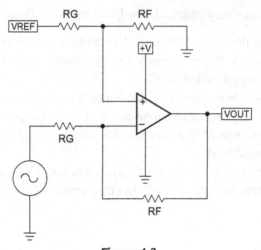

Figure 4.3
Inverting op amp with DC offset.

divider to ground, therefore the DC operating point is set by that voltage divider. The DC gain on V_{REF} is determined by that voltage divider, and by R_F and R_G on the inverting side (remembering that you can short the independent voltage source for the analysis). The inverting AC gain, meanwhile, is set by R_F and R_G in the bottom leg of the circuit. Putting these thoughts down algebraically:

Eq. (4.1) is written with the aid of superposition, and simplified algebraically, to acquire Eq. (4.2).

$$V_{OUT} = V_{REF}\left(\frac{R_F}{R_G + R_F}\right)\left(\frac{R_F + R_G}{R_G}\right) - V_{IN}\frac{R_F}{R_G} \tag{4.1}$$

$$V_{OUT} = (V_{REF} - V_{IN})\frac{R_F}{R_G} \tag{4.2}$$

For $V_{REF} = V_{IN}$, one obtains Eq. (4.3), and there is no output voltage from the circuit regardless of the input voltage.

$$V_{OUT} = (V_{REF} - V_{IN})\frac{R_F}{R_G} = (V_{IN} - V_{IN})\frac{R_F}{R_G} = 0 \tag{4.3}$$

When $V_{REF} = 0$, $V_{OUT} = -V_{IN}(R_F/R_G)$, there are two possible solutions to Eq. (4.2). First, when V_{IN} is any positive voltage, V_{OUT} should be negative voltage. The circuit cannot achieve a negative voltage with a positive supply, so the output saturates at the lower power supply rail. Second, when V_{IN} is any negative voltage, the output spans the normal range according to Eq. (4.5).

$$V_{IN} \geq 0, \quad V_{OUT} = 0 \tag{4.4}$$

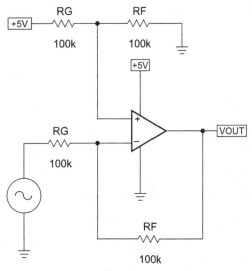

Figure 4.4
Inverting op amp with V_{CC} bias.

$$V_{IN} \leq 0, \quad V_{OUT} = |V_{IN}| \frac{R_F}{R_G} \tag{4.5}$$

When V_{REF} equals the supply voltage, V_{CC}, we obtain Eq. (4.6). In Eq. (4.6), when V_{IN} is negative, V_{OUT} should exceed V_{CC}; that is impossible, so the output saturates. When V_{IN} is positive, the circuit acts as an inverting amplifier.

$$V_{OUT} = (V_{CC} - V_{IN}) \frac{R_F}{R_G} \tag{4.6}$$

The transfer curve for the circuit shown in Fig. 4.4 is shown in Fig. 4.5.

Four op amps were tested in the circuit configuration shown in Fig. 4.4. Three op amps, LM358, TL07X, and TLC272, had output voltage spans of 2.3−3.75 V. This performance does not justify the ideal op amp assumption that was made in the previous chapter unless the output voltage swing is severely limited. Limited output or input voltage swing is one of the worst deficiencies a single-supply op amp can have because the limited voltage swing limits the circuit's dynamic range. Also, limited voltage swing frequently results in distortion of large signals. The fourth op amp tested was the TLV247X, which was designed for rail-to-rail operation in single-supply circuits. The TLV247X plotted a perfect curve (results limited by the instrumentation) and performance that justifies the use of ideal assumptions. Some of the older op amps must limit their transfer equation as shown in Eq. (4.7).

$$V_{OUT} = (V_{CC} - V_{IN}) \frac{R_F}{R_G} \quad \text{for} \quad V_{OH} \geq V_{OUT} \geq V_{OL} \tag{4.7}$$

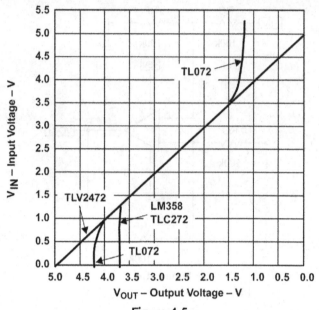

Figure 4.5
Transfer curve for inverting op amp with V_{CC} bias.

The noninverting op amp circuit is shown in Fig. 4.6.

The only difference between Figs. 4.6 and 4.3 is that the relative positions of the input signal and V_{REF} have been swapped. Therefore, the gain on V_{REF} is determined by R_F and R_G in the bottom leg of the circuit, and the signal gain is determined by voltage divider composed of R_F and R_G in the top leg of the circuit, times gain provided by R_F and R_G in the bottom leg.

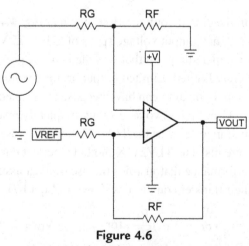

Figure 4.6
Noninverting op amp.

Eq. (4.8) is written with the aid of superposition, and simplified algebraically, to acquire Eq. (4.9).

$$V_{OUT} = V_{IN}\left(\frac{R_F}{R_G + R_F}\right)\left(\frac{R_F + R_G}{R_G}\right) - V_{REF}\frac{R_F}{R_G} \tag{4.8}$$

$$V_{OUT} = (V_{IN} - V_{REF})\frac{R_F}{R_G} \tag{4.9}$$

When $V_{REF} = 0$, $V_{OUT} = V_{IN}\frac{R_F}{R_G}$, there are two possible circuit solutions. First, when V_{IN} is a negative voltage, V_{OUT} must be a negative voltage. The circuit cannot achieve a negative output voltage with a positive supply, so the output saturates at the lower power supply rail. Second, when V_{IN} is a positive voltage, the output spans the normal range as shown by Eq. (4.11).

$$V_{IN} \leq 0, \quad V_{OUT} = 0 \tag{4.10}$$

$$V_{IN} \geq 0, \quad V_{OUT} = V_{IN} \tag{4.11}$$

The noninverting op amp circuit shown in Fig. 4.6 is constructed with $V_{CC} = 5$ V, $R_G = R_F = 100$ kΩ. The transfer curve for this circuit is shown in Fig. 4.7; a TLV247X serves as the op amp.

There are many possible variations of inverting and noninverting circuits. At this point many designers analyze these variations hoping to stumble upon the one that solves the circuit problem. A design methodology is needed, one that is guaranteed to provide the

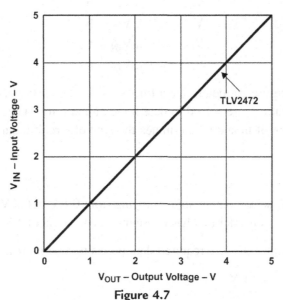

Figure 4.7

Transfer curve for noninverting op amp.

correct circuit configuration each and every time. Fortunately, there is such a methodology, and it comes from the realm of the Cartesian coordinate system. This uniform design methodology starts by employing simultaneous equations to render specified data into equation form. When the form of the desired equation is known, a circuit that fits the equation can be chosen to solve the problem. The resulting equation must be a straight line, thus there are only four possible solutions.

4.4 *Simultaneous Equations*

Taking an orderly path to developing a circuit that works the first time starts here; follow these steps until the equation of the op amp is determined. Use the specifications given for the circuit coupled with simultaneous equations to determine what form the op amp equation must have. Go to the section that illustrates that equation form (called a case), solve the equation to determine the resistor values, and you have a working solution.

A linear op amp transfer function is limited to the equation of a straight line (Eq. 4.12).

$$y = \pm mx \pm b \qquad (4.12)$$

The equation of a straight line has four possible solutions depending upon the sign of the slope "m," and the intercept "b"; thus simultaneous equations yield solutions in four forms. Four circuits must be developed; one for each form of the equation of a straight line. The four equations, cases, or forms of a straight line are given in Eqs. (4.13) through (4.16), where electronic terminology has been substituted for math terminology.

$$V_{OUT} = +mV_{IN} + b \qquad (4.13)$$
$$V_{OUT} = +mV_{IN} - b \qquad (4.14)$$
$$V_{OUT} = -mV_{IN} + b \qquad (4.15)$$
$$V_{OUT} = -mV_{IN} - b \qquad (4.16)$$

Only two points are required to determine a line, so given two data points for V_{OUT} and V_{IN}, simultaneous equations are solved to determine m and b for the equation that satisfies the given data. The sign of m and b determines the type of circuit required to implement the solution.

An example:

Circuit requirement: "A sensor output signal ranging from 0.1 V to 0.2 V must be interfaced into an analog-to-digital converter that has an input voltage range of 1 V to 4 V."

This requirement generates the two required data points to determine a straight line:

1. $V_{OUT} = 1$ V at $V_{IN} = 0.1$ V
2. $V_{OUT} = 4$ V at $V_{IN} = 0.2$ V

This is all we need to generate the simultaneous equations and solve for the straight line, circuit topology, and component values!

The data points are inserted into Eq. (4.13), as shown in Eqs. (4.17) and (4.18), to obtain m and b for the specifications.

$$1 = m(0.1) + b \tag{4.17}$$
$$4 = m(0.2) + b \tag{4.18}$$

Multiply Eq. (4.17) by 2 and subtract it from Eq. (4.18).

$$2 = m(0.2) + 2b \tag{4.19}$$
$$b = -2 \tag{4.20}$$

After algebraic manipulation of Eq. (4.17), substitute Eq. (4.20) into Eq. (4.17) to obtain Eq. (4.21).

$$m = \frac{2+1}{0.1} = 30 \tag{4.21}$$

Now m and b are substituted back into Eq. (4.13) yielding Eq. (4.22).

$$V_{OUT} = 30V_{IN} - 2 \tag{4.22}$$

Notice, although Eq. (4.13) was the starting point, the form of Eq. (4.22) is identical to the format of Eq. (4.14). The specifications or given data determine the sign of m and b, and starting with Eq. (4.13), the final equation form is discovered after m and b are calculated. The next step required to complete the problem solution is to develop a circuit that has an $m = 30$ and $b = -2$. If we so desired, we could now go to Section 4.4.2—the Case 2 section, to complete the design. But this example was only meant to illustrate how to determine the case, Section 4.4.2 describes a solution for a very similar case.

Circuits were developed for Eqs. (4.13) through (4.16), and they are given under the headings Case 1 through Case 4 respectively. There are different circuits that will yield the same equations, but these circuits were selected because they do not require negative references.

4.4.1 Case 1: $V_{OUT} = +mV_{IN} + b$

The circuit configuration that yields a solution for Case 1 is shown in Fig. 4.8. If you look carefully at the circuit, hopefully something stands out! Both the input signal and the voltage reference are connected (through resistors) to the noninverting "+" input. Both "m" and "b" in the Case 1 equation are positive. This should be an "aha" moment, because this will hold true for the other cases. However, there is much analysis to be done before you can determine the exact circuit schematic and the exact component values. But your understanding should be enhanced at this point.

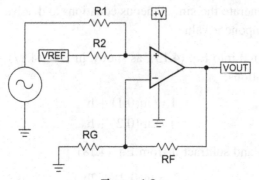

Figure 4.8
Schematic for Case 1: $V_{OUT} = +mV_{IN} + b$.

Another thing that should stand out to you is that the noninverting input configuration is nothing more than a summer circuit, with a noninverting gain determined by R_F and R_G. But be careful, it would be easy to underestimate the task at this point! R_1 and R_2 are also voltage dividers on the input signal and V_{REF}, respectively!

The circuit equation is written using the voltage divider rule and superposition.

$$V_{OUT} = V_{IN}\left(\frac{R_2}{R_1 + R_2}\right)\left(\frac{R_F + R_G}{R_G}\right) + V_{REF}\left(\frac{R_1}{R_1 + R_2}\right)\left(\frac{R_F + R_G}{R_G}\right) \qquad (4.23)$$

The equation of a straight line (case 1) is repeated in Eq. (4.24) below so comparisons can be made between it and Eq. (4.23).

$$V_{OUT} = mV_{IN} + b \qquad (4.24)$$

Equating coefficients yields Eqs. (4.25) and (4.26).

$$m = \left(\frac{R_2}{R_1 + R_2}\right)\left(\frac{R_F + R_G}{R_G}\right) \qquad (4.25)$$

$$b = V_{REF}\left(\frac{R_1}{R_1 + R_2}\right)\left(\frac{R_F + R_G}{R_G}\right) \qquad (4.26)$$

A Case 1 Example:

The circuit specifications are as follows:

1. $V_{OUT} = 1$ V at $V_{IN} = 0.01$ V
2. $V_{OUT} = 4.5$ V at $V_{IN} = 1$ V

As before, this is all we need to complete the design. Except for two details—the supply and reference voltage. Sometimes, no reference voltage is available, and it is necessary to use the supply voltage. A reference voltage source is left out of the design as a space and cost savings measure, and it sacrifices noise performance, accuracy, and stability performance. Cost is an important specification, but the V_{CC} supply must be specified well enough to do the job. Assume for this example that $+5$ V is used for both supply and reference.

Each step in the subsequent design procedure is included in this analysis to ease. Many steps will be skipped when subsequent cases are analyzed.

The data are substituted into simultaneous equations.

$$1 = m(0.01) + b \qquad (4.27)$$
$$4.5 = m(1.0) + b \qquad (4.28)$$

Eq. (4.27) is multiplied by 100 (Eq. 4.29) and Eq. (4.28) is subtracted from Eq. (4.29) to obtain Eq. (4.30).

$$100 = m(1.0) + 100b \qquad (4.29)$$
$$b = \frac{95.5}{99} = 0.9646 \qquad (4.30)$$

The slope of the transfer function, m, is obtained by substituting b into Eq. (4.27).

$$m = \frac{1 - b}{0.01} = \frac{1 - 0.9646}{0.01} = 3.535 \qquad (4.31)$$

Now that b and m are calculated, the resistor values can be calculated. Eqs. (4.25) and (4.26) are solved for the quantity $(R_F + R_G)/R_G$, and then they are set equal in Eq. (4.32) thus yielding Eq. (4.33).

$$\frac{R_F + R_G}{R_G} = m\left(\frac{R_1 + R_2}{R_2}\right) = \frac{b}{V_{CC}}\left(\frac{R_1 + R_2}{R_1}\right) \qquad (4.32)$$
$$R_2 = \frac{3.535}{\dfrac{0.9646}{5}}R_1 = 18.316R_1 \qquad (4.33)$$

Choose $R_1 = 10 \text{ k}\Omega$, and that sets the value of $R_2 = 183.16 \text{ k}\Omega$. The closest 5% resistor value to $183.16 \text{ k}\Omega$ is $180 \text{ k}\Omega$; therefore, select $R_1 = 10 \text{ k}\Omega$ and $R_2 = 180 \text{ k}\Omega$. Being forced to yield to reality by choosing standard resistor values means that there is an error in the circuit transfer function, because m and b are not exactly the same as calculated. The real world constantly forces compromises into circuit design, but the good circuit designer accepts the challenge and throws money or brains at the challenge. Using 10 cent resistors with a 10 cent op amp is hard to justify except in precision circuits; however, the price of 1% resistors has plummeted in recent years, and there is seldom a need to make such a compromise today.

The left half of Eq. (4.32) is used to calculate R_F and R_G.

$$\frac{R_F + R_G}{R_G} = m\left(\frac{R_1 + R_2}{R_2}\right) = 3.535\left(\frac{180 + 10}{180}\right) = 3.73 \tag{4.34}$$

$$R_F = 2.73 R_G \tag{4.35}$$

The resulting circuit equation is given below.

$$V_{OUT} = 3.5 V_{IN} + 0.97 \tag{4.36}$$

The gain setting resistor, R_G, is selected as 10 kΩ, and 27 kΩ, the closest 5% standard value is selected for the feedback resistor, R_F. Again, there is a slight error involved with standard resistor values. This circuit must have an output voltage swing from 1 to 4.5 V. The circuit with the selected component values is shown in Fig. 4.9 and the transfer curve is shown in Fig. 4.10.

The transfer curve shown is a straight line, and that means that the circuit is linear. The V_{OUT} intercept is about 0.98 V rather than 1 V as specified, and this is excellent performance considering that the components were selected from 5% resistor values. The output voltage measured 4.53 V when the input voltage was 1 V. Considering the low and high input voltage errors, it is safe to conclude that the resistor tolerances have skewed the gain slightly, but this is still excellent performance for 5% components. Often lab data similar to that shown here are more accurate than the 5% resistor tolerance, but do not fall into the trap of expecting this performance, because you will be disappointed if you do.

The resistors were selected in the kΩ range arbitrarily. The gain and offset specifications determine the resistor ratios, but supply current, frequency response, and op amp drive capability determine their absolute values. The resistor value selection in this design is high because modern op amps do not have input current offset problems, and they yield

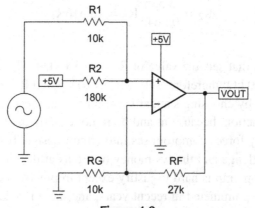

Figure 4.9

Case 1 example circuit.

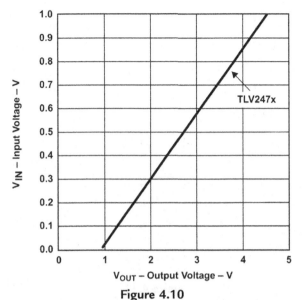

Figure 4.10

Case 1 example circuit measured transfer curve.

reasonable frequency response. If higher frequency response is demanded, the resistor values must decrease, and resistor value decreases reduce input current errors, while supply current increases. When the resistor values get low enough, it becomes hard for another circuit, or possibly the op amp, to drive the resistors.

4.4.2 Case 2: $V_{OUT} = +mV_{IN} - b$

The circuit shown in Fig. 4.11 yields a solution for Case 2. Before delving into the math, let us take a broad view of what is going on. The slope m is positive, so the input signal is applied to the noninverting input. The intercept b is negative, so it is applied to the inverting input. Noninverting gain for the signal input is provided by R_F and R_G, and inverting gain for the reference is also provided by R_F and R_G. At this point, though, you should be noticing a fundamental limitation of this circuit. There will be some error introduced by the resistors R_1 and R_2, which form a voltage divider off of $+V_{REF}$. This error can be mitigated somewhat by making the scale of the resistors R_G and R_F large in comparison with R_1 and R_2—say 100 times the value. This is the most problematic solution for the four cases presented in this chapter. If you were concerned only with AC response on the input signal, you could bypass R_2 with a capacitor, but we are assuming in this chapter that you are concerned with DC response—so that will not work. The best solution would be to use a separate op amp to buffer R_1 and R_2, thus presenting a low impedance to R_G. Indeed, the price of dual op amps, and the packages they are offered in, may make this a viable solution. But for the sake of this analysis, we will proceed with the

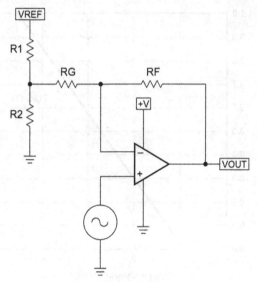

Figure 4.11
Schematic for Case 2: $V_{OUT} = +mV_{IN} - b$.

assumption that only a single op amp is available. The solution it yields is surprisingly satisfactory!

The circuit equation is obtained by taking the Thevenin equivalent circuit looking into the junction of R_1 and R_2. After the R_1, R_2 circuit is replaced with the Thevenin equivalent circuit, the gain is calculated with the ideal gain equation (Eq. 4.37).

$$V_{OUT} = V_{IN}\left(\frac{R_F + R_G + R_1\|R_2}{R_G + R_1\|R_2}\right) - V_{REF}\left(\frac{R_2}{R_1 + R_2}\right)\left(\frac{R_F}{R_G + R_1\|R_2}\right) \qquad (4.37)$$

Comparing terms in Eqs. (4.37) and (4.14) enables the extraction of m and b.

$$m = \frac{R_F + R_G + R_1\|R_2}{R_G + R_1\|R_2} \qquad (4.38)$$

$$|b| = V_{REF}\left(\frac{R_2}{R_1 + R_2}\right)\left(\frac{R_F}{R_G + R_1\|R_2}\right) \qquad (4.39)$$

The specifications for an example design are $V_{OUT} = 1.5$ V at $V_{IN} = 0.2$ V, $V_{OUT} = 4.5$ V at $V_{IN} = 0.5$ V, and $V_{REF} = V_{CC} = 5$ V. The simultaneous equations, Eqs. 4.40 and 4.41, are written below.

$$1.5 = 0.2m + b \qquad (4.40)$$
$$4.5 = 0.5m + b \qquad (4.41)$$

From these equations we find that b = -0.5 and m = 10. Making the assumption that $R_1 \| R_2 \ll R_G$ simplifies the calculations of the resistor values.

$$m = 10 = \frac{R_F + R_G}{R_G} \tag{4.42}$$

$$R_F = 9R_G \tag{4.43}$$

Let $R_G = 20$ kΩ, and then $R_F = 180$ kΩ.

$$b = V_{CC}\left(\frac{R_F}{R_G}\right)\left(\frac{R_2}{R_1 + R_2}\right) = 5\left(\frac{180}{20}\right)\left(\frac{R_2}{R_1 + R_2}\right) \tag{4.44}$$

$$R_1 = \frac{1 - 0.01111}{0.01111}R_2 = 89R_2 \tag{4.45}$$

Select $R_2 = 820$ Ω, and R_1 equals 72.98 kΩ. Since 72.98 kΩ is not a standard 5% resistor value, R_1 is selected as 75 kΩ. The difference between the selected and calculated value of R_1 has about a 3% effect on b, and this error shows up in the transfer function as an intercept rather than a slope error. The parallel resistance of R_1 and R_2 is approximately 820 Ω and this is much less than R_G, which is 20 kΩ, thus the earlier assumption that $R_G \gg R_1 \| R_2$ is justified. The final circuit is shown in Fig. 4.12 and the measured transfer curve for this circuit is shown in Fig. 4.13. Notice that I went ahead and added the capacitor across R_2. This is *not* for AC gain, although it would certainly have that effect. It is merely there for bypassing on the DC potential that generates the intercept b, to reduce noise (since we are using the power supply as $+V_{REF}$).

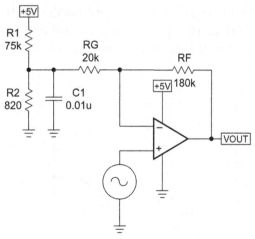

Figure 4.12
Case 2 example circuit.

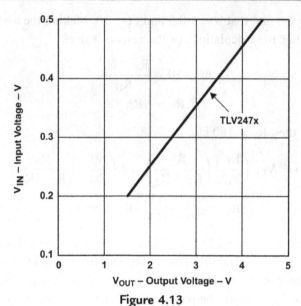

Figure 4.13
Case 2 example circuit measured transfer curve.

The TLV247X was used to build the test circuit because of its wide dynamic range. The transfer curve plots are very close to the theoretical curve; the direct result of using a high performance op amp.

4.4.3 Case 3: $V_{OUT} = -mV_{IN} + b$

The circuit shown in Fig. 4.14 yields the transfer function desired for Case 3. This is really straightforward. The intercept is positive; it is applied to the noninverting input. The slope is negative; it is applied to the inverting input. The inverting gain on the signal is

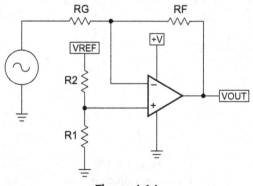

Figure 4.14
Schematic for Case 3: $V_{OUT} = -mV_{IN} + b$.

simple, being determined by R_F and R_G. The reference voltage goes through a voltage divider and then has noninverting gain determined by R_F and R_G.

The circuit equation is obtained with superposition.

$$V_{OUT} = -V_{IN}\left(\frac{R_F}{R_G}\right) + V_{REF}\left(\frac{R_1}{R_1 + R_2}\right)\left(\frac{R_F + R_G}{R_G}\right) \qquad (4.46)$$

Comparing terms between Eqs. (4.45) and (4.15) enables the extraction of m and b.

$$|m| = \frac{R_F}{R_G} \qquad (4.47)$$

$$b = V_{REF}\left(\frac{R_1}{R_1 + R_2}\right)\left(\frac{R_F + R_G}{R_G}\right) \qquad (4.48)$$

The design specifications for an example circuit are $V_{OUT} = 1$ V at $V_{IN} = -0.1$ V, $V_{OUT} = 6$ V at $V_{IN} = -1$ V, and $V_{REF} = V_{CC} = 10$ V.

Up until now, we have been using a TLV247 family op amp for our real world circuits. Let us complicate things a bit! The supply voltage available for this circuit is 10 V, and this exceeds the maximum allowable supply voltage for the TLV247X. Also, assume this circuit must drive a back-terminated cable that looks like two 50 Ω resistors connected in series, thus the op amp must be able to drive 6/100 = 60 mA. The stringent op amp selection criteria limits the of op amps if ideal op amp equations are going to be used. The TLC07X has excellent single-supply input performance coupled with high output current drive capability, so it is selected for this circuit.

The simultaneous equations (Eqs. 4.49 and 4.50) are written below.

$$1 = (-0.1)m + b \qquad (4.49)$$

$$6 = (-1)m + b \qquad (4.50)$$

From these equations we find that b = 0.444 and m = -5.6.

$$|m| = 5.56 = \frac{R_F}{R_G} \qquad (4.51)$$

$$R_F = 5.56 R_G \qquad (4.52)$$

Let $R_G = 10$ kΩ, and then $R_F = 56.6$ kΩ, which is not a standard 5% value, hence R_F is selected as 56 kΩ.

$$b = V_{CC}\left(\frac{R_F + R_G}{R_G}\right)\left(\frac{R_1}{R_1 + R_2}\right) = 10\left(\frac{56 + 10}{10}\right)\left(\frac{R_1}{R_1 + R_2}\right) \qquad (4.53)$$

$$R_2 = \frac{66 - 0.4444}{0.4444}R_1 = 147.64 R_1 \qquad (4.54)$$

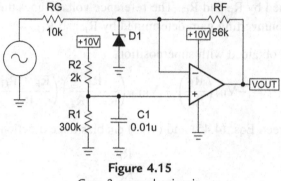

Figure 4.15
Case 3 example circuit.

The final equation for the example is given below.

$$V_{OUT} = -5.56V_{IN} + 0.444 \qquad (4.55)$$

Select $R_1 = 2$ kΩ and $R_2 = 295.28$ kΩ. Since 295.28 kΩ is not a standard 5% resistor value, R_1 is selected as 300 kΩ. The difference between the selected and calculated value of R_1 has a nearly insignificant effect on b. The final circuit is shown in Fig. 4.15, and the measured transfer curve for this circuit is shown in Fig. 4.16.

There could be an issue that would destroy the op amp. As long as the circuit works normally, there are no problems handling the negative voltage input to the circuit, because

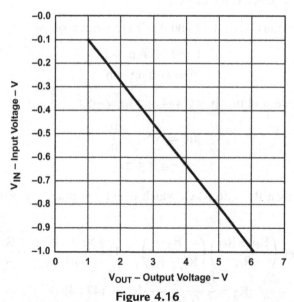

Figure 4.16
Case 3 example circuit measured transfer curve.

the inverting lead of the TLC07X is at a positive voltage. The positive op amp input lead is at a voltage of approximately 65 mV, and normal op amp operation keeps the inverting op amp input lead at the same voltage because of the assumption that the error voltage is zero. However, when V_{CC} is powered down while there is a negative voltage on the input circuit, most of the negative voltage appears on the inverting op amp input lead.

The most prudent solution is to connect the diode, D_1, with its cathode on the inverting op amp input lead and its anode at ground. If a negative voltage gets on the inverting op amp input lead, it is clamped to ground by the diode. Select the diode type as Schottky, so the voltage drop across the diode is about 200 mV; this small voltage does not harm most op amp inputs. As a further precaution, R_G can be split into two resistors with the diode inserted at the junction of the two resistors. This places a current limiting resistor between the diode and the inverting op amp input lead.

4.4.4 Case 4: $V_{OUT} = -mV_{IN} - b$

The circuit shown in Fig. 4.17 yields a solution for Case 4. As you can see, because both the slope m and intercept b are negative, both are applied to the inverting input. Furthermore, Case 4 reduces to that of the simple summation circuit described in an earlier chapter. The final solution, however, must take some other things into account to avoid turnoff problems like those described in Case 3.

The circuit equation is obtained by using superposition to calculate the response to each input. The individual responses to V_{IN} and V_{REF} are added to obtain Eq. (4.56).

$$V_{OUT} = -V_{IN}\frac{R_F}{R_{G1}} - V_{REF}\frac{R_F}{R_{G2}} \qquad (4.56)$$

Comparing terms in Eqs. (4.56) and (4.16) enables the extraction of m and b.

$$|m| = \frac{R_F}{R_{G1}} \qquad (4.57)$$

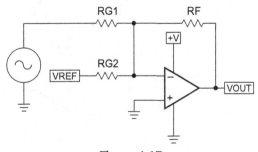

Figure 4.17
Schematic for Case 4: $V_{OUT} = -mV_{IN} - b$.

$$|b| = V_{REF}\frac{R_F}{R_{G2}} \tag{4.58}$$

The design specifications for an example circuit are: $V_{OUT} = 1$ V at $V_{IN} = -0.1$ V, $V_{OUT} = 5$ V at $V_{IN} = -0.3$ V, $V_{REF} = V_{CC} = 5$ V, $R_L = 10$ kΩ, and 5% resistor tolerances. The simultaneous equations, Eqs. (4.59) and (4.60), are written below.

$$1 = (-0.1)m + b \tag{4.59}$$
$$5 = (-0.3)m + b \tag{4.60}$$

From these equations we find that $b = -1$ and $m = -20$. Setting the magnitude of m equal to Eq. (4.57) yields Eq. (4.61).

$$|m| = 20 = \frac{R_F}{R_{G1}} \tag{4.61}$$

$$R_F = 20R_{G1} \tag{4.62}$$

Let $R_{G1} = 1$ kΩ, and then $R_F = 20$ kΩ.

$$|b| = V_{CC}\left(\frac{R_F}{R_{G1}}\right) = 5\left(\frac{R_F}{R_{G2}}\right) = 1 \tag{4.63}$$

$$R_{G2} = \frac{R_F}{0.2} = \frac{20}{0.2} = 100 \text{ k}\Omega \tag{4.64}$$

The final equation for this example is given in Eq. (4.63).

$$V_{OUT} = -20V_{IN} - 1 \tag{4.65}$$

The final circuit is shown in Fig. 4.18 and the measured transfer curve for this circuit is shown in Fig. 4.19.

The TLV247X was used to build the test circuit because of its wide dynamic range. The transfer curve plots very close to the theoretical curve.

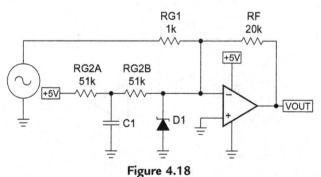

Figure 4.18
Case 4 example circuit.

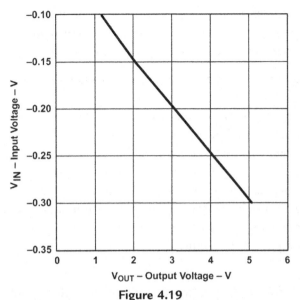

Figure 4.19

Case 4 example circuit measured transfer curve.

As in Case 3, as long as the circuit works normally there are no problems handling the negative voltage input to the circuit because the inverting lead of the TLV247X is at a positive voltage. The positive op amp input lead is grounded, and normal op amp operation keeps the inverting op amp input lead at ground because of the assumption that the error voltage is zero. When V_{CC} is powered down while there is a negative voltage on the inverting op amp input lead there is a possibility of circuit damage.

The most prudent solution is to connect the diode, D_1, with its cathode on the inverting op amp input lead and its anode at ground. If a negative voltage gets on the inverting op amp input lead it is clamped to ground by the diode. Select the diode type as germanium or Schottky, so the voltage drop across the diode is about 200 mV; this small voltage does not harm most op amp inputs. R_{G2} is split into two resistors ($R_{G2A} = R_{G2B} = 51\ k\Omega$) with a capacitor inserted at the junction of the two resistors. This decouples V_{CC}.

4.5 Summary

Single-supply op amp design is more complicated than split-supply op amp design, but with a logical design approach excellent results are achieved. Single-supply design used to be considered technically limiting because older op amps had limited capability. Op amps such as the TLC247X, TLC07X, and TLC08X have excellent single-supply parameters; thus when used in the correct applications these op amps yield rail-to-rail performance equal to their split-supply counterparts.

Single-supply op amp design usually involves some form of biasing, and this requires more thought, so single-supply op amp design needs discipline and a procedure. The recommended design procedure for single-supply op amp design is as follows:

- Substitute the specification data into simultaneous equations to obtain m and b (the slope and intercept of a straight line).
- Let m and b determine the form of the circuit.
- Choose the circuit configuration that fits the form.
- Using the circuit equations for the circuit configuration selected, calculate the resistor values.
- Build the circuit, take data, and verify performance.
- Test the circuit for nonstandard operating conditions (circuit power off while interface power is on, over/under range inputs, etc.).
- Add protection components as required.
- Retest.

When this procedure is followed, good results follow. As single-supply circuit designers expand their horizon, new challenges require new solutions. Remember, the only equation a linear op amp can produce is the equation of a straight line. That equation only has four forms. The new challenges may consist of multiple inputs, common-mode voltage rejection, or something different, but this method can be expanded to meet these challenges.

CHAPTER 5

On Beyond Case 4

5.1 A Continuum of Applications

The previous chapter presented four cases. Other configurations like inverting and noninverting gain with no offset have also been covered in previous chapters. These comprise the vast majority of applications. During my tenure as an applications engineer, I supported customers with real world requirements, and there are some applications that were not addressed in the preceding material. In particular, applications which required attenuation, I had thought this would be a little used application. This proved not to be the case! An attendee to our initial book launch lecture series related a tale of woe—where a unity gain circuit worked perfectly, a gain of 1/10 was unstable, and a gain of 1/100 oscillated uncontrollably! As two of the authors of this book contemplated this dilemma, we began to realize that the cases with no offset (inverting and noninverting gain) are nothing more than cases of slopes with intercepts (b) of zero. The cases 1—4 in the previous chapter define the cases with gains >1, but not cases of attenuation with or without an offset. We also realized that even voltage regulators belong in a continuum of op amp applications, because they are offsets (b) with no slope (m).

This is best shown in the following table, which lists what has been covered thus far, and what has yet to be discussed (Table 5.1). For the sake of brevity, cases with a negative reference will not be covered; there are very few negative voltage references manufactured

Table 5.1: The Gain and Offset Matrix

		b < 0	b = 0	b > 0
Noninverting	m > 1	Case 2 (Section 4.4.2)	Noninverting gain (Section 2.3)	Case 1 (Section 4.4.1)
	m = 1	Section 5.4	Noninverting buffer (Section 5.7)	
	m < 1		Section 5.2	Section 5.3
	m = 0	Negative reference or regulator (Chapter 21)	Ground	Positive reference or regulator (Chapter 20)
Inverting	m < −1	Section 5.7	Section 5.5	Section 5.6
	m ≥ −1	Case 4 (Section 4.4.4)	Inverting gain (Section 2.4)	Case 3 (Section 4.4.3)

Op Amps for Everyone. http://dx.doi.org/10.1016/B978-0-12-811648-7.00005-4

or utilized. Also, for the sake of brevity, the equations will not be derived as they were in Chapter 4. They were all developed, however, using the same voltage divider, superposition, and other laws covered in previous chapters.

Clearly there is some work yet to be done, particularly in the cases of attenuation. For the designer to have a complete set of tools to understand every combination of gain and offset, they need to know more than the and applications that they are most familiar with (inverting and noninverting gain, and noninverting buffers). The four cases presented in Chapter 4 have supplemented the familiar basic circuits, and this chapter presents the rest of the cases.

5.2 Noninverting Attenuator With Zero Offset

The simplest of all the new cases is that of noninverting attenuation. It is done by building on the voltage divider principle (Fig. A.5), and adding a unity gain op amp buffer (Fig. 5.1).

5.3 Noninverting Attenuation With Positive Offset

The case of noninverting attenuation with positive offset is a minor variation of the noninverting attenuator, in this case adding a second input for the reference, which is also attenuated by the voltage divider law (Fig. 5.2).

5.4 Noninverting Attenuation With Negative Offset

The case of noninverting attenuation with negative offset is another slight variation of the noninverting buffer. In this case, instead of applying the reference by superposition to the noninerting input, the reference is applied to the inverting input through an inverting gain stage. The only limitation is that the gain on the reference must be equal to or greater than the stable bandwidth of the op amp (Fig. 5.3).

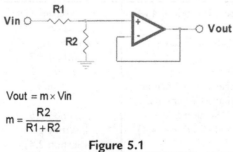

$$Vout = m \times Vin$$
$$m = \frac{R2}{R1+R2}$$

Figure 5.1
Noninverting attenuator.

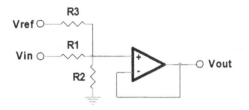

$$Vout = m \times Vin + b$$

$$m = \frac{1/R1}{1/R1 + 1/R2 + 1/R3}$$

$$b = Vref \times \frac{1/R3}{1/R1 + 1/R2 + 1/R3}$$

Figure 5.2
Noninverting attenuation with positive offset.

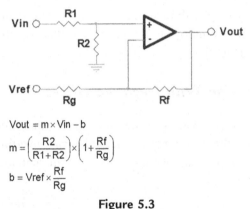

$$Vout = m \times Vin - b$$

$$m = \left(\frac{R2}{R1 + R2}\right) \times \left(1 + \frac{Rf}{Rg}\right)$$

$$b = Vref \times \frac{Rf}{Rg}$$

Figure 5.3
Noninverting attenuation with negative offset.

5.5 Inverting Attenuation With Zero Offset

This is the most often misdesigned circuit of all the cases. Many inexperienced designers create an unstable stage by attempting to extrapolate an inverting gain stage to the attenuation case by making R_G greater than R_F. The easiest fix for this problem is to use a voltage divider followed by a unity gain buffer as described in Section 5.2. If inverting gain is absolutely required, then a similar solution can be implemented by adding a voltage divider to the input of an inverting gain stage.

The balance here is simple enough to understand. R_{IN} is split into $R_{IN}A$ and $R_{IN}B$, the sum of which cannot be greater than R_F. By the addition of Ratten, the effective attenuation of the stage can be any value desired, while the gain of the stage from $R_{IN}B$ and R_F is always between a gain of 1 and 2 (Fig. 5.4).

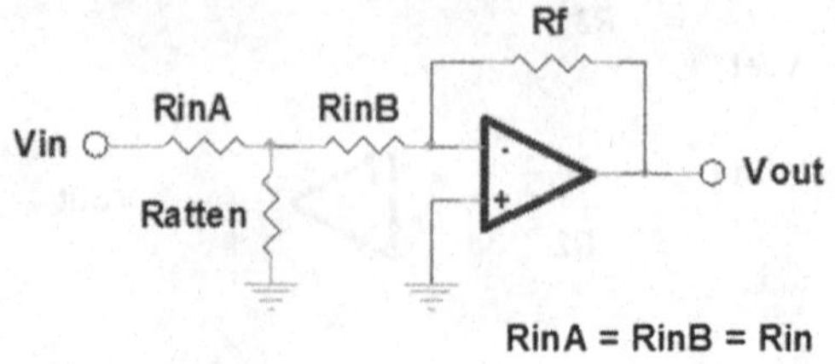

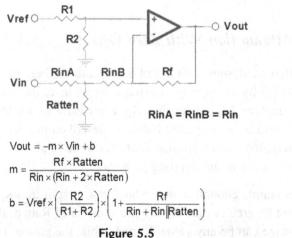

$$\text{Vout} = -m \times \text{Vin}$$

$$m = \frac{\text{Rf} \times \text{Ratten}}{\text{Rin} \times (\text{Rin} + 2 \times \text{Ratten})}$$

Figure 5.4

Inverting attenuation with zero offset.

5.6 Inverting Attenuation With Positive Offset

If positive offset is needed along with inverting attenuation, a combination of the previous section and Section 5.2 can be employed. Just remember that the offset is also attenuated by the same factor as V_{IN} (Fig. 5.5).

5.7 Inverting Attenuation With Negative Offset

If negative offset is needed along with inverting attenuation, the reference can be added to the inverting input using a voltage summation method at the inverting input. Again, it is important to have a gain of more than one on the V_{REF} gain channel, or instability may result (Fig. 5.6).

$$\text{Vout} = -m \times \text{Vin} + b$$

$$m = \frac{\text{Rf} \times \text{Ratten}}{\text{Rin} \times (\text{Rin} + 2 \times \text{Ratten})}$$

$$b = \text{Vref} \times \left(\frac{\text{R2}}{\text{R1} + \text{R2}} \right) \times \left(1 + \frac{\text{Rf}}{\text{Rin} + \text{Rin} \| \text{Ratten}} \right)$$

Figure 5.5

Noninverting attenuation with positive offset.

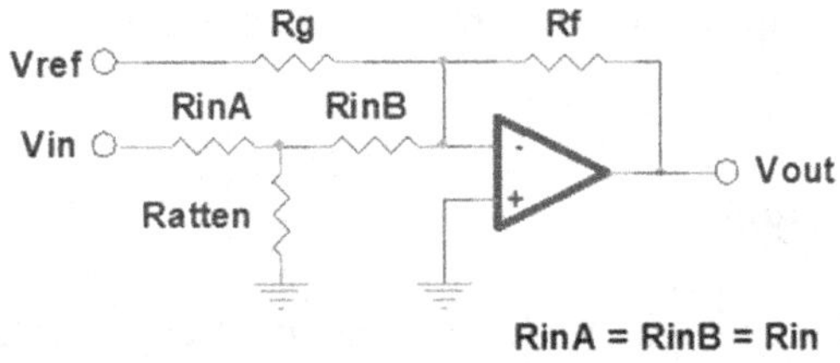

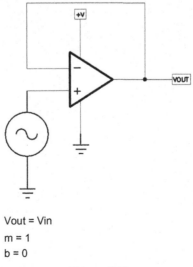

$$Vout = -m \times Vin - b$$

$$m = \dfrac{Rf \times Ratten}{Rin \times (Rin + 2 \times Ratten)}$$

$$b = Vref \times \dfrac{Rf}{Rg}$$

Figure 5.6
Noninverting attenuation with negative offset.

5.8 Noninverting Buffer

The case of a noninverting buffer is a special case of the noninverting (with zero offset) gain circuit. When the gain is determined by $1 + R_F/R_G$, the lowest gain you can possibly achieve out of a noninverting gain circuit is one. You would obtain this by making $R_G \gg R_F$, in effect infinite. This implies open, or not present in the circuit. The value of R_F becomes unimportant, so it can just be made zero. There are reasons, however, why you might want to put a resistor across R_F. The primary reason being so that, in the future, if there is a change to the circuit, an R_G can be easily added to the board. Other reasons might include power consumption or stability (Fig. 5.7). In the

$$Vout = Vin$$

$$m = 1$$

$$b = 0$$

Figure 5.7
Inverting unity gain buffer.

case of current feedback amplifiers—covered in a later chapter—R_F is absolutely required!

5.9 Signal Chain Design

The circuits above should give a way to design just about any interface circuit that the designer needs. The primary use of these gain/attenuation circuits, with or without offset, is the interface between an input voltage and a data converter. The primary use of negative reference/regulation circuits is to provide a steady and clean source of power. Do not worry; however, our old friend, the op amp, plays a prominent role in the humble voltage regulator!

CHAPTER 6

Feedback and Stability Theory

6.1 Introduction to Feedback Theory

With this chapter, we begin a quest for understanding. There is a specific question that comes up over and over again in interaction with customers and other engineers: "Why can't I make an inverting attenuator by making $R_G > R_F$?" The answer to this question is technical. It is difficult. It requires a solid knowledge of the topics to follow this in the next chapter. Most importantly, it requires knowledge of feedback theory. The inverting attenuator is not only the topic covered in these chapters, however. Inverting attenuators are just one aspect of the larger topic of op amp stability. This where we begin to introduce the characteristics of real-world op amps in earnest!

The gain of all op amps decreases as frequency increases, and the decreasing gain results in decreasing accuracy as the ideal op amp assumption ($a \Rightarrow \infty$) breaks down. In most real op amps the open-loop gain starts to decrease before 10 Hz, so an understanding of feedback is required to predict the closed-loop performance of the op amp. The real-world application of op amps is feedback controlled and depends on op amp open-loop gain at a given frequency. A designer must know the theory to be able to predict the circuit response regardless of frequency or open-loop gain.

Ideal op amp circuits can be designed without knowledge of feedback analysis tools, but these circuits are limited to low frequencies. Also, an understanding of feedback analysis tools is required to understand why effects like ringing and oscillations occur.

6.2 Block Diagram Math and Manipulations

Electronic systems and circuits are often represented by block diagrams, and block diagrams have a unique algebra and set of transformations [1]. Block diagrams are used because they are a shorthand pictorial representation of the cause-and-effect relationship between the input and output in a real system. They are a convenient method for characterizing the functional relationships between components. It is not necessary to understand the functional details of a block to manipulate a block diagram.

Op Amps for Everyone. http://dx.doi.org/10.1016/B978-0-12-811648-7.00006-6

The input impedance of each block is assumed to be infinite to preclude loading. Also, the output impedance of each block is assumed to be zero to enable high fan-out. The systems designer sets the actual impedance levels, but the fan-out assumption is valid because the block designers adhere to the system designer's specifications. All blocks multiply the input times the block quantity (see Fig. 6.1) unless otherwise specified within the block. The quantity within the block can be a constant as shown in Fig. 6.1C, or it can be a complex math function involving Laplace transforms. The blocks can perform time-based operations such as differentiation and integration.

Adding and subtracting are done in special blocks called summing points. Fig. 6.2 gives several examples of summing points. Summing points can have unlimited inputs, can add or subtract, and can have mixed signs yielding addition and subtraction within a single summing point. Fig. 6.3 defines the terms in a typical control system, and Fig. 6.4 defines the terms in a typical electronic feedback system. Multiloop feedback systems (Fig. 6.5) are intimidating, but they can be reduced to a single loop feedback system, as shown in the figure, by writing equations and solving for V_{OUT}/V_{IN}. An easier method for reducing multiloop feedback systems to single-loop feedback systems is to follow the rules and use the transforms given in Fig. 6.6.

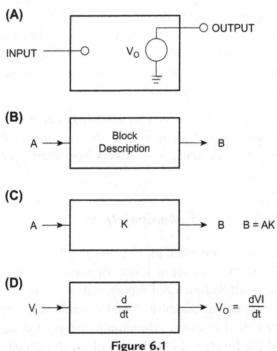

Figure 6.1

Definition of blocks. (A) Input/output impedance (B) signal flow arrows (C) block multiplication (D) blocks perform functions as indicated.

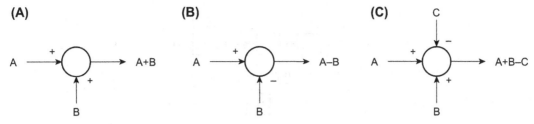

Figure 6.2

Summary points. (A) Additive summary point (B) subtractive summary point (C) multiple input summary points.

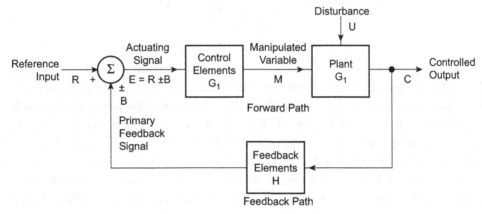

Figure 6.3

Definition of control system terms.

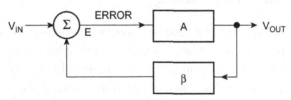

Figure 6.4

Definition of an electronic feedback circuit.

The following are block diagram reduction rules:

- Combine cascade blocks
- Combine parallel blocks
- Eliminate interior feedback loops
- Shift summing points to the left
- Shift takeoff points to the right
- Repeat until canonical form is obtained

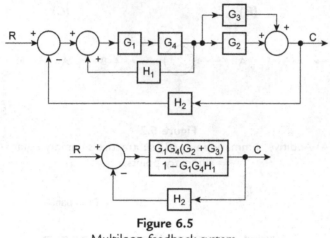

Figure 6.5
Multiloop feedback system.

Fig. 6.6 gives the block diagram transforms. The idea is to reduce the diagram to its canonical form because the canonical feedback loop is the simplest form of a feedback loop, and its analysis is well documented. All feedback systems can be reduced to the canonical form, so all feedback systems can be analyzed with the same math. A canonical loop exists for each input to a feedback system; although the stability dynamics are independent of the input, the output results are input dependent. The response of each input of a multiple input feedback system can be analyzed separately and added through superposition.

6.3 Feedback Equation and Stability

Fig. 6.7 shows the canonical form of a feedback loop with control system and electronic system terms. The terms make no difference except that they have meaning to the system engineers, but the math does have meaning, and it is identical for both types of terms. The electronic terms and negative feedback sign are used in this analysis, because subsequent chapters deal with electronic applications. The output equation is written in Eq. (6.1).

$$V_{OUT} = EA \tag{6.1}$$

The error equation is written in Eq. (6.2).

$$E = V_{IN} - \beta V_{OUT} \tag{6.2}$$

Combining Eqs. (6.1) and (6.2) yields Eq. (6.3).

$$\frac{V_{OUT}}{A} = V_{IN} - \beta V_{OUT} \tag{6.3}$$

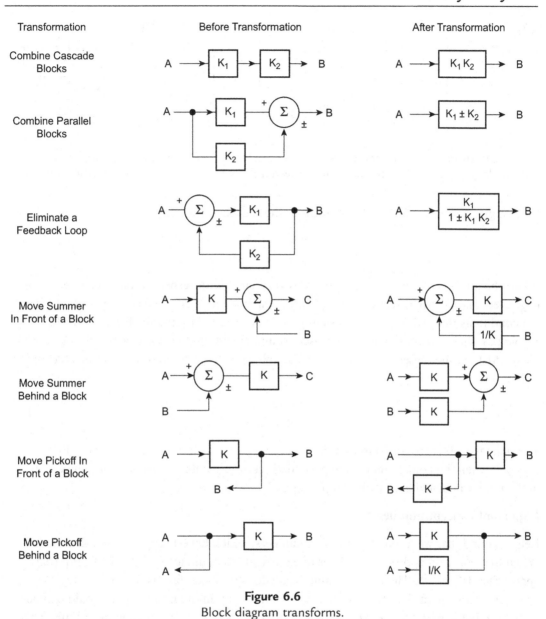

Figure 6.6
Block diagram transforms.

Collecting terms yields Eq. (6.4).

$$V_{OUT}\left(\frac{1}{A} + \beta\right) = V_{IN} \tag{6.4}$$

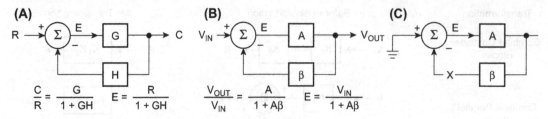

Figure 6.7
Comparison of control and electronic canonical feedback systems. (A) Control system terminology (B) electronics terminology (C) feedback loop is broken to calculate the loop gain.

Rearranging terms yields the classic form of the feedback Eq. (6.5).

$$\frac{V_{OUT}}{V_{IN}} = \frac{A}{1 + A\beta} \tag{6.5}$$

When the quantity $A\beta$ in Eq. (6.5) becomes very large with respect to one, the one can be neglected, and Eq. (6.5) reduces to Eq. (6.6), which is the ideal feedback equation. Under the conditions that $A\beta \gg 1$, the system gain is determined by the feedback factor β. Stable passive circuit components are used to implement the feedback factor, thus in the ideal situation, the closed-loop gain is predictable and stable because β is predictable and stable.

$$\frac{V_{OUT}}{V_{IN}} = \frac{1}{\beta} \tag{6.6}$$

The quantity $A\beta$ is so important that it has been given a special name: loop gain. In Fig. 6.7, when the voltage inputs are grounded (current inputs are opened) and the loop is broken, the calculated gain is the loop gain, $A\beta$.

Important concept number 1:

Keep in mind that we are using complex numbers, which have magnitude and direction. When the loop gain approaches -1, or to express it mathematically $1\angle -180°$, Eq. (6.5) approaches $1/0 \Rightarrow \infty$. The circuit output heads for infinity as fast as it can use the equation of a straight line. If the output were not energy limited, the circuit would explode the world, but happily, it is energy limited, so somewhere it comes up against a limit. This limit is either the voltage rail of the op amp or an uncontrolled oscillation—as the circuit has plenty of energy to work with out of the power supply. Many an op amp has burned up as it sucks more and more power from the power supply, exceeding the power rating of its output transistors! What makes the loop gain $A\beta$ have a magnitude of $1\angle -180°$? It is the presence of capacitors—internal and external to the op amp! Be patient, we will explain these capacitors more later.

Active devices in electronic circuits exhibit nonlinear phenomena when their output approaches a power supply rail, and the nonlinearity reduces the gain to the point where the loop gain no longer equals $1 \angle -180°$. Now the circuit can do two things: first, it can become stable at the power supply limit, or second, it can reverse direction (because stored charge keeps the output voltage changing) and head for the negative power supply rail.

The first state where the circuit becomes stable at a power supply limit is named lockup; the circuit will remain in the locked up state until power is removed and reapplied. The second state where the circuit bounces between power supply limits is named oscillatory. Remember, the loop gain, $A\beta$, is the sole factor determining stability of the circuit or system. Inputs are grounded or disconnected, so they have no bearing on stability.

Eqs. (6.1) and (6.2) are combined and rearranged to yield Eq. (6.7), which is the system or circuit error equation.

$$E = \frac{V_{IN}}{1 + A\beta} \tag{6.7}$$

First, notice that the error is proportional to the input signal. This is the expected result because a bigger input signal results in a bigger output signal, and bigger output signals require more drive voltage. As the loop gain increases, the error decreases, thus large loop gains are attractive for minimizing errors.

6.4 Bode Analysis of Feedback Circuits

H.W. Bode developed a quick, accurate, and easy method of analyzing feedback amplifiers, and he published a book about his techniques in 1945 [2]. Operational amplifiers were in their infancy and still have limited applications when Bode published his book, but they fall under the general classification of feedback amplifiers and are easily analyzed with Bode techniques. The mathematical manipulations required to analyze a feedback circuit are complicated because they involve multiplication and division. Bode developed the Bode plot, which simplifies the analysis through the use of graphical techniques.

The Bode equations are log equations that take the form $20 \, \mathrm{Log}(F(t)) = 20 \, \mathrm{Log}(|F(t)|) +$ phase angle. Terms that are normally multiplied and divided can now be added and subtracted because they are log equations. The addition and subtraction is done graphically, thus easing the calculations and giving the designer a pictorial representation of circuit performance. Eq. (6.8) is written for the low-pass filter shown in Fig. 6.8.

$$\frac{V_{OUT}}{V_{IN}} = \frac{\frac{1}{Cs}}{R + \frac{1}{Cs}} = \frac{1}{1 + RCs} = \frac{1}{1 + \tau s} \tag{6.8}$$

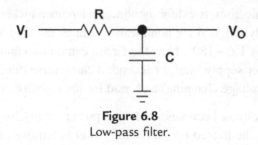

Figure 6.8
Low-pass filter.

where: $s = j\omega$, $j = \sqrt{(-1)}$, and $RC = \tau$.

The magnitude of this transfer function is $|V_{OUT}/V_{IN}| = 1\bigg/\sqrt{1^2 + (\tau\omega)^2}$. This

magnitude, $|V_{OUT}/V_{IN}| \cong 1$ when $\tau = 0.1/\tau$, it equals 0.707 when $\tau = 1/\tau$, and it is approximately $= 0.1$ when $\tau = 10/\tau$. These points are plotted in Fig. 6.9 using straight line approximations. The negative slope is -20 dB/decade or -6 dB/octave. The magnitude curve is plotted as a horizontal line until it intersects the breakpoint where $\tau = 1/\tau$. The negative slope begins at the breakpoint because the magnitude starts decreasing at that point. The gain is equal to 1 or 0 dB at very low frequencies, equal to 0.707 or -3 dB at the break frequency, and it keeps falling with a -20 dB/decade slope for higher frequencies.

The phase shift for the low-pass filter or any other transfer function is calculated with the aid of Eq. (6.9).

$$\phi = \tangent^{-1}\left(\frac{\text{Real}}{\text{Imaginary}}\right) = -\tangent^{-1}\left(\frac{\omega\tau}{1}\right) \qquad (6.9)$$

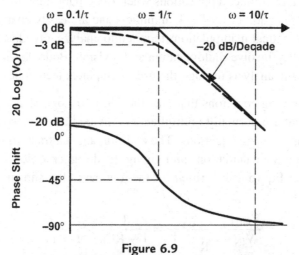

Figure 6.9
Bode plot of low-pass filter transfer function.

The phase shift is much harder to approximate because the tangent function is nonlinear. Normally the phase information is only required around the 0 dB intercept point for an active circuit, so the calculations are minimized. The phase is shown in Fig. 6.9, and it is approximated by remembering that the tangent of 90 degrees is 1, the tangent of 60 degrees is $\sqrt{3}$, and the tangent of 30 degrees is $\sqrt{3}/3$.

A breakpoint occurring in the denominator is called a pole, and it slopes down. Conversely, a breakpoint occurring in the numerator is called a zero, and it slopes up. When the transfer function has multiple poles and zeros, each pole or zero is plotted independently, and the individual poles/zeros are added graphically. If multiple poles, zeros, or a pole/zero combination have the same breakpoint, they are plotted on top of each other. Multiple poles or zeros cause the slope to change by multiples of 20 dB/ decade.

An example of a transfer function with multiple poles and zeros is a band reject filter (see Fig. 6.10). The transfer function of the band reject filter is given in Eq. (6.10).

$$G = \frac{V_{OUT}}{V_{IN}} = \frac{(1 + \tau s)(1 + \tau s)}{2\left(1 + \frac{\tau s}{0.44}\right)\left(1 + \frac{\tau s}{4.56}\right)} \tag{6.10}$$

The pole zero plot for each individual pole and zero is shown in Fig. 6.11, and the combined pole zero plot is shown in Fig. 6.12.

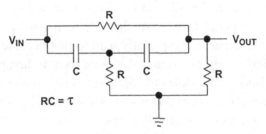

Figure 6.10
Band reject filter.

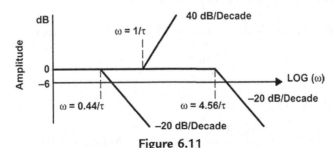

Figure 6.11
Individual pole zero plot of band reject filter.

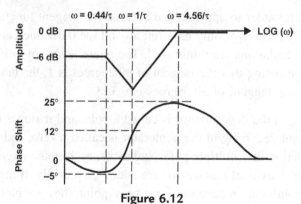

Figure 6.12

Combined pole zero plot of band reject filter.

The individual pole zero plots show the DC gain of 1/2 plotting as a straight line from the −6 dB intercept. The two zeros occur at the same break frequency, thus they add to a 40-dB/decade slope. The two poles are plotted at their breakpoints of $\tau = 0.44/\tau$ and $\tau = 4.56/\tau$. The combined amplitude plot intercepts the amplitude axis at −6 dB because of the DC gain, and then breaks down at the first pole. When the amplitude function gets to the double zero, the first zero cancels out the first pole, and the second zero breaks up. The upward slope continues until the second pole cancels out the second zero, and the amplitude is flat from that point out in frequency.

When the separation between all the poles and zeros is great, a decade or more in frequency, it is easy to draw the Bode plot. As the poles and zeros get closer together, the plot gets harder to make. The phase is especially hard to plot because of the tangent function, but picking a few salient points and sketching them in first gets a pretty good approximation [3]. The Bode plot enables the designer to get a good idea of pole zero placement, and it is valuable for fast evaluation of possible compensation techniques. When the situation gets critical, accurate calculations must be made and plotted to get an accurate result.

6.5 Bode Analysis Applied to Op Amps

First, let us apply Bode analysis to an ideal op amp. Consider Eq. (6.11).

$$\frac{V_{OUT}}{V_{IN}} = \frac{A}{1 + A\beta} \tag{6.11}$$

Taking the log of Eq. (6.11) yields Eq. (6.12).

$$20 \, Log\left(\frac{V_{OUT}}{V_{IN}}\right) = 20 \, Log(A) - 20 \, Log(1 + A\beta) \tag{6.12}$$

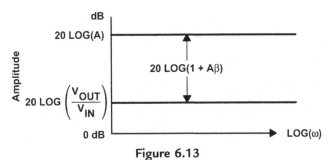

Figure 6.13
The ideal op amp Bode plot, when no pole exists in Eq. (6.12).

If A and β do not contain any poles or zeros, there will be no breakpoints. Then the Bode plot of Eq. (6.12) looks like that shown in Fig. 6.13, and because there are no poles to contribute negative phase shift, the circuit cannot oscillate.

Now, will add in the characteristics of real-world amplifiers. All real amplifiers have many poles, caused by multiple internal capacitances. The original IC operational amplifier, the μA709, as described in Chapter 1, was of this type, and as the internal parasitic poles accumulated, the forced the loop gain Aβ to quickly assume the value $1 \angle -180°$, which led to instability. Surprisingly, an uncompensated amplifier can have a fairly high bandwidth, but the designers using it were forced to navigate the minefield of instability to take advantage of that bandwidth.

To create a "user-friendly" op amp, IC designers knew they could not eliminate parasitic internal capacitances. If you cannot beat them, join them! The solution was to intentionally create a pole in the op amp response—a pole large enough to swamp (and mask) the effects of the internal poles. This created an internally compensated amplifier that appears to have a single pole. The first such op amp was the μA741, and it was a commercial success! Even inexperienced designers could successfully design inverting and noninverting gain stages—the bulk of the applications for op amps. The world was safe for inexperienced analog designers once more!

A compensated op amp has an equation similar to that given in Eq. (6.13).

$$A = \frac{a}{1 + j\dfrac{\omega}{\omega_a}} \tag{6.13}$$

The Bode plot for this single compensation pole op amp is shown in Fig. 6.14.

The amplifier gain, A, intercepts the amplitude axis at 20 Log(A), and it breaks down at a slope of −20 dB/decade at $\omega = \omega_a$. Ideally, the negative slope continues for all frequencies greater than the breakpoint, $\omega = \omega_a$. Unfortunately, this is not the case.

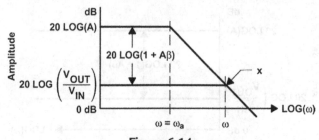

Figure 6.14
When Eq. (6.12) has a single pole.

Important concept number 2:

Let us take a break and take stock of where we are for a moment. We have a real-world op amp, such as the μA741, with an intentional, single, internal pole that dominates the response, given the Bode plot a downward turn at 20 dB/decade. This slope intercepts the 0 dB point on the horizontal axis, determining what is known as the unity gain point of the op amp, and ultimately this is the advertised bandwidth of the op amp. Now, the meat of this important concept: the effects of the intentional dominant pole go away after the unity gain intercept on the Bode plot. At gains less than unity, internal parasitic capacitances accumulate rapidly and soon cause the loop gain Aβ to have a $1 \angle -180°$ in the denominator!

This is where most designers run into problems. If you have one dominant pole at unity gain, the phase shift at unity gain is only $\angle -90°$ (the effect of one capacitor), and the circuit is stable. But it is headed toward instability, with unity gain being the LEAST stable operation point of the op amp. As the Bode plot of Fig. 6.14 continues, you are not guaranteed a single pole. Other poles come into play, thus gains less than unity become unstable, and rapidly!

Exceptions to the rule:

There are a few op amps on the market that are undercompensated, that is the dominant pole crosses at a point above the 0 dB line of the Bode Plot. This is done to extend the bandwidth of the amplifier, but at the cost of unity gain stability.

Let us move on and get a bit more detailed. The closed-loop circuit gain intercepts the amplitude axis at 20 Log(V_{OUT}/V_{IN}), and because ω does not have any poles or zeros, it is constant until its projection intersects the amplifier gain at point X. After intersection with the amplifier gain curve, the closed-loop gain follows the amplifier gain because the amplifier is the controlling factor.

Actually, the closed-loop gain starts to roll off earlier, and it is down 3 dB at point X. At point X the difference between the closed-loop gain and the amplifier gain is −3 dB, thus according to Eq. (6.12) the term $-20 \, \text{Log}(1 + A\beta) = -3$ dB. The magnitude of 3 dB is $\sqrt{2}$, hence $\sqrt{1 + (A\beta)^2} = \sqrt{2}$, and elimination of the radicals shows that $A\beta = 1$. There is a method [4] of relating phase shift and stability to the slope of the closed-loop gain curves, but only the Bode method is covered here. An excellent discussion of poles, zeros, and their interaction is given by M.E. Van Valkenberg [5], and he also includes some excellent prose to liven the discussion.

6.6 Loop Gain Plots Are the Key to Understanding Stability

Stability is determined by the loop gain, and when $A\beta = -1 = |1| \angle -180°$ instability or oscillation occurs. If the magnitude of the gain exceeds one, it is usually reduced to one by circuit nonlinearities, so oscillation generally results for situations where the gain magnitude exceeds one.

Consider oscillator design, which depends on nonlinearities to decrease the gain magnitude; if the engineer designed for a gain magnitude of one at nominal circuit conditions, the gain magnitude would fall below one under worst-case circuit conditions causing oscillation to cease. Thus, the prudent engineer designs for a gain magnitude of one under worst-case conditions knowing that the gain magnitude is much more than one under optimistic conditions. The prudent engineer depends on circuit nonlinearities to reduce the gain magnitude to the appropriate value, but this same engineer pays a price of poorer distortion performance. Sometimes a design compromise is reached by putting a nonlinear component, such as a lamp, in the feedback loop to control the gain without introducing distortion.

Some high gain control systems always have a gain magnitude greater than one, but they avoid oscillation by manipulating the phase shift. The amplifier designer, who pushes the amplifier for superior frequency performance, has to be careful not to let the loop gain phase shift accumulate to 180 degrees. Problems with overshoot and ringing pop up before the loop gain reaches 180 degrees phase shift, thus the amplifier designer must keep a close eye on loop dynamics. Ringing and overshoot are handled in the next section, so preventing oscillation is emphasized in this section. Eq. (6.14) has the form of many loop gain transfer functions or circuits, so it is analyzed in detail.

$$(A)\beta = \frac{(K)}{(1 + \tau_1(s))(1 + \tau_2(s))} \qquad (6.14)$$

The quantity, K, is the DC gain, and it plots as a straight line with an intercept of 20 Log(K). The Bode plot of Eq. (6.14) is shown in Fig. 6.15. The two breakpoints, $\tau = \tau_1 = 1/\tau_1$ and $\tau = \tau_2 = 1/\tau_2$, are plotted in the Bode plot. Each breakpoint adds -20 dB/decade slope to the plot, and 45 degrees phase shift accumulates at each

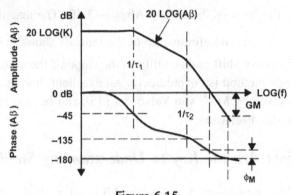

Figure 6.15
Magnitude and phase plot of Eq. (6.14).

breakpoint. This transfer function is referred to as a two slope because of the two breakpoints. The slope of the curve, when it crosses the 0 dB intercept, indicates phase shift and the ability to oscillate. Notice that a one-slope system can only accumulate 90 degrees phase shift, so when a transfer function passes through 0 dB with a one slope, it cannot oscillate. Furthermore, a two-slope system can accumulate 180 degrees phase shift; therefore, a transfer function with a two or greater slope is capable of oscillation.

A one slope crossing the 0 dB intercept is stable, whereas a two or greater slope crossing the 0 dB intercept may be stable or unstable depending upon the accumulated phase shift. Fig. 6.15 defines two stability terms: the phase margin, ϕ_M, and the gain margin, G_M. Of these two terms the phase margin is much more important because phase shift is critical for stability. Phase margin is a measure of the difference in the actual phase shift and the theoretical 180 degrees required for oscillation, and the phase margin measurement or calculation is made at the 0 dB crossover point. The gain margin is measured or calculated at the 180 degrees phase crossover point. Phase margin is expressed mathematically in Eq. (6.15).

$$\phi_M = 180 - \text{tangent}^{-1}(A\beta) \tag{6.15}$$

The phase margin in Fig. 6.15 is very small, 20 degrees, so it is hard to measure or predict from the Bode plot. A designer probably does not want a 20-degree phase margin because the system overshoots and rings badly, but this case points out the need to calculate small phase margins carefully. The circuit is stable, and it does not oscillate because the phase margin is positive. Also, the circuit with the smallest phase margin has the highest frequency response and bandwidth.

Increasing the loop gain to $(K + C)$ as shown in Fig. 6.16 shifts the magnitude plot up. If the pole locations are kept constant, the phase margin reduces to zero as shown, and the

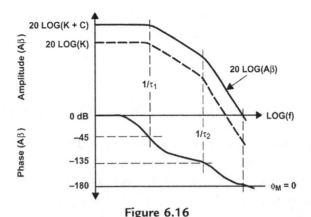

Figure 6.16
Magnitude and phase plot of the loop gain increased to $(K + C)$.

circuit will oscillate. The circuit is not good for much in this condition because production tolerances and worst case conditions ensure that the circuit will oscillate when you want it to amplify, and vice versa.

The circuit poles are spaced closer in Fig. 6.17, and this results in a faster accumulation of phase shift. The phase margin is zero because the loop gain phase shift reaches 180 degrees before the magnitude passes through 0 dB. This circuit oscillates, but it is not a very stable oscillator because the transition to 180 degrees phase shift is very slow. Stable oscillators have a very sharp transition through 180 degrees.

When the closed-loop gain is increased the feedback factor, β, is decreased because $V_{OUT}/V_{IN} = 1/\beta$ for the ideal case. This, in turn, decreases the loop gain, $A\beta$, thus the stability

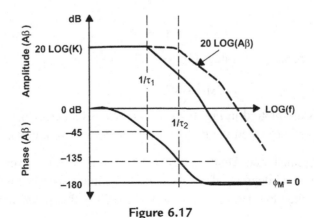

Figure 6.17
Magnitude and phase plot of the loop gain with pole spacing reduced.

increases. In other words, increasing the closed-loop gain makes the circuit more stable. Stability is not important except to oscillator designers because overshoot and ringing become intolerable to linear amplifiers long before oscillation occurs. The overshoot and ringing situation is investigated next.

6.7 The Second-Order Equation and Ringing/Overshoot Predictions

The second-order equation is a common approximation used for feedback system analysis because it describes a two-pole circuit, which is the most common approximation used. All real circuits are more complex than two poles, but except for a small fraction, they can be represented by a two-pole equivalent. The second-order equation is extensively described in electronic and control literature [6].

$$(1 + A\beta) = 1 + \frac{K}{(1 + \tau_1 s)(1 + \tau_2 s)} \tag{6.16}$$

After algebraic manipulation Eq. (6.16) is presented in the form of Eq. (6.17).

$$s^2 + s\frac{\tau_1 + \tau_2}{\tau_1 \tau_2} + \frac{1 + K}{\tau_1 \tau_2} = 0 \tag{6.17}$$

Eq. (6.17) is compared to the second-order control Eq. (6.18), and the damping ratio, ζ, and natural frequency, w_N, are obtained through like term comparisons.

$$s^2 + 2\zeta\omega_N s + \omega_N^2 \tag{6.18}$$

Comparing these equations yields formulas for the phase margin and percent overshoot as a function of damping ratio.

$$\omega_N = \sqrt{\frac{1 + K}{\tau_1 \tau_2}} \tag{6.19}$$

$$\xi = \frac{\tau_1 + \tau_2}{2\omega_N \tau_1 \tau_2} \tag{6.20}$$

When the two poles are well separated, Eq. (6.21) is valid.

$$\phi_M\_M\_tangent^{-2}(2\xi) \tag{6.21}$$

The salient equations are plotted in Fig. 6.18, which enables a designer to determine the phase margin and overshoot when the gain and pole locations are known.

Enter Fig. 6.18 at the calculated damping ratio, say 0.4, and read the overshoot at 25% and the phase margin at 42 degrees. If a designer had a circuit specification of 5% maximum overshoot, then the damping ratio must be 0.78 with a phase margin of 62 degrees.

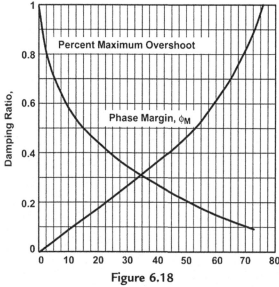

Figure 6.18

Phase margin and overshoot versus damping ratio.

References

[1] DiStefano, Stubberud, Williams, Theory and Problems of Feedback and Control Systems, Schaum's Outline Series, Mc Graw Hill Book Company, 1967.
[2] H.W. Bode, Network Analysis and Feedback Amplifier Design, D. Van Nostrand, Inc., 1945.
[3] T. Frederickson, Intuitive Operational Amplifiers, McGraw Hill Book Company, 1988.
[4] J.L. Bower, P.M. Schultheis, Introduction to the Design of Servomechanisms, Wiley, 1961.
[5] M.E. Van Valkenberg, Network Analysis, Prentice-Hall, 1964.
[6] V. Del Toro, S. Parker, Principles of Control Systems Engineering, McGraw–Hill, 1960.

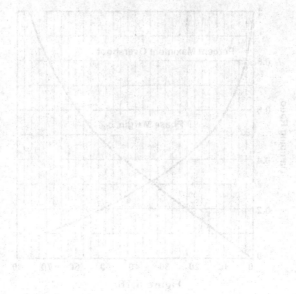

Figure 3.10c.

CHAPTER 7

Development of the Nonideal Op Amp Equations

7.1 Introduction

There are two types of error sources in op amps, and they fall under the general classification of DC and AC errors. Examples of DC errors are input offset voltage and input bias current. The DC errors stay constant over the usable op amp frequency range; therefore, the input bias current is 10 pA at 1 kHz and it is 10 pA at 10 kHz. Because of their constant and controlled behavior, DC errors are not considered until later chapters.

AC errors are flighty, so we address them here by developing a set of nonideal equations that account for AC errors. The AC errors may show up under DC conditions, but they get worse as the operating frequency increases. A good example of an AC error is common-mode rejection ratio (CMRR). Most op amps have a guaranteed CMRR specification, but this specification is only valid at DC or very low frequencies. Further inspection of the data sheet reveals that CMRR decreases as operating frequency increases. Several other specifications that fall into the category of AC specifications are output impedance, power supply rejection ratio, peak-to-peak output voltage, differential gain, differential phase, and phase margin.

Differential gain is the most important AC specification because the other AC specifications are derived from the differential gain. Until now, differential gain has been called op amp gain or op amp open-loop gain, and we shall continue with that terminology. Let the data sheet call it differential gain.

As shown in prior chapters, when frequency increases, the op amp gain decreases and errors increase. This chapter develops the equations that illustrate the effects of the gain changes. We start with a review of the basic canonical feedback system stability because the op amp equations are developed using the same techniques.

Amplifiers are built with active components such as transistors. Pertinent transistor parameters such as transistor gain are subject to drift and initial inaccuracies from many sources, so amplifiers being built from these components are subject to drift and inaccuracies. The drift and inaccuracy is minimized or eliminated by using negative feedback. The op amp circuit configuration employs feedback to make the transfer equation of the circuit independent of the amplifier parameters (well almost), and while doing this, the circuit transfer function is made dependent on external passive components. The external passive components can be purchased to meet almost any drift or accuracy specification; only the cost and size of the passive components limit their use.

Once feedback is applied to the op amp, it is possible for the op amp circuit to become unstable. Certain amplifiers belong to a family called internally compensated op amps; they contain internal capacitors that are sometimes advertised as precluding instabilities. Although internally compensated op amps should not oscillate when operated under specified conditions, many have relative stability problems that manifest themselves as poor phase response, ringing, and overshoot. The only absolutely stable internally compensated op amp is the one lying on the workbench without power applied! All other internally compensated op amps oscillate under some external circuit conditions.

Noninternally compensated or *externally* compensated op amps are unstable without the addition of external stabilizing components. This situation is a disadvantage in many cases because they require additional components, but the lack of internal compensation enables the top-drawer circuit designer to squeeze the last drop of performance from the op amp. You have two options: op amps internally compensated by the IC manufacturer or op amps externally compensated by you. Compensation, except that done by the op amp manufacturer, must be done external to the IC. Surprisingly enough, internally compensated op amps require external compensation for demanding applications.

Compensation is achieved by adding external components that modify the circuit transfer function so that it becomes unconditionally stable. There are several different methods of compensating an op amp, and as you might suspect, there are pros and cons associated with each method of compensation. After the op amp circuit is compensated, it must be analyzed to determine the effects of compensation. The modifications that compensation have on the closed-loop transfer function often determine which compensation scheme is most profitably employed.

7.2 Review of the Canonical Equations

A block diagram for a generalized feedback system is repeated in Fig. 7.1. This simple block diagram is sufficient to determine the stability of any system.

The output and error equation development is repeated below:

$$V_{OUT} = EA \tag{7.1}$$

$$E = V_{IN} = V_{OUT} \tag{7.2}$$

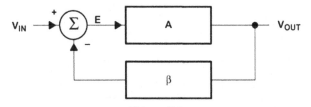

Figure 7.1
Feedback system block diagram.

Combining Eqs. (7.1) and (7.2) yields Eq. (7.3):

$$\frac{V_{OUT}}{A} = V_{IN} - \beta V_{OUT} \tag{7.3}$$

Collecting terms yields Eq. (7.4):

$$V_{OUT}\left(\frac{1}{A} + \beta\right) = V_{IN} \tag{7.4}$$

Rearranging terms yields the classic form of the feedback equation.

$$\frac{V_{OUT}}{V_{IN}} = \frac{A}{1 + A\beta} \tag{7.5}$$

Notice that Eq. (7.5) reduces to Eq. (7.6) when the quantity $A\beta$ in Eq. (7.5) becomes very large with respect to one. Eq. (7.6) is called the ideal feedback equation because it depends on the assumption that $A\beta \gg 1$, and it finds extensive use when amplifiers are assumed to have ideal qualities. Under the conditions that $A\beta \gg 1$, the system gain is determined by the feedback factor β. Stable passive circuit components are used to implement the feedback factor, thus the ideal closed-loop gain is predictable and stable because β is predictable and stable.

$$\frac{V_{OUT}}{V_{IN}} = \frac{1}{\beta} \tag{7.6}$$

The quantity $A\beta$ is so important that it has been given a special name, loop gain. Consider Fig. 7.2; when the voltage inputs are grounded (current inputs are opened) and the loop is broken, the calculated gain is the loop gain, $A\beta$. Now, keep in mind that this is a mathematics of complex numbers, which have magnitude and direction. When the loop gain approaches -1, or to express it mathematically $1 \angle -180$ degrees, Eq. (7.5) approaches infinity because $1/0 \Rightarrow \infty$. The circuit output heads for infinity as fast as it can use the equation of a straight line. If the output were not energy limited the circuit would explode the world, but it is energy limited by the power supplies so the world stays intact.

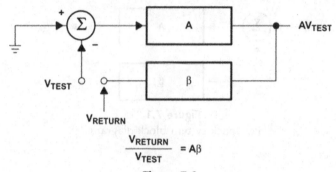

$$\frac{V_{RETURN}}{V_{TEST}} = A\beta$$

Figure 7.2
Feedback loop broken to calculate loop gain.

Active devices in electronic circuits exhibit nonlinear behavior when their output approaches a power supply rail, and the nonlinearity reduces the amplifier gain until the loop gain no longer equals 1 $\angle -180$ degrees. Now the circuit can do two things: first, it could become stable at the power supply limit or second, it can reverse direction (because stored charge keeps the output voltage changing) and head for the negative power supply rail.

The first state where the circuit becomes stable at a power supply limit is named lockup; the circuit will remain in the locked up state until power is removed. The second state where the circuit bounces between power supply limits is named oscillatory. Remember, the loop gain, $A\beta$, is the sole factor that determines stability for a circuit or system. Inputs are grounded or disconnected when the loop gain is calculated, so they have no effect on stability. The loop gain criteria are analyzed in depth later.

Eqs. (7.1) and (7.2) are combined and rearranged to yield Eq. (7.7), which gives an indication of system or circuit error.

$$E = \frac{V_{IN}}{1 + A\beta} \tag{7.7}$$

First, notice that the error is proportional to the input signal. This is the expected result because a bigger input signal results in a bigger output signal, and bigger output signals require more drive voltage. Second, the loop gain is inversely proportional to the error. As the loop gain increases the error decreases, thus large loop gains are attractive for minimizing errors. Large loop gains also decrease stability, thus there is always a trade-off between error and stability.

7.3 Noninverting Op Amps

A noninverting op amp is shown in Fig. 7.3. The dummy variable, V_B, is inserted to make the calculations easier, and a is the op amp gain.

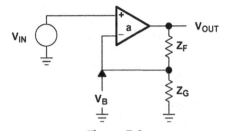

Figure 7.3
Noninverting op amp.

Eq. (7.8) is the amplifier transfer equation.

$$V_{OUT} = a(V_{IN} \pm V_B) \qquad (7.8)$$

The output equation is developed with the aid of the voltage divider rule. Using the voltage divider rule assumes that the op amp impedance is low.

$$V_B = \frac{V_{OUT}Z_G}{Z_F + Z_G} \quad \text{for } I_B = 0 \qquad (7.9)$$

Combining Eqs. (7.8) and (7.9) yields Eq. (7.10).

$$V_{OUT} = aV_{IN} - \frac{aZ_G V_{OUT}}{Z_G + Z_F} \qquad (7.10)$$

Rearranging terms in Eq. (7.10) yields Eq. (7.11), which describes the transfer function of the circuit.

$$\frac{V_{OUT}}{V_{IN}} = \frac{a}{1 + \dfrac{aZ_G}{Z_G + Z_F}} \qquad (7.11)$$

Eq. (7.5) is repeated as Eq. (7.12) to make a term by term comparison of the equations easy.

$$\frac{V_{OUT}}{V_{IN}} = \frac{A}{1 + A\beta} \qquad (7.12)$$

By virtue of the comparison we get Eq. (7.13), which is the loop gain equation for the noninverting op amp. The loop gain equation determines the stability of the circuit. The comparison also shows that the open-loop gain, A, is equal to the op amp open-loop gain, a, for the noninverting circuit.

$$A\beta = \frac{aZ_G}{Z_G + Z_F} \qquad (7.13)$$

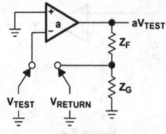

Figure 7.4
Open-loop noninverting op amp.

Eq. (7.13) is also derived with the aid of Fig. 7.4, which shows the open-loop noninverting op amp.

The test voltage, V_{TEST}, is multiplied by the op amp open-loop gain to obtain the op amp output voltage, aV_{TEST}. The voltage divider rule is used to calculate Eq. (7.15), which is identical to Eq. (7.14) after some algebraic manipulation.

$$V_{RETURN} = \frac{aV_{TEST}Z_G}{Z_F + Z_G} \qquad (7.14)$$

$$\frac{V_{RETURN}}{V_{TEST}} = A\beta = \frac{aZ_G}{Z_F + Z_G} \qquad (7.15)$$

7.4 Inverting Op Amps

The inverting op amp circuit is shown in Fig. 7.5. The dummy variable (VA) is inserted to make the calculations easier, and a is the op amp open-loop gain.

The transfer equation is given in Eq. (7.16):

$$V_{OUT} = -aV_A \qquad (7.16)$$

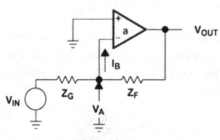

Figure 7.5
Inverting op amp.

The node voltage (Eq. 7.17) is obtained with the aid of superposition and the voltage divider rule. Eq. (7.18) is obtained by combining Eqs. (7.16) and (7.17).

$$V_A = \frac{V_{IN}Z_F}{Z_G + Z_F} + \frac{V_{OUT}Z_G}{Z_G + Z_F} \quad \text{for } I_B = 0 \tag{7.17}$$

$$\frac{V_{OUT}}{V_{IN}} = \frac{\dfrac{-aZ_F}{Z_G + Z_F}}{1 + \dfrac{aZ_G}{Z_G + Z_F}} \tag{7.18}$$

Eq. (7.16) is the transfer function of the inverting op amp. By virtue of the comparison between Eqs. (7.18) and (7.14), we get Eq. (7.15) again, which is also the loop gain equation for the inverting op amp circuit. The comparison also shows that the open-loop gain (A) is different from the op amp open-loop gain (a) for the noninverting circuit.

The inverting op amp with the feedback loop broken is shown in Fig. 7.6, and this circuit is used to calculate the loop gain given in Eq. (7.19).

$$\frac{V_{RETURN}}{V_{TEST}} = \frac{aZ_G}{Z_G + Z_F} = A\beta \tag{7.19}$$

Several things must be mentioned at this point in the analysis. First, the transfer functions for the noninverting and inverting Eqs. (7.13) and (7.18), are different. For a common set of Z_G and Z_F values, the magnitude and polarity of the gains are different. Second, the loop gain of both circuits, as given by Eqs. (7.15) and (7.19), is identical. Thus, the stability performance of both circuits is identical although their transfer equations are different. This makes the important point that *stability is not dependent on the circuit inputs*. Third, the A gain block shown in Fig. 7.1 is different for each op amp circuit. By comparison of Eqs. (7.5), (7.11), and (7.18), we see that $A_{NON-INV} = a$ and $A_{INV} = aZ_F \div (Z_G + Z_F)$.

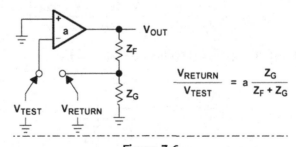

Figure 7.6

Inverting op amp: feedback loop broken for loop gain calculation.

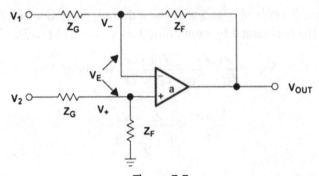

Figure 7.7
Differential amplifier circuit.

7.5 Differential Op Amps

The differential amplifier circuit is shown in Fig. 7.7. The dummy variable, V_E, is inserted to make the calculations easier, and a is the open-loop gain.

Eq. (7.20) is the circuit transfer equation.

$$V_{OUT} = aV_E = V_+ \pm V_- \tag{7.20}$$

The positive input voltage, V_+, is written in Eq. (7.21) with the aid of superposition and the voltage divider rule.

$$V_+ = V_2 \frac{Z_F}{Z_F + Z_G} \tag{7.21}$$

The negative input voltage, V_-, is written in Eq. (7.22) with the aid of superposition and the voltage divider rule.

$$V_- = V_1 \frac{Z_F}{Z_F + Z_G} - V_{OUT} \frac{Z_G}{Z_F + Z_G} \tag{7.22}$$

Combining Eqs. (7.20)–(7.22) yields Eq. (7.23).

$$V_{OUT} = a \left[\frac{V_2 Z_F}{Z_F + Z_G} - \frac{V_1 Z_F}{Z_F + Z_G} - \frac{V_{OUT} Z_G}{Z_F + Z_G} \right] \tag{7.23}$$

After algebraic manipulation, Eq. (7.23) reduces to Eq. (7.24).

$$\frac{V_{OUT}}{V_2 - V_1} = \frac{\dfrac{aZ_F}{Z_F + Z_G}}{1 + \dfrac{aZ_G}{Z_F + Z_G}} \tag{7.24}$$

The comparison method reveals that the loop gain as shown in Eq. (7.25) is identical to that shown in Eqs. (7.13) and (7.19).

$$A\beta = \frac{aZ_G}{Z_G + Z_F} \qquad (7.25)$$

Again, the loop gain, which determines stability, is only a function of the closed loop and independent of the inputs.

7.6 Are You Smarter Than an Op Amp?

The discussion above is important enough that I need to bring a point to your attention again before we leave this chapter. The real-world equations for the noninverting and inverting gain stages (and the differential as well) impose some limitations on how you apply op amps. You must be aware of these limitations or you will not get the gain you are expecting!

Let me reintroduce you to the open-loop Bode response of a typical voltage-feedback op amp (Fig. 7.8).

Notice that there is no frequency specified on the horizontal axis—this discussion applies to low-speed as well as high-speed op amps. You should consider three elements of the Bode plot.

- There is the open-loop response starting on the vertical gain axis, and sloping down to intercept the frequency axis. Consider this the op amp's "speed limit" at any frequency.

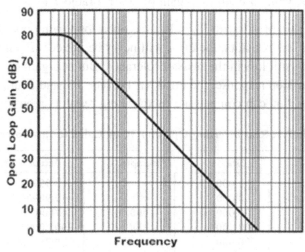

Figure 7.8
Bode response of a typical op amp.

- There is the region below and to the left of the open-loop response. This is the operating region of the op amp.
- There is the region above and to the right of the open-loop response. This is a region where you cannot operate an op amp.

Let me anger some semiconductor manufacturers and data sheet authors. If an open-loop response plot is not included in a data sheet, there is something to hide, i.e., not much of an operating region!

Repeating Eq. (7.11) for a noninverting op amp:

$$\frac{V_{OUT}}{V_{IN}} = \frac{a}{1 + \frac{aZ_G}{Z_G + Z_F}} \qquad (7.11)$$

And Eq. (7.18) for an inverting op amp stage:

$$\frac{V_{OUT}}{V_{IN}} = \frac{\frac{-aZ_F}{Z_G + Z_F}}{1 + \frac{aZ_G}{Z_G + Z_F}} \qquad (7.18)$$

Those equations are *ugly*! But when a ≫ the closed-loop gain at a given frequency, the approximations you are used to work well enough. In all practicality, you almost never have to worry about this limitation. But if you try to make a high-gain and/or a high-frequency amplifier stage, these equations explain why you will not get the gain you expected.

For the inverting case, with an open-loop gain of 80 dB (Table 7.1).

When you attempt a gain of −1 with $R_F = R_G$, the open-loop gain only contributes 0.02% error. If you are using 1% resistors, the error will never be noticed. Even at a gain of −10, the error is only 0.1%, still lost in the resistor tolerance. But at a gain of −100, fully 40 dB below the open-loop response, the error has become 1% and could be noticeable in precision applications. You can compensate by tweaking resistors a bit, but do not

Table 7.1: Real Inverting Op Amp Stage Gains for a = 80 dB

a	R_G	R_F	Attempted	Actual	Error (%)
10,000	100,000	100,000	−1	−0.9998	−0.0200
10,000	10,000	100,000	−10	−9.9890	−0.1099
10,000	1000	100,000	−100	−99.0001	−0.9999
10,000	100	100,000	−1000	−909.0083	−9.0992
10,000	10	100,000	−10,000	−4999.7500	−50.0025
10,000	1	100,000	−100,000	−9090.8264	−90.9092
10,000	1	1.00E + 12	−1E + 12	−9999.9999	−100

Table 7.2: Real Noninverting Op Amp Stage Gains for a = 80 dB

a	R_G	R_F	Attempted	Actual	Error (%)
10,000	100,000	100,000	2	1.9996	−0.0200
10,000	10,000	100,000	11	10.9879	−0.1099
10,000	1000	100,000	101	99.9901	−0.9999
10,000	100	100,000	1001	909.9173	−9.0992
10,000	10	100,000	10,001	5000.2500	−50.0025
10,000	1	100,000	100,001	9090.9174	−90.9092
10,000	1	1.00E + 12	1E + 12	9999.9999	−100

outsmart yourself! Any thermal drift at all will amplify the errors, which will be dominated by the lower value resistor R_G. The op amp bandwidth also has some thermal drift, changing the gain. By the time you attempt a gain of 10,000, the error balloons to 50% and no reasonable amount of tweaking will help. If you do something ridiculous like using 1 Ω for R_G and 1 TΩ for R_F, the highest gain the stage will provide is still less than −10,000. Just like the speed of light, you can theoretically get as close as you want to the open-loop response plot, but you can never achieve it.

This is a similar table for the noninverting case (Table 7.2).

This may all sound esoteric and theoretical, far removed from your experience, but keep it in the back of your mind when you are designing gain stages. This is the way op amps really behave, and simple gain expressions you are used to are only approximations. They can mislead you. To give you a simple guide, think of open-loop Bode plots in the following way (Fig. 7.9).

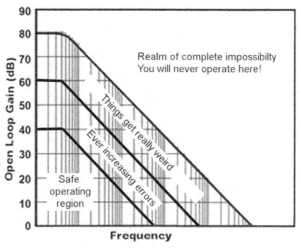

Figure 7.9
Bode response representation of safe operating region.

Notice, there is not a lot of safe (errors $\leq 1\%$) operating region. All of the sudden, that high bandwidth amplifier does not have such a high bandwidth, does it? Just remember that log frequency plot above extends indefinitely in decades to the left, so you may just have to choose a higher bandwidth amplifier. Also, the starting open-loop gain on this op loop plot is 80 dB. You can find op amps with open-loop gains of 120 dB or more. Another thing to remember is the safe operating region is for gain stages only! Things get even stranger for filter circuits. The author recently discovered that a 1 GHz op amp was completely inadequate for a 1 MHz band-pass filter with a high Q! That is three decades—a factor of 1000—in frequency!

CHAPTER 8

Voltage-Feedback Op Amp Compensation

8.1 Introduction

Voltage-feedback amplifiers (VFAs) have been with us for about 60 years, and they have been problems for circuit designers since the first day. You see, the feedback that makes them versatile and accurate also has a tendency to make them unstable. The operational amplifier (op amp) circuit configuration uses a high-gain amplifier whose parameters are determined by external feedback components. The amplifier gain is so high that without these external feedback components, the slightest input signal would saturate the amplifier output. The op amp is in common usage, so this configuration is examined in detail, but the results are applicable to many other voltage-feedback circuits. Current-feedback amplifiers (CFAs) are similar to VFAs, but the differences are important enough to warrant CFAs being handled separately.

Stability as used in electronic circuit terminology is often defined as achieving a nonoscillatory state. This is a poor, inaccurate definition of the word. Stability is a relative term, and this situation makes people uneasy because relative judgments are exhaustive. It is easy to draw the line between a circuit that oscillates and one that does not oscillate, so we can understand why some people believe that oscillation is a natural boundary between stability and instability.

Feedback circuits exhibit poor phase response, overshoot, and ringing long before oscillation occurs, and these effects are considered undesirable by circuit designers. This chapter is not concerned with oscillators; thus, relative stability is defined in terms of performance. By definition, when designers decide what trade-offs are acceptable, they determine what the relative stability of the circuit is. A relative stability measurement is the damping ratio (ζ), and the damping ratio. The damping ratio is related to phase margin, hence phase margin is another measure of relative stability. The most stable circuits have the longest response times, lowest bandwidth, highest accuracy, and least overshoot. The least stable circuits have the fastest response times, highest bandwidth, lowest accuracy, and some overshoot.

Op amps left in their native state oscillate without some form of compensation. The first IC op amps were very hard to stabilize, but there were a lot of good analog designers around in the 1960s, so we used them. Internally compensated op amps were introduced in the late 1960s in an attempt to make op amps easy for everyone to use. Unfortunately, internally compensated op amps sacrifice a lot of bandwidth and still oscillate under some conditions, so an understanding of compensation is required to apply op amps.

Internal compensation provides a worst-case trade-off between stability and performance. Uncompensated op amps require more attention, but they can do more work. Both are covered here.

Compensation is a process of applying a judicious patch in the form of an RC network to make up for a less than perfect op amp or circuit. There are many different problems that can introduce instability, thus there are many different compensation schemes.

8.2 Internal Compensation

Op amps are internally compensated to save external components and to enable their use by less knowledgeable people. It takes some measure of analog knowledge to compensate an analog circuit. Internally compensated op amps normally are stable when they are used in accordance with the applications instructions. Internally compensated op amps are not unconditionally stable. They are multiple pole systems, but they are internally compensated such that they appear as a single pole system over much of the frequency range. Internal compensation severely decreases the possible closed-loop bandwidth of the op amp.

Internal compensation is accomplished in several ways, but the most common method is to connect a capacitor across the collector–base junction of a voltage gain transistor (see Fig. 8.1). The Miller effect multiplies the capacitor value by an amount approximately equal to the stage gain, thus the Miller effect uses small value capacitors for compensation.

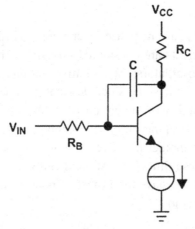

Figure 8.1
Miller effect compensation.

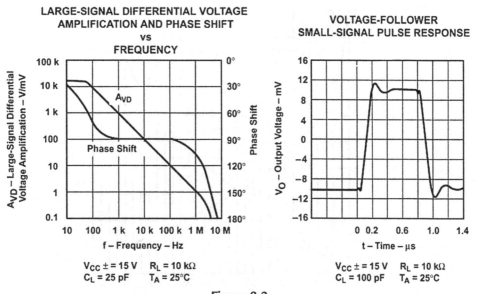

LARGE-SIGNAL DIFFERENTIAL VOLTAGE AMPLIFICATION AND PHASE SHIFT vs FREQUENCY

VOLTAGE-FOLLOWER SMALL-SIGNAL PULSE RESPONSE

$V_{CC}\pm = 15\,V \quad R_L = 10\,k\Omega$
$C_L = 25\,pF \quad T_A = 25°C$

$V_{CC}\pm = 15\,V \quad R_L = 10\,k\Omega$
$C_L = 100\,pF \quad T_A = 25°C$

Figure 8.2
TL03X frequency- and time-response plots.

Fig. 8.2 shows the gain/phase diagram for an older op amp (TL03X). When the gain crosses the 0 dB axis (gain equal to one) the phase shift is approximately 108 degrees, thus the op amp must be modeled as a second-order system because the phase shift is more than 90 degrees.

This yields a phase margin of $\phi = 180$ degrees−108 degrees = 72 degrees, thus the circuit should be very stable. Referring to Fig. 8.3, the damping ratio is one and the expected overshoot is zero. Fig. 8.2 shows approximately 10% overshoot, which is unexpected, but inspecting Fig. 8.2 further reveals that the loading capacitance for the two plots is different. The pulse response is loaded with 100 pF rather than 25 pF shown for the gain/phase plot, and this extra loading capacitance accounts for the loss of phase margin.

Why does the loading capacitance make the op amp unstable? Look closely at the gain/phase response between 1 and 9 MHz, and observe that the gain curve changes slope drastically while the rate of phase change approaches 120 degrees/decade. The radical gain/phase slope change proves that several poles are located in this area. The loading capacitance works with the op amp output impedance to form another pole, and the new pole reacts with the internal op amp poles. As the loading capacitor value is increased, its pole migrates down in frequency, causing more phase shift at the 0 dB crossover frequency. The proof of this is given in the TL03X data sheet where plots of ringing and oscillation versus loading capacitance are shown.

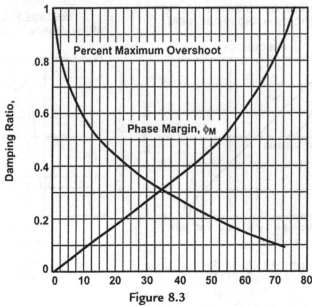

Figure 8.3

Phase margin and percent overshoot versus damping ratio.

Fig. 8.4 shows similar plots for the TL07X, which is the newer family of op amps. Notice that the phase shift is approximately 100 degrees when the gain crosses the 0 dB axis. This yields a phase margin of 80 degrees, which is close to unconditionally stable. The slope of the phase curve changes to 180 degrees/decade about one decade from the 0 dB crossover point. The radical slope change causes suspicion about the 90 degrees phase margin; furthermore the gain curve must be changing radically when the phase is changing radically. The gain/phase plot may not be totally false, but it sure is overly optimistic.

The TL07X pulse response plot shows approximately 20% overshoot. There is no loading capacitance indicated on the plot to account for a seemingly unconditionally stable op amp exhibiting this large an overshoot. Something is wrong here: the analysis is wrong, the plots are wrong, or the parameters are wrong. Fig. 8.5 shows the plots for the TL08X family of op amps, which are sisters to the TL07X family. The gain/phase curve and pulse response is virtually identical, but the pulse response lists a 100 pF loading capacitor. This little exercise illustrates three valuable points: first, if the data seem wrong it probably is wrong; second, even the factory people make mistakes; and third, the loading capacitor makes op amps ring, overshoot, or oscillate.

The frequency- and time-response plots for the TLV277X family of op amps is shown in Figs. 8.6 and 8.7. First, notice that the information is more sophisticated because the phase response is given in degrees of phase margin; second, both gain/phase plots are done with substantial loading capacitors (600 pF), so they have some practical value; and third, the phase margin is a function of power supply voltage.

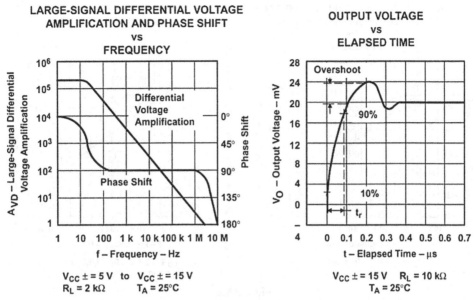

Figure 8.4

TL07X frequency- and time-response plots.

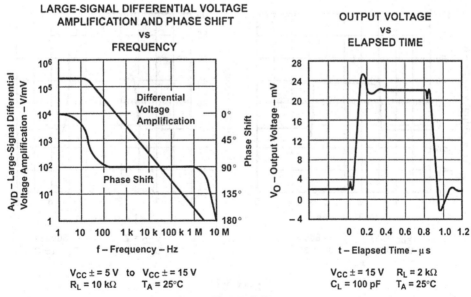

Figure 8.5

TL08X frequency- and time-response plots.

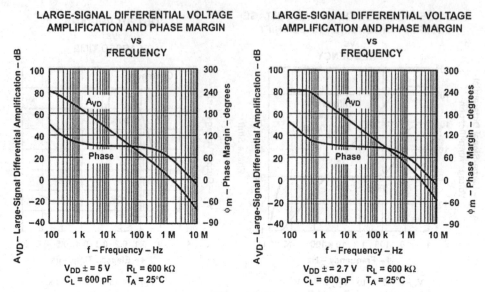

Figure 8.6
TLV277X frequency-response plots.

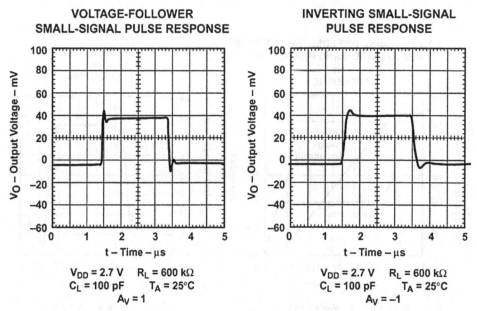

Figure 8.7
TLV227X time-response plots.

At $V_{CC} = 5$ V, the phase margin at the 0 dB crossover point is 60 degrees, while it is 30 degrees at $V_{CC} = 2.7$ V. This translates into an expected overshoot of 18% at $V_{CC} = 5$ V, and 28% at $V_{CC} = 2.7$ V. Unfortunately the time-response plots are done with 100 pF loading capacitance, hence we cannot check our figures very well. The $V_{CC} = 2.7$ V overshoot is approximately 2%, and it is almost impossible to figure out what the overshoot would have been with a 600 pF loading capacitor. The small-signal pulse response is done with mV signals, and that is a more realistic measurement than using the full signal swing.

Internally compensated op amps are very desirable because they are easy to use, and they do not require external compensation components. Their drawback is that the bandwidth is limited by the internal compensation scheme. The op amp open-loop gain eventually (when it shows up in the loop gain) determines the error in an op amp circuit. In a noninverting buffer configuration, the TL277X is limited to 1% error at 50 kHz ($V_{CC} = 2.7$ V) because the op amp gain is 40 dB at that point. Circuit designers can play tricks such as bypassing the op amp with a capacitor to emphasize the high-frequency gain, but the error is still 1%. Keep Eq. (8.1) in mind because it defines the error. If the TLV277X were not internally compensated, it could be externally compensated for a lower error at 50 kHz because the gain would be much higher.

$$E = \frac{V_{IN}}{1 + A\beta} \tag{8.1}$$

8.3 External Compensation, Stability, and Performance

Nobody compensates an op amp just because it is there; they have a reason to compensate the op amp, and that reason is usually stability. They want the op amp to perform a function in a circuit where it is potentially unstable. Internally and noninternally compensated op amps are compensated externally because certain circuit configurations do cause oscillations. Several potentially unstable circuit configurations are analyzed in this section, and the reader can extend the external compensation techniques as required.

Other reasons for externally compensating op amps are noise reduction, flat amplitude response, and obtaining the highest bandwidth possible from an op amp. An op amp generates noise, and noise is generated by the system. The noise contains many frequency components, and when a high-pass filter is incorporated in the signal path, it reduces high-frequency noise. Compensation can be employed to roll off the op amp's high frequency, closed-loop response, thus causing the op amp to act as a noise filter. Internally compensated op amps are modeled with a second-order equation, and this means that the output voltage can overshoot in response to a step input. When this overshoot (or peaking) is undesirable, external compensation can increase the phase margin to 90 degrees where there is no peaking.

An uncompensated op amp has the highest bandwidth possible. External compensation is required to stabilize uncompensated op amps, but the compensation can be tailored to the specific circuit, thus yielding the highest possible bandwidth consistent with the pulse response requirements.

8.4 Dominant-Pole Compensation

We saw that capacitive loading caused potential instabilities, thus an op amp loaded with an output capacitor is a circuit configuration that must be analyzed. This circuit is called dominant-pole compensation because if the pole formed by the op amp output impedance and the loading capacitor is located close to the zero frequency axis, it becomes dominant. The op amp circuit is shown in Fig. 8.8, and the open-loop circuit used to calculate the loop gain (Aβ) is shown in Fig. 8.9.

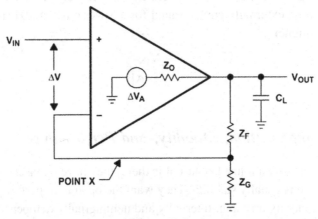

Figure 8.8
Capacitively loaded op amp.

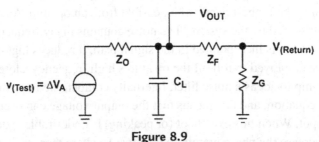

Figure 8.9
Capacitively loaded op amp with loop broken for loop gain (Aβ) calculation.

The analysis starts by looking into the capacitor and taking the Thevenin equivalent circuit.

$$V_{TH} = \frac{\Delta Va}{Z_O C_L s + 1} \tag{8.2}$$

$$Z_{TH} = \frac{Z_O}{Z_O C_L s + 1} \tag{8.3}$$

Then the output equation is written.

$$V_{RETURN} = \frac{V_{TH} Z_G}{Z_G + Z_F + Z_{TH}} = \frac{\Delta Va}{Z_O C_L s + 1} \left(\frac{Z_G}{Z_F + Z_G + \dfrac{Z_O}{Z_O C_L s + 1}} \right) \tag{8.4}$$

Rearranging terms yields Eq. (8.5).

$$\frac{V_{RETURN}}{V_{TEST}} = A\beta = \frac{\dfrac{aZ_G}{Z_F + Z_G + Z_O}}{\dfrac{(Z_F + Z_G) Z_O C_L s}{Z_F + Z_G + Z_O} + 1} \tag{8.5}$$

When the assumption is made that $(Z_F + Z_G) \gg Z_O$, Eq. (8.5) reduces to Eq. (8.6).

$$A\beta = \frac{aZ_G}{Z_F + Z_G} \left(\frac{1}{Z_O C_L s + 1} \right) \tag{8.6}$$

Eq. (8.7) models the op amp as a second-order system. Hence, substituting the second-order model for a in Eq. (8.6) yields Eq. (8.8), which is the stability equation for the dominant pole compensation circuit.

$$a = \frac{K}{(s + \tau_1)(s + \tau_2)} \tag{8.7}$$

$$A\beta = \frac{K}{(s + \tau_1)(s + \tau_2)} \frac{Z_G}{Z_F + Z_G} \frac{1}{Z_O C_L s + 1} \tag{8.8}$$

Several conclusions can be drawn from Eq. (8.8) depending on the location of the poles. If the Bode plot of Eq. (8.7), the op amp transfer function, looks like that shown in Fig. 8.10, it only has 25 degrees phase margin, and there is approximately 48% overshoot. When the pole introduced by Z_O and C_L moves toward the zero frequency axis, it comes close to the τ_2 pole, and it adds phase shift to the system. Increased phase shift increases peaking and decreases stability. In the real world, many loads, especially cables, are capacitive, and an op amp like the one pictured in Fig. 8.10 would ring while driving a capacitive load. The load capacitance causes peaking and instability in internally compensated op amps when the op amps do not have enough phase margin to allow for the phase shift introduced by the load.

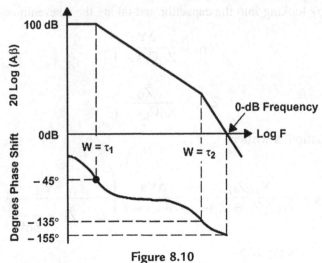

Figure 8.10
Possible Bode plot of the op amp described in Eq. (8.7).

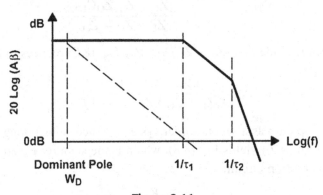

Figure 8.11
Dominant-pole compensation plot.

Prior to compensation, the Bode plot of an uncompensated op amp looks like that shown in Fig. 8.11. Notice that the breakpoints are located close together thus accumulating about 180 degrees of phase shift before the 0 dB crossover point; the op amp is not usable and probably unstable. Dominant-pole compensation is often used to stabilize these op amps. If a dominant pole, in this case ω_D, is properly placed, it rolls off the gain so that τ_1 introduces 45 degrees phase at the 0 dB crossover point. After the dominant pole is introduced the op amp is stable with 45 degrees phase margin, but the op amp gain is drastically reduced for frequencies higher than ω_D. This procedure works well for internally compensated op amps, but is seldom used for externally compensated op amps because inexpensive discrete capacitors are readily available.

Assuming that $Z_O \ll Z_F$, the closed-loop transfer function is easy to calculate because C_L is enclosed in the feedback loop. The ideal closed-loop transfer equation is the same as Eq. (6.11) for the noninverting op amp and is repeated below as Eq. (8.9).

$$\frac{V_{OUT}}{V_{IN}} = \frac{a}{1 + \dfrac{aZ_G}{Z_G + Z_F}} \tag{8.9}$$

When $a \Rightarrow \infty$ Eq. (8.9) reduces to Eq. (8.10)

$$\frac{V_{OUT}}{V_{IN}} = \frac{Z_F + Z_G}{Z_G} \tag{8.10}$$

As long as the op amp has enough compliance and current to drive the capacitive load, and Z_O is small, the circuit functions as though the capacitor was not there. When the capacitor becomes large enough, its pole interacts with the op amp pole causing instability. When the capacitor is huge, it completely kills the op amp's bandwidth, thus lowering the noise while retaining a large low-frequency gain.

8.5 Gain Compensation

When the closed-loop gain of an op amp circuit is related to the loop gain, as it is in voltage-feedback op amps, the closed-loop gain can be used to stabilize the circuit. This type of compensation cannot be used in current-feedback op amps because the mathematical relationship between the loop gain and ideal closed-loop gain does not exist. The loop gain equation is repeated as Eq. (8.11). Notice that the closed-loop gain parameters Z_G and Z_F are contained in Eq. (8.11); hence the stability can be controlled by manipulating the closed-loop gain parameters.

$$A\beta = \frac{aZ_G}{Z_G + Z_F} \tag{8.11}$$

The original loop gain curve for a closed-loop gain of one is shown in Fig. 8.12, and it is or comes very close to being unstable. If the closed-loop noninverting gain is changed to 9, then K changes from K/2 to K/10. The loop gain intercept on the Bode plot (Fig. 8.12) moves down 14 dB, and the circuit is stabilized.

Gain compensation works for inverting or noninverting op amp circuits because the loop gain equation contains the closed-loop gain parameters in both cases. When the closed-loop gain is increased, the accuracy and the bandwidth decrease. As long as the application can stand the higher gain, gain compensation is the best type of compensation to use. Uncompensated versions of normally internally compensated op amps are offered for sale as stable op amps with minimum gain restrictions. As long as gain in the circuit you design exceeds the gain specified on the data sheet, this is economical and a safe mode of operation.

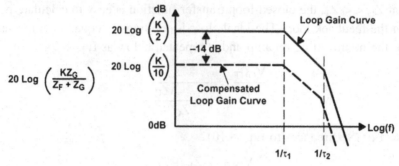

Figure 8.12
Gain compensation.

8.6 Lead Compensation

Sometimes lead compensation is forced on the circuit designer because of the parasitic capacitance associated with packaging and wiring op amps. Fig. 8.13 shows the circuit for lead compensation; notice the capacitor in parallel with R_F. That capacitor is often made by parasitic wiring and the ground plane, and high-frequency circuit designers go to great lengths to minimize or eliminate it. What is good in one sense is bad in another, because adding the parallel capacitor is a good way to stabilize the op amp and reduce noise. Let us analyze the stability first, and then we will analyze the closed-loop performance.

The loop equation for the lead-compensation circuit is given by Eq. (8.12).

$$A\beta = \left(\frac{R_G}{R_G + R_F}\right)\left(\frac{R_F Cs + 1}{R_G \| R_F Cs + 1}\right)\left(\frac{K}{(s + \tau_1)(s + \tau_2)}\right) \tag{8.12}$$

The compensation capacitor introduces a pole and zero into the loop equation. The zero always occurs before the pole because $R_F > R_F \| R_G$. When the zero is properly placed, it cancels out the τ_2 pole along with its associated phase shift. The original transfer function is

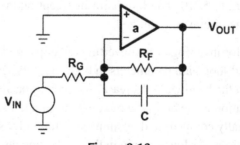

Figure 8.13
Lead-compensation circuit.

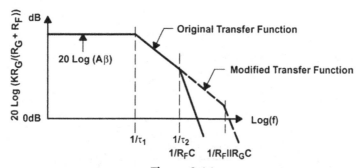

Figure 8.14
Lead-compensation Bode plot.

shown in Fig. 8.14 drawn in solid lines. When the R_FC zero is placed at $\omega = 1/\tau_2$, it cancels out the τ_2 pole causing the Bode plot to continue on a slope of -20 dB/decade. When the frequency gets to $\omega = 1/(R_F\|R_G)C$, this pole changes the slope to -40 dB/decade. Properly placed, the capacitor aids stability, but what does it do to the closed-loop transfer function? The equation for the inverting op amp closed-loop gain is repeated below:

$$\frac{V_{OUT}}{V_{IN}} = \frac{\dfrac{-aZ_F}{Z_G + Z_F}}{1 + \dfrac{aZ_G}{Z_G + Z_F}} \tag{8.13}$$

When a approaches infinity, Eq. (8.13) reduces to Eq. (8.14).

$$\frac{V_{OUT}}{V_{IN}} = -\frac{Z_F}{Z_{IN}} \tag{8.14}$$

Substituting $R_F\|C$ for Z_F and R_G for Z_G in Eq. (8.14) yields Eq. (8.15), which is the ideal closed-loop gain equation for the lead-compensation circuit.

$$\frac{V_{OUT}}{V_{IN}} = -\frac{R_F}{R_G}\left(\frac{1}{R_FCs + 1}\right) \tag{8.15}$$

The forward gain for the inverting amplifier is given by Eq. (8.16). Compare Eq. (8.13) with Eq. (6.5) to determine A.

$$A = \frac{aZ_F}{Z_G + A_F} = \left(\frac{aR_F}{R_G + R_F}\right)\left(\frac{1}{R_F\|R_GCs + 1}\right) \tag{8.16}$$

The op amp gain (a), the forward gain (A), and the ideal closed-loop gain are plotted in Fig. 8.15. The op amp gain is plotted for reference only. The forward gain for the inverting op amp is not the op amp gain. Notice that the forward gain is reduced by the factor $R_F/(R_G + R_F)$, and it contains a high-frequency pole. The ideal closed-loop gain

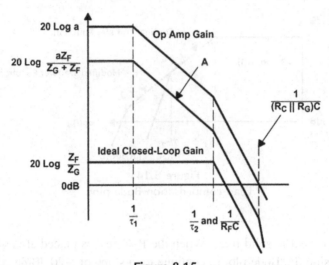

Figure 8.15
Inverting op amp with lead compensation.

follows the ideal curve until the $1/R_FC$ breakpoint (same location as $1/\tau_2$ breakpoint), and then it slopes down at -20 dB/decade. Lead compensation sacrifices the bandwidth between the $1/R_FC$ breakpoint and the forward gain curve. The location of the $1/R_FC$ pole determines the bandwidth sacrifice, and it can be much greater than shown here. The pole caused by R_F, R_G, and C does not appear until the op amp's gain has crossed the 0 dB axis, thus it does not affect the ideal closed-loop transfer function.

The forward gain for the noninverting op amp is a; compare Eqs. (6.11) to (6.5). The ideal closed-loop gain is given by Eq. (8.17).

$$\frac{V_{OUT}}{V_{IN}} = \frac{Z_F + Z_G}{Z_G} = \left(\frac{R_F + R_G}{R_G}\right)\left(\frac{R_F\|R_GCs + 1}{R_FCs + 1}\right) \tag{8.17}$$

The plot of the noninverting op amp with lead compensation is shown in Fig. 8.16. There is only one plot for both the op amp gain (a) and the forward gain (A), because they are identical in the noninverting circuit configuration. The ideal starts out as a flat line, but it slopes down because its closed-loop gain contains a pole and a zero. The pole always occurs closer to the low-frequency axis because $R_F > R_F\|R_G$. The zero flattens the ideal closed-loop gain curve, but it never does any good because it cannot fall on the pole. The pole causes a loss in the closed-loop bandwidth by the amount separating the closed-loop and forward gain curves.

Although the forward gain is different in the inverting and noninverting circuits, the closed-loop transfer functions take very similar shapes. This becomes truer as the closed-loop gain increases because the noninverting forward gain approaches the op amp gain. This

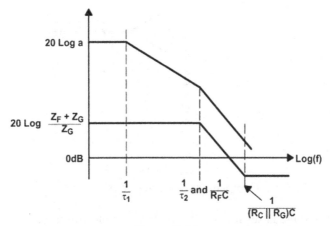

Figure 8.16
Noninverting op amp with lead compensation.

relationship cannot be relied on in every situation, and each circuit must be checked to determine the closed-loop effects of the compensation scheme.

8.7 Compensated Attenuator Applied to Op Amp

Stray capacitance on op amp inputs is a problem that circuit designers are always trying to get away from because it decreases stability and causes peaking. The circuit shown in Fig. 8.17 has some stray capacitance (C_G) connected from the inverting input to ground. Eq. (8.18) is the loop gain equation for the circuit with input capacitance.

$$A\beta = \left(\frac{R_G}{R_G + R_F}\right)\left(\frac{1}{R_G\|R_FCs + 1}\right)\left(\frac{K}{(\tau_1 s + 1)(\tau_2 s + 1)}\right) \qquad (8.18)$$

Op amps having high input and feedback resistors are subject to instability caused by stray capacitance on the inverting input. Referring to Eq. (8.18), when the $1/(R_F\|R_GC_G)$ pole

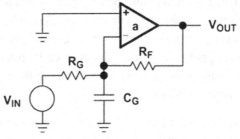

Figure 8.17
Op amp with stray capacitance on the inverting input.

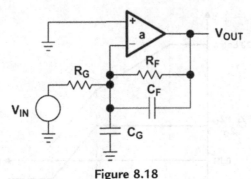

Figure 8.18
Compensated attenuator circuit.

moves close to τ_2 the stage is set for instability. Reasonable component values for a complementary metal–oxide–semiconductor (CMOS) op amp are $R_F = 1\ M\Omega$, $R_G = 1\ M\Omega$, and $C_G = 10\ pF$. The resulting pole occurs at 318 kHz, and this frequency is lower than the breakpoint of τ_2 for many op amps. There is 90 degrees of phase shift resulting from τ_1, the $1/(R_F\|R_GC)$ pole adds 45 degrees phase shift at 318 kHz, and τ_2 starts to add another 45 degrees phase shift at about 600 kHz. This circuit is unstable because of the stray input capacitance. The circuit is compensated by adding a feedback capacitor as shown in Fig. 8.18.

The loop gain with C_F added is given by Eq. (8.19).

$$A\beta = \left[\frac{\dfrac{R_G}{R_GC_Gs + 1}}{\dfrac{R_G}{R_GC_Gs + 1} + \dfrac{R_F}{R_FC_Fs + 1}} \right] \left(\frac{K}{(\tau_1s + 1)(\tau_2s + 1)} \right) \qquad (8.19)$$

If $R_GC_G = R_FC_F$ Eq. (8.19) reduces to Eq. (8.20).

$$A\beta = \left[\frac{R_G}{R_G + R_F} \right] \left(\frac{K}{(\tau_1s + 1)(\tau_2s + 1)} \right) \qquad (8.20)$$

The compensated attenuator Bode plot is shown in Fig. 8.19. Adding the correct $1/R_FC_F$ breakpoint cancels out the $1/R_GC_G$ breakpoint; the loop gain is independent of the capacitors. Now is the time to take advantage of the stray capacitance. C_F can be formed by running a wide copper strip from the output of the op amp over the ground plane under R_F; do not connect the other end of this copper strip. The circuit is tuned by removing some copper (a razor works well) until all peaking is eliminated. Then measure the copper and have an identical trace put on the printed-circuit board.

The inverting and noninverting closed-loop gain equations are a function of frequency. Eq. (8.21) is the closed-loop gain equation for the inverting op amp. When $R_FC_F = R_GC_G$, Eq. (8.21) reduces to Eq. (8.22), which is independent of the breakpoint. This also

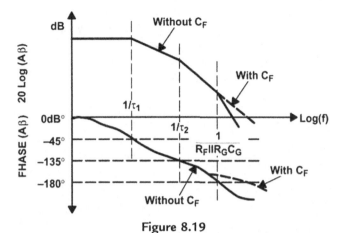

Figure 8.19
Compensated attenuator bode plot.

happens to the noninverting op amp circuit. This is one of the few occasions when the compensation does not affect the closed-loop gain frequency response.

$$\frac{V_{OUT}}{V_{IN}} = \frac{\dfrac{R_F}{R_F C_F s + 1}}{\dfrac{R_G}{R_G C_G s + 1}} \tag{8.21}$$

$$\text{When } R_F C_F = R_G C_G \quad \frac{V_{OUT}}{V_{IN}} = -\left(\frac{R_F}{R_G}\right) \tag{8.22}$$

8.8 Lead-Lag Compensation

Lead-lag compensation stabilizes the circuit without sacrificing the closed-loop gain performance. It is often used with uncompensated op amps. This type of compensation provides excellent high-frequency performance. The circuit schematic is shown in Fig. 8.20, and the loop gain is given by Eq. (8.23).

$$A\beta = \frac{K}{(\tau_1 s + 1)(\tau_2 s + 1)} \frac{R_G}{R_G + R_F} \frac{RCs + 1}{\dfrac{(RR_G + RR_F + R_G R_F)}{(R_G + R_F)}Cs + 1} \tag{8.23}$$

Referring to Fig. 8.21, a pole is introduced at $\omega = 1/RC$, and this pole reduces the gain 3 dB at the breakpoint. When the zero occurs prior to the first op amp pole, it cancels out the phase shift caused by the $\omega = 1/RC$ pole. The phase shift is completely canceled before the second op amp pole occurs, and the circuit reacts as if the pole was never introduced. Nevertheless, $A\beta$ is reduced by 3 dB or more, so the loop gain crosses the 0 dB axis at a lower frequency. The beauty of lead-lag compensation is that the closed-loop ideal gain is

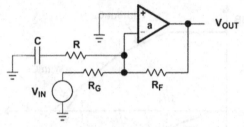

Figure 8.20
Lead-lag compensated op amp.

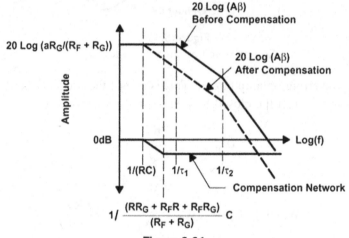

Figure 8.21
Bode plot of lead-lag compensated op amp.

not affected as shown below. The Thevenin equivalent of the input circuit is calculated in Eq. (8.24), the circuit gain in terms of Thevenin equivalents is calculated in Eq. (8.25), and the ideal closed-loop gain is calculated in Eq. (8.26).

$$V_{TH} = V_{IN} \frac{R + \frac{1}{Cs}}{R + R_G + \frac{1}{Cs}} \qquad R_{TH} \frac{R_G \left(R + \frac{1}{Cs} \right)}{R + R_G + \frac{1}{Cs}} \tag{8.24}$$

$$V_{OUT} = -V_{TH} \frac{R_F}{R_{TH}} \tag{8.25}$$

$$-\frac{V_{OUT}}{V_{IN}} = \frac{R + \frac{1}{Cs}}{R + R_G + \frac{1}{Cs}} \frac{R_F}{\frac{R_G \left(R + \frac{1}{Cs} \right)}{R + R_G + \frac{1}{Cs}}} = \frac{R_F}{R_G} \tag{8.26}$$

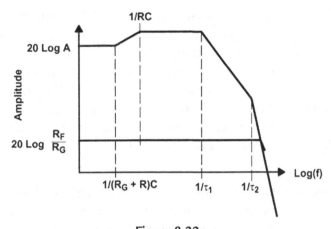

Figure 8.22
Closed-loop plot of lead-lag compensated op amp.

Eq. (8.26) is intuitively obvious because the RC network is placed across a virtual ground. As long as the loop gain, Aβ, is large, the feedback will null out the closed-loop effect of RC, and the circuit will function as if it were not there. The closed-loop log plot of the lead-lag-compensated op amp is given in Fig. 8.22. Notice that the pole and zero resulting from the compensation occur and are gone before the first amplifier poles come on the scene. This prevents interaction, but it is not required for stability.

8.9 Comparison of Compensation Schemes

Internally compensated op amps can, and often do, oscillate under some circuit conditions. Internally compensated op amps need an external pole to get the oscillation or ringing started, and circuit stray capacitances often supply the phase shift required for instability. Loads, such as cables, often cause internally compensated op amps to ring severely.

Dominant-pole compensation is often used in IC design because it is easy to implement. It rolls off the closed-loop gain early; thus, it is seldom used as an external form of compensation unless filtering is required. Load capacitance, depending on its pole location, usually causes the op amp to ring. Large load capacitance can stabilize the op amp because it acts as dominant-pole compensation.

The simplest form of compensation is gain compensation. High closed-loop gains are reflected in lower loop gains, and in turn, lower loop gains increase stability. If an op amp circuit can be stabilized by increasing the closed-loop gain, do it.

Stray capacitance across the feedback resistor tends to stabilize the op amp because it is a form of lead compensation. This compensation scheme is useful for limiting the circuit bandwidth, but it decreases the closed-loop gain.

Stray capacitance on the inverting input works with the parallel combination of the feedback and gain setting resistors to form a pole in the Bode plot, and this pole decreases the circuit's stability. This effect is normally observed in high-impedance circuits built with CMOS op amps. Adding a feedback capacitor forms a compensated attenuator scheme that cancels out the input pole. The cancellation occurs when the input and feedback RC time constants are equal. Under the conditions of equal time constants, the op amp functions as though the stray input capacitance was not there. An excellent method of implementing a compensated attenuator is to build a stray feedback capacitor using the ground plane and a trace of the output node.

Lead-lag compensation stabilizes the op amp, and it yields the best closed-loop frequency performance. Contrary to some published opinions, no compensation scheme will increase the bandwidth beyond that of the op amp. Lead-lag compensation just gives the best bandwidth for the compensation.

8.10 Conclusions

The stability criteria often is not oscillation, rather it is circuit performance as exhibited by peaking and ringing.

The circuit bandwidth can often be increased by connecting an external capacitor in parallel with the op amp. Some op amps have hooks that enable a parallel capacitor to be connected in parallel with a portion of the input stages. This increases bandwidth because it shunts high frequencies past the low-bandwidth gm stages (see Appendix B), but this method of compensation depends on the op amp type and manufacturer.

The compensation techniques given here are adequate for the majority of applications. When the new and challenging application presents itself, use the procedure outlined here to invent your own compensation techniques.

CHAPTER 9

Current-Feedback Op Amps

9.1 Introduction

When I first heard the term "current-feedback amplifier (CFA)," I was initially confused. I had visions of something really complicated, perhaps associated with the little used Norton current theorem, or even the esoteric Norton op amps of days past. I had nothing to worry about! So that you do not have to go through any apprehension at all, I will present a summary right here of the main points:

- Your standard gain circuits, noninverting and inverting, use the same exact formulas for design. Inverting is still $-R_F/R_G$, noninverting is still $1 + R_F/R_G$.
- You have to use the value of R_F recommended in the data sheet. You can use R_G to set your gain.
- Use noninverting gain circuit or you will load your source.
- Never put a capacitor directly across the feedback path from output to inverting input.
- You *might* get some more gain out of a circuit using a CFA. Probably not as much as the hype suggests.
- If you are trying to make a DC gain system, go back to voltage-feedback op amps. Current-feedback op amps have lousy DC performance.
- If you are trying to make analog filters, use a voltage-feedback op amp. Current-feedback op amps are not particularly good at analog filtering because of the "no capacitor across the feedback" rule.
- Most current-feedback op amps are terrific for high-speed and high-current drive applications. This is where they shine.

That is it in a nutshell — now we can delve into the internals.

CFAs do not have the traditional differential amplifier input structure, thus they sacrifice the parameter matching inherent to that structure. The CFA circuit configuration prevents them from obtaining the precision of voltage-feedback amplifiers (VFAs), but the circuit configuration that sacrifices precision results in increased bandwidth and slew rate. The higher bandwidth is relatively independent of closed-loop gain, so the constant gain-bandwidth restriction applied to VFAs is removed for CFAs. The slew rate of CFAs is much improved from their counterpart VFAs because their structure enables the output stage to supply slewing current until the output reaches its final value. In general, VFAs

Op Amps for Everyone. http://dx.doi.org/10.1016/B978-0-12-811648-7.00009-1

are used for precision and general purpose applications, whereas CFAs are restricted to high-frequency applications above 100 MHz.

CFAs, unlike previous generation high-frequency amplifiers, have eliminated the AC coupling requirement; they are usually DC-coupled while they operate in the gigahertz range. CFAs have much faster slew rates than VFAs, so they have faster rise/fall times and less intermodulation distortion.

9.2 Current-Feedback Amplifier Model

The CFA model is shown in Fig. 9.1. The noninverting input of a CFA connects to the input of the input buffer, so it has very high impedance similar to that of a bipolar transistor noninverting VFA input. The inverting input connects to the input buffer's output, so the inverting input impedance is equivalent to a buffer's output impedance, which is very low. Z_B models the input buffer's output impedance, and it is usually less than 50 Ω. The input buffer gain, G_B, is as close to one as IC design methods can achieve, and it is small enough to neglect in the calculations.

The output buffer provides low output impedance for the amplifier. Again, the output buffer gain, G_{OUT}, is very close to one, so it is neglected in the analysis. The output impedance of the output buffer is ignored during the calculations. This parameter may influence the circuit performance when driving very low impedance or capacitive loads, but this is usually not the case. The input buffer's output impedance cannot be ignored because affects stability at high frequencies.

The current-controlled current source, Z, is a transimpedance. The transimpedance in a CFA serves the same function as gain in a VFA; it is the parameter that makes the performance of the op amp dependent only on the passive parameter values. Usually the

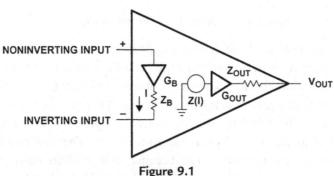

Figure 9.1
Current-feedback amplifier model.

transimpedance is very high, in the MΩ range, so the CFA gains accuracy by closing a feedback loop in the same manner that the VFA does.

9.3 Development of the Stability Equation

The stability equation is developed with the aid of Fig. 9.2. Remember, stability is independent of the input, and stability depends solely on the loop gain, $A\beta$. Breaking the loop at point X, inserting a test signal, V_{TI}, and calculating the return signal V_{TO} develop the stability equation.

The circuit used for stability calculations is shown in Fig. 9.3 where the model of Fig. 9.1 is substituted for the CFA symbol. The input and output buffer gain, and output buffer output impedance have been deleted from the circuit to simplify calculations. This approximation is valid for almost all applications.

The transfer equation is given in Eq. (9.1), and the Kirchoff's law is used to write Eqs. (9.2) and (9.3).

$$V_{TO} = I_1 Z \tag{9.1}$$
$$V_{TI} = I_2 (Z_F + Z_G \parallel Z_B) \tag{9.2}$$
$$I_2 (Z_G \parallel Z_B) = I_1 Z_B \tag{9.3}$$

Eqs. (9.2) and (9.3) are combined to yield Eq. (9.4).

$$V_{TI} = I_1 (Z_F + Z_G \parallel Z_B)\left(1 + \frac{Z_B}{Z_G}\right) = I_1 Z_F \left(1 + \frac{Z_B}{Z_F \parallel Z_G}\right) \tag{9.4}$$

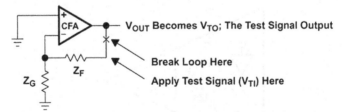

Figure 9.2
Stability analysis circuit.

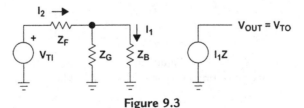

Figure 9.3
Stability analysis circuit.

Dividing Eq. (9.1) by Eq. (9.4) yields Eq. (9.5), and this is the open-loop transfer equation. This equation is commonly known as the loop gain.

$$A\beta = \frac{V_{TO}}{V_{TI}} = \frac{Z}{\left(Z_F\left(1 + \frac{Z_B}{Z_F \| Z_G}\right)\right)} \tag{9.5}$$

9.4 The Noninverting Current-Feedback Amplifier

The closed-loop gain equation for the noninverting CFA is developed with the aid of Fig. 9.4, where external gain setting resistors have been added to the circuit. The buffers are shown in Fig. 9.4, but because their gains equal one and they are included within the feedback loop, the buffer gain does not enter into the calculations.

Eq. (9.6) is the transfer equation, Eq. (9.7) is the current equation at the inverting node, and Eq. (9.8) is the input loop equation. These equations are combined to yield the closed-loop gain equation, Eq. (9.9).

$$V_{OUT} = I_Z \tag{9.6}$$

$$I = \left(\frac{V_A}{Z_G}\right) - \left(\frac{V_{OUT} - V_A}{Z_F}\right) \tag{9.7}$$

$$V_A = V_{IN} - IZ_B \tag{9.8}$$

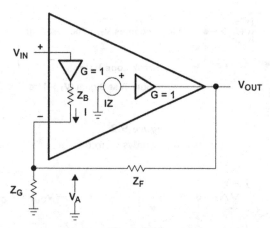

Figure 9.4
Noninverting current-feedback amplifier.

$$\frac{V_{OUT}}{V_{IN}} = \frac{Z\left(1 + \frac{Z_F}{Z_G}\right)}{Z_F\left(1 + \frac{Z_B}{Z_F\|Z_G}\right)}{1 + \frac{Z}{Z_F\left(1 + \frac{Z_B}{Z_F\|Z_G}\right)}} \tag{9.9}$$

When the input buffer output impedance, Z_B, approaches zero, Eq. (9.9) reduces to Eq. (9.10).

$$\frac{V_{OUT}}{V_{IN}} = \frac{\frac{Z\left(1 + \frac{Z_F}{Z_G}\right)}{Z_F}}{1 + \frac{Z}{Z_F}} = \frac{1 + \frac{Z_F}{Z_G}}{1 + \frac{Z_F}{Z}} \tag{9.10}$$

When the transimpedance, Z, is very high, the term Z_F/Z in Eq. (9.10) approaches zero, and Eq. (9.10) reduces to Eq. (9.11); the ideal closed-loop gain equation for the CFA. The ideal closed-loop gain equations for the CFA and VFA are identical, and the degree to which they depart from ideal is dependent on the validity of the assumptions. The VFA has one assumption that the direct gain is very high, whereas the CFA has two assumptions that the transimpedance is very high and that the input buffer output impedance is very low. As would be expected, two assumptions are much harder to meet than one, thus the CFA departs from the ideal more than the VFA does.

$$\frac{V_{OUT}}{V_{IN}} = 1 + \frac{Z_F}{Z_G} \tag{9.11}$$

9.5 The Inverting Current-Feedback Amplifier

The inverting CFA configuration is seldom used because the inverting input impedance is very low ($Z_B \| Z_F + Z_G$). When Z_G is made dominant by selecting it as a high-resistance value it overrides the effect of Z_B. Z_F must also be selected as a high value to achieve at least unity gain, and high values for Z_F result in poor bandwidth performance, as we will see in the next section. If Z_G is selected as a low value the frequency sensitive Z_B causes the gain to increase as frequency increases. These limitations restrict inverting applications of the inverting CFA (Fig. 9.5).

The current equation for the input node is written as Eq. (9.12). Eq. (9.13) defines the dummy variable, V_A, and Eq. (9.14) is the transfer equation for the CFA. These equations

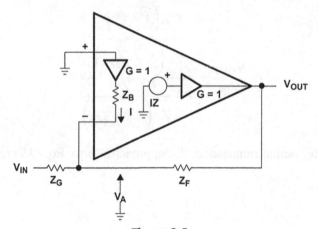

Figure 9.5
Inverting current-feedback amplifier.

are combined and simplified leading to Eq. (9.15), which is the closed-loop gain equation for the inverting CFA.

$$1 + \frac{V_{IN} - V_A}{Z_G} = \frac{V_A - V_{OUT}}{Z_F} \tag{9.12}$$

$$IZ_B = -V_A \tag{9.13}$$

$$IZ = V_{OUT} \tag{9.14}$$

$$\frac{V_{OUT}}{V_{IN}} = -\frac{\dfrac{Z}{Z_G\left(1 + \dfrac{Z_B}{Z_F\|Z_G}\right)}}{1 + \dfrac{Z}{Z_F\left(1 + \dfrac{Z_B}{Z_F\|Z_G}\right)}} \tag{9.15}$$

When Z_B approaches zero, Eq. (9.15) reduces to Eq. (9.16).

$$\frac{V_{OUT}}{V_{IN}} = -\frac{\dfrac{1}{Z_G}}{\dfrac{1}{Z} + \dfrac{1}{Z_F}} \tag{9.16}$$

When Z is very large, Eq. (9.16) becomes Eq. (9.17), which is the ideal closed-loop gain equation for the inverting CFA.

$$\frac{V_{OUT}}{V_{IN}} = -\frac{Z_F}{Z_G} \tag{9.17}$$

The ideal closed-loop gain equation for the inverting VFA and CFA op amps are identical. Both configurations have lower input impedance than the noninverting configuration has, but the VFA has one assumption, whereas the CFA has two assumptions. Again, as was the case with the noninverting counterparts, the CFA is less ideal than the VFA because of the two assumptions. The zero Z_B assumption always breaks down in bipolar junction transistors as is shown later. The CFA is almost never used in the differential amplifier configuration because of the CFA's gross input impedance mismatch.

9.6 Stability Analysis

The stability equation is repeated as Eq. (9.18).

$$A\beta = \frac{V_{TO}}{V_{TI}} = \frac{Z}{\left(Z_F\left(1 + \frac{Z_B}{Z_F\|Z_G}\right)\right)} \tag{9.18}$$

Comparing Eqs. (9.9) and (9.15) to Eq. (9.18) reveals that the inverting and noninverting CFA op amps have identical stability equations. This is the expected result because stability of any feedback circuit is a function of the loop gain, and the input signals have no affect on stability. The two op amp parameters affecting stability are the transimpedance, Z, and the input buffer's output impedance, Z_B. The external components affecting stability are Z_G and Z_F. The designer controls the external impedance, although stray capacitance that is a part of the external impedance sometimes seems to be uncontrollable. Stray capacitance is the primary cause of ringing and overshoot in CFAs. Z and Z_B are CFA op amp parameters that cannot be controlled by the circuit designer, so he has to live with them.

Prior to determining stability with a Bode plot, we take the log of Eq. (9.18), and plot the logs (Eqs. 9.19 and 9.20) in Fig. 9.6.

$$20\ LOG|A\beta| = 20\ LOG|Z| - 20\ LOG\left|Z_F\left(1 + \frac{Z_B}{Z_F\|Z_B}\right)\right| \tag{9.19}$$

$$\phi = TANGENT^{-1}(A\beta) \tag{9.20}$$

This enables the designer to add and subtract components of the stability equation graphically.

The plot in Fig. 9.6 assumes typical values for the parameters:

$$Z = \frac{1M\Omega}{(1 + \tau_1 S)(1 + \tau_2 S)} \tag{9.21}$$

$$Z_B = 70\Omega \tag{9.22}$$

$$Z_G = Z_F\ 1\ k\Omega \tag{9.23}$$

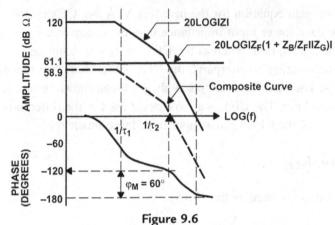

Figure 9.6
Bode plot of stability equation.

The transimpedance has two poles, and the plot shows that the op amp will be unstable without the addition of external components because 20 LOG|Z| crosses the 0-dB axis after the phase shift is 180 degrees. Z_F, Z_B, and Z_G reduce the loop gain 61.1 dB, so the circuit is stable because it has 60 degrees phase margin. Z_F is the component that stabilizes the circuit. The parallel combination of Z_F and Z_G contributes little to the phase margin because Z_B is very small, so Z_B and Z_G have little effect on stability.

The manufacturer determines the optimum value of R_F during the characterization of the IC. Referring to Fig. 9.6, it is seen that when R_F exceeds the optimum value recommended by the IC manufacturer, stability increases. The increased stability has a price called decreased bandwidth. Conversely, when R_F is less than the optimum value recommended by the IC manufacturer, stability decreases, and the circuit response to step inputs is overshoot or possibly ringing. Sometimes the overshoot associated with less than optimum R_F is tolerated because the bandwidth increases as R_F decreases. The peaked response associated with less than optimum values of R_F can be used to compensate for cable droop caused by cable capacitance.

When $Z_B = 0 \ \Omega$ and $Z_F = R_F$ the loop gain equation is $A\beta = Z/R_F$. Under these conditions, Z and R_F determine stability, and a value of R_F can always be found to stabilize the circuit. The transimpedance and feedback resistor have a major impact on stability, and the input buffer's output impedance has a minor effect on stability. Since Z_B increases with an increase in frequency, it tends to increase stability at higher frequencies. Eq. (9.18) is rewritten as Eq. (9.24), but it has been manipulated so that the ideal closed-loop gain is readily apparent.

$$A\beta = \frac{Z}{Z_F + Z_B\left(1 + \dfrac{R_F}{R_G}\right)} \tag{9.24}$$

The closed-loop ideal gain equation (inverting and noninverting) shows up in the denominator of Eq. (9.24), so the closed-loop gain influences the stability of the op amp. When Z_B approaches zero, the closed-loop gain term also approaches zero, and the op amp becomes independent of the ideal closed-loop gain. Under these conditions, R_F determines stability, and the bandwidth is independent of the closed-loop gain. Many people claim that the CFA bandwidth is independent of the gain, and that claim's validity is dependent on the ratios Z_B/Z_F being very low.

Z_B is important enough to warrant further investigation, so the equation for Z_B is given below.

$$Z_B \cong h_{ib} + \frac{R_B}{\beta_0 + 1} \left[\frac{1 + \dfrac{s\beta_0}{\omega_T}}{1 + \dfrac{s\beta_0}{(\beta_0 + 1)\omega_T}} \right] \tag{9.25}$$

At low frequencies $h_{ib} = 50\ \Omega$ and $R_B/(\beta_0 + 1) = 25$, so $Z_B = 75\ \Omega$. Z_B varies in accordance with Eq. (9.25) at high frequencies. Also, the transistor parameters in Eq. (9.25) vary with transistor type; they are different for NPN and PNP transistors. Because Z_B is dependent on the output transistors being used, and this is a function of the quadrant the output signal is in, Z_B has an extremely wide variation. Z_B is a small factor in the equation, but it adds a lot of variability to the current-feedback op amp.

9.7 Selection of the Feedback Resistor

The feedback resistor determines stability, and it affects closed-loop bandwidth, so it must be selected very carefully. Most CFA IC manufacturers employ applications and product engineers who spend a great deal of time and effort selecting R_F. They measure each noninverting gain with several different feedback resistors to gather data. Then they pick a compromise value of R_F that yields stable operation with acceptable peaking, and that value of R_F is recommended on the data sheet for that specific gain. This procedure is repeated for several different gains in anticipation of the various gains their customer applications require (often $G = 1$, 2, or 5). When the value of R_F or the gain is changed from the values recommended on the data sheet, bandwidth and/or stability is affected.

When the circuit designer must select a different R_F value from that recommended on the data sheet, he gets into stability or low bandwidth problems. Lowering R_F decreases stability, and increasing R_F decreases bandwidth. What happens when the designer needs to operate at a gain not specified on the data sheet? The designer must select a new value of R_F for the new gain, but there is no guarantee that new value of R_F is an optimum value. One solution to the R_F selection problem is to assume that the loop gain, $A\beta$, is a linear function. Then the assumption can be made that $(A\beta)_1$ for a gain of one equals

$(A\beta)_N$ for a gain of N, and that this is a linear relationship between stability and gain. Eqs. (9.26) and (9.27) are based on the linearity assumption.

$$\frac{Z}{Z_{F1} + Z_B\left(1 + \frac{Z_{F1}}{Z_{G1}}\right)} = \frac{Z}{Z_{FN} + Z_B\left(1 + \frac{Z_{FN}}{Z_{GN}}\right)} \tag{9.26}$$

$$Z_{FN} = Z_{F1} + Z_B\left(\left(1 + \frac{Z_{F1}}{Z_{G1}}\right) - \left(1 + \frac{Z_{FN}}{Z_{GN}}\right)\right) \tag{9.27}$$

Eq. (9.27) leads one to believe that a new value Z_F can easily be chosen for each new gain. This is not the case in the real world; the assumptions do not hold up well enough to rely on them. When you change to a new gain not specified on the data sheet, Eq. (9.27), at best, supplies a starting point for R_F, but you must test to determine the final value of R_F.

When the R_F value recommended on the data sheet cannot be used, an alternate method of selecting a starting value for R_F is to use graphical techniques. The graph shown in Fig. 9.7 is a plot of the typical 300 MHz CFA data given in Table 9.1.

Enter the graph at the new gain, say $A_{CL} = 6$, and move horizontally until you reach the intersection of the gain versus feedback resistance curve. Then drop vertically to the resistance axis and read the new value of R_F (500 Ω in this example). Enter the graph at

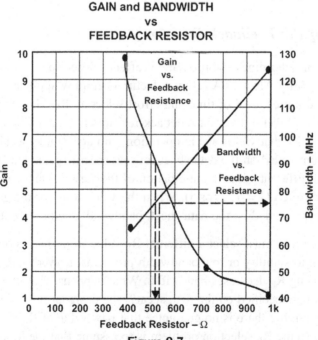

Figure 9.7

Plot of current-feedback amplifier R_F, G, and BW.

Table 9.1: Data Set for Curves in Fig. 9.7

Gain (A_{CL})	R_F (Ω)	Bandwidth (MHz)
+1	1000	125
+2	681	95
+10	383	65

the new value of R_F, and travel vertically until you intersect the bandwidth versus feedback resistance curve. Now move to the bandwidth axis to read the new bandwidth (75 MHz in this example). As a starting point you should expect to get approximately 75 MHz bandwidth with a gain of 6 and $R_F = 500\ \Omega$. Although this technique yields more reliable solutions than Eq. (9.27) does, op amp peculiarities, circuit board stray capacitances, and wiring make extensive testing mandatory. The circuit must be tested for performance and stability at each new operating point.

9.8 Stability and Input Capacitance

When designer lets the circuit board introduce stray capacitance on the inverting input node to ground, it causes the impedance Z_G to become reactive. The new impedance, Z_G, is given in Eq. (9.28), and Eq. (9.29) is the stability equation that describes the situation.

$$Z_G = \frac{R_G}{1 + R_G C_G s} \tag{9.28}$$

$$A\beta = \frac{Z}{Z_B + \dfrac{Z_F}{Z_G^2 + Z_B Z_G}} \tag{9.29}$$

$$A\beta = \frac{2}{R_F\left(1 + \dfrac{R_B}{R_F \| R_G}\right)(1 + R_B \| R_F \| R_G C_G s)} \tag{9.30}$$

Eq. (9.29) is the stability equation when Z_G consists of a resistor in parallel with stray capacitance between the inverting input node and ground. The stray capacitance, C_G, is a fixed value because it is dependent on the circuit layout. The pole created by the stray capacitance is dependent on R_B because it dominates R_F and R_G. R_B fluctuates with manufacturing tolerances, so the $R_B C_G$ pole placement is subject to IC manufacturing tolerances. As the $R_B C_G$ combination becomes larger, the pole moves toward the zero frequency axis, lowering the circuit stability. Eventually it interacts with the pole contained in Z, $1/\tau_2$, and instability results.

The effects of stray capacitance on CFA closed-loop performance are shown in Fig. 9.8.

Notice that the introduction of C_G causes more than 3 dB peaking in the CFA frequency response plot, and it increases the bandwidth about 18 MHz. Two picofarads are not a lot

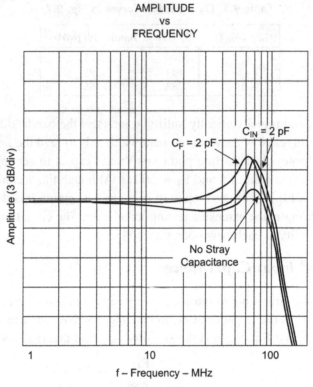

AMPLITUDE
vs
FREQUENCY

Figure 9.8
Effects of stray capacitance on current-feedback amplifiers.

of capacitance because a sloppy layout can easily add four or more picofarads to the circuit.

9.9 Stability and Feedback Capacitance

When a stray capacitor is formed across the feedback resistor, the feedback impedance is given by Eq. (9.31). Eq. (9.32) gives the loop gain when a feedback capacitor has been added to the circuit.

$$Z_F = \frac{R_F}{1 + R_F C_F s} \tag{9.31}$$

$$A\beta = \frac{Z(1 + R_F C_F s)}{R_F\left(1 + \dfrac{R_B}{R_F \| R_G}\right)(1 + R_B \| R_F \| R_G C_F s)} \tag{9.32}$$

This loop gain transfer function contains a pole and zero, thus, depending on the pole/zero placement, oscillation can result. The Bode plot for this case is shown in Fig. 9.9. The

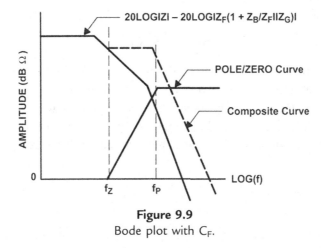

Figure 9.9
Bode plot with C_F.

original and composite curves cross the 0-dB axis with a slope of -40 dB/decade, so either curve can indicate instability. The composite curve crosses the 0-dB axis at a higher frequency than the original curve, hence the stray capacitance has added more phase shift to the system. The composite curve is surely less stable than the original curve. Adding capacitance to the inverting input node or across the feedback resistor usually results in instability. R_B largely influences the location of the pole introduced by C_F, thus here is another case where stray capacitance leads to instability.

Fig. 9.8 shows that $C_F = 2$ pF adds about 4 dB of peaking to the frequency response plot. The bandwidth increases about 10 MHz because of the peaking. C_F and C_G are the major causes of overshoot, ringing, and oscillation in CFAs, and the circuit board layout must be carefully done to eliminate these stray capacitances.

9.10 Compensation of C_F and C_G

When C_F and C_G both are present in the circuit they may be adjusted to cancel each other out. The stability equation for a circuit with C_F and C_G is Eq. (9.33).

$$A\beta = \frac{Z(1 + R_F C_F s)}{R_F\left(1 + \dfrac{R_B}{R_F \| R_G}\right)(R_B \| R_F \| R_G (C_F + C_G)s + 1)} \tag{9.33}$$

If the zero and pole in Eq. (9.33) are made to cancel each other, the only poles remaining are in Z. Setting the pole and zero in Eq. (9.33) equal yields Eq. (9.34) after some algebraic manipulation.

$$R_F C_F = C_G (R_G \| R_B) \tag{9.34}$$

R_B dominates the parallel combination of R_B and R_G, so Eq. (9.34) is reduced to Eq. (9.35).

$$R_F C_F = R_B C_G \qquad\qquad (9.35)$$

R_B is an IC parameter, so it is dependent on the IC process. R_B is an important IC parameter, but it is not important enough to be monitored as a control variable during the manufacturing process. R_B has widely spread, unspecified parameters, thus depending on R_B for compensation is risky. Rather, the prudent design engineer assures that the circuit will be stable for any reasonable value of R_B, and that the resulting frequency response peaking is acceptable.

9.11 Summary

Constant gain-bandwidth is not a limiting criterion for the CFA, so the feedback resistor is adjusted for maximum performance. Stability is dependent on the feedback resistor; as R_F is decreased, stability is decreased, and when R_F goes to zero the circuit becomes unstable. As R_F is increased stability increases, but the bandwidth decreases.

The inverting input impedance is very high, but the noninverting input impedance is very low. This situation precludes CFAs from operation in the differential amplifier configuration. Stray capacitance on the inverting input node or across the feedback resistor always leads to peaking, usually to ringing, and sometimes to oscillations. A prudent circuit designer scans the PC board layout for stray capacitances, and he eliminates them. Breadboarding and lab testing are a must with CFAs. The CFA performance can be improved immeasurably with a good layout, good decoupling capacitors, and low inductance components.

CHAPTER 10

Voltage- and Current-Feedback Op Amp Comparison

10.1 Introduction

The name, operational amplifier, was given to voltage-feedback amplifiers (VFAs) when they were the only op amps in existence. These amplifiers could be programmed with external components to perform various math operations on a signal; thus, they were nicknamed op amps. Current-feedback amplifiers (CFAs) have been around for a few decades, but their popularity has only increased in the last several years. Two factors limiting the popularity of CFAs are their application difficulty and lack of precision—you do *not* use a CFA in applications where DC response is included in the output, such as the cases developed in previous chapters.

So where is a CFA used? The answer is in very high-speed AC-coupled applications. The VFA is familiar component, and there are several variations of internally compensated VFAs that can be used with little applications work. Because of its long history, the VFA comes in many varieties and packages, so there are VFAs applicable to almost any job. But VFA bandwidth is limited, so it cannot function as well at high signal frequencies as the CFA can. For now, the signal frequency and precision separates the applications of the two op amp configurations.

The VFA has some other redeeming virtues, such as excellent precision, that makes it the desirable amplifier in low-frequency applications. Many functions other than signal amplification are accomplished at low frequencies, and functions like level-shifting a signal require precision. Fortunately, precision is not required in most high-frequency applications where amplification or filtering of a signal is predominant, so CFAs are suitable to high-frequency applications. The lack of precision coupled with the application difficulties prevents the CFA from replacing the VFA.

10.2 Precision

The long-tailed pair input structure gives the VFA its precision; the long-tailed pair is shown in Fig. 10.1.

Op Amps for Everyone. http://dx.doi.org/10.1016/B978-0-12-811648-7.00010-8

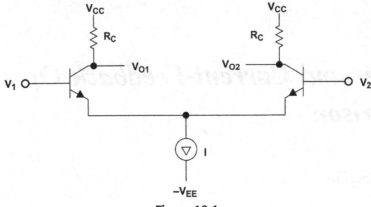

Figure 10.1
Long-tailed pair.

The transistors, Q_1 and Q_2, are very carefully matched for initial and drift tolerances. Careful attention is paid to detail in the transistor design to insure that parameters such as current gain, β, and base-emitter voltage, V_{BE}, are matched between the input transistors, Q_1 and Q_2. When $V_{B1} = V_{B2}$, the current, I, splits equally between the transistors, and $V_{O1} = V_{O2}$. As long as the transistor parameters are matched, the collector currents stay equal. The slightest change of V_{B1} with respect to V_{B2} causes a mismatch in the collector currents and a differential output voltage $|V_{B1} - V_{B2}|$.

When temperature or other outside influences change transistor parameters such as current gain or base-emitter voltage, as long as the change is equal, it causes no change in the differential output voltage. IC designers go to great lengths to ensure that transistor parameter changes due to external influences do not cause a differential output voltage change. The slightest change in either base voltage causes a differential output voltage change, and gross changes in external conditions do not cause a differential output voltage change. This is the formula for a precision amplifier because it can amplify small input changes while ignoring changes in the parameters or ambient conditions.

This is a simplified explanation, and there are many different techniques used to ensure transistor matching. Some of the techniques used to match input transistors are parameter trimming, special layout techniques, thermal balancing, and symmetrical layouts. The long-tailed pair is an excellent circuit configuration for obtaining precision in the input circuit, but the output circuit has one fault. The output circuit collector impedance has to be high to achieve high gain in the first stage. High impedance coupled with the Miller capacitance discussed previously forms a quasidominant pole compensation circuit that has poor high-frequency response.

The noninverting input of the CFA (see Fig. 10.2) connects to a buffer input inside the op amp. The inverting input of the CFA connects to a buffer output inside the CFA. Buffer

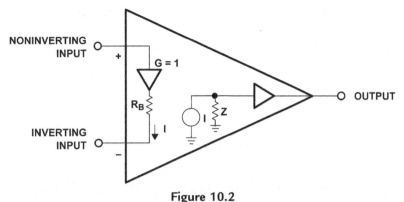

Figure 10.2
Ideal current-feedback amplifier.

inputs and outputs have dramatically different impedance levels, so any matching becomes a moot point. The buffer cannot reject common-mode voltages introduced by parameter drifts because it has no common-mode rejection capability. The input current causes a voltage drop across the input buffer's output impedance, RB, and there is no way that this voltage drop can be distinguished from an input signal.

The CFA circuit configuration was selected for high-frequency amplification because it has current-controlled gain and a current-dominant input. Being a current device, the CFA does not have the Miller effect problem that the VFA has. The input structure of the CFA sacrifices precision for bandwidth, and CFAs achieve usable bandwidths ten times the usable VFA bandwidth.

10.3 Bandwidth

The bandwidth of a circuit is defined by high-frequency errors. When the gain falls off at high frequencies unequal frequency amplification causes the signal to become distorted. The signal loses its high-frequency components; an example of high-frequency signal degradation is a square wave with sharp corners that is amplified and turned into slump cornered semi-sine wave. The error equation for any feedback circuit is repeated in Eq. (10.1).

$$E = \frac{V_{IN}}{1 + A\beta} \tag{10.1}$$

This equation is valid for any feedback circuit, so it applies equally to a VFA or a CFA. The loop gain equation for any VFA is repeated as Eq. (10.2).

$$A\beta = \frac{aR_G}{R_F + R_G} \tag{10.2}$$

Eq. (10.2) is rewritten below as Eqs. (10.3) and (10.4) for the noninverting and inverting circuits, respectively. In each case, the symbol G_{CLNI} and G_{CLI} represent the closed-loop gain for the noninverting and inverting circuits, respectively.

$$A\beta = \frac{a}{\frac{R_F + R_G}{R_G}} = \frac{a}{G_{CLNI}} \tag{10.3}$$

$$A\beta = \frac{a}{\frac{R_F + R_G}{R_G}} = \frac{a}{G_{CLI} + 1} \tag{10.4}$$

In both cases the loop gain decreases as the closed-loop gain increases, thus all VFA errors increase as the closed-loop gain increases. The error increase is mathematically coupled to the closed-loop gain equation, so there is no working around this fact. For the VFA, effective bandwidth decreases as the closed-loop gain increases because the loop gain decreases as the closed-loop gain increases.

A plot of the VFA loop gain, closed-loop gain, and error is given in Fig. 10.3. Referring to Fig. 10.3, the direct gain, A, is the op amp open loop gain, a, for a noninverting op amp. The direct gain for an inverting op amp is $(a(Z_F/(Z_G + Z_F)))$. The Miller effect causes the direct gain to fall off at high frequencies, thus error increases as frequency increases because the effective loop gain decreases. At a given frequency, the error also increases when the closed-loop gain is increased.

The CFA is a current operated device; hence, it not nearly as subject to the Miller effect resulting from stray capacitance as the VFA is. The absence of the Miller effect enables the CFA's frequency response to hold up far better than the VFA's does. A plot of the CFA loop

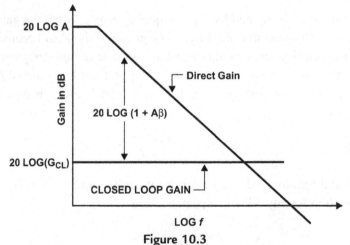

Figure 10.3
Voltage-feedback amplifier gain versus frequency.

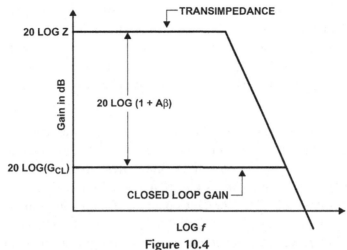

Figure 10.4
Current-feedback amplifier gain versus frequency.

gain, transimpedance, and error is given in Fig. 10.4. Notice that the transimpedance stays at the large low-frequency intercept value until much higher frequencies than the VFA does.

The loop gain equation for the CFA is repeated here as Eq. (10.5).

$$A\beta = \frac{Z}{R_F\left(1 + \dfrac{R_B}{R_F \parallel R_G}\right)} \tag{10.5}$$

When the input buffer output resistance approaches zero, Eq. (10.5) reduces to Eq. (10.6).

$$A\beta = \frac{Z}{R_F} \tag{10.6}$$

Eq. (10.6) shows that the closed-loop gain has no effect on the loop gain when $R_B = 0$, so under ideal conditions one would expect the transimpedance to fall off with a zero slope. Fig. (10.4) shows that there is a finite slope, but much less than that of a VFA, and the slope is caused by R_B not being equal to zero. For example, R_B is usually 50 Ω when $R_F = 1000$ Ω at $A_{CL} = 1$. If we let $R_F = R_G$, then $R_F \parallel R_G = 500$ Ω, and $R_B/R_F \parallel R_G = 50/500 = 0.1$.

Substituting this value into Eq. (10.6) yields Eq. (10.7), and Eq. (10.7) is almost identical to Eq. (10.6). RB does cause some interaction between the loop gain and the transimpedance, but because the interaction is secondary the CFA gain falls off with a faster slope.

$$A\beta = \frac{Z}{1.1R_F} \tag{10.7}$$

The direct gain of a VFA starts falling off early, often at 10 or 100 Hz, but the transimpedance of a CFA does not start falling off until much higher frequencies. The VFA is constrained by the gain-bandwidth limitation imposed by the closed-loop gain being incorporated within the loop gain. The CFA, with the exception of the effects of R_B, does not have this constraint. This adds up to the CFA being the superior high-frequency amplifier.

Remember the discussion at the end of Chapter 7—are you smarter than an op amp? This is your way to exceed that speed limit (at least the speed limit you would get with a similar VFA), with some constraints on your ambition. Real open loop plots of CFAs in real data sheets are seldom as dramatic as Fig. 10.4, but offer at least some improvement over VFAs. Whether it is with a VFA or CFA op amps, when it comes to open loop bandwidth, a truism from extreme sports applies here: "go big or go home." Within constraints of cost and power consumption in your design, if precision in amplitude is a requirement, throw as much open loop bandwidth at the application as you can!

10.4 Stability

Stability in a feedback system is defined by the loop gain, and no other factor, including the inputs or type of inputs, affects stability. The loop gain for a VFA is given in Eq. (10.2). Examining Eq. (10.2), we see that the stability of a VFA depends on two items; the op amp transfer function, a, and the gain setting components, Z_F/Z_G.

The op amp contains many poles, and if it is not internally compensated, it requires external compensation. The op amp always has at least one dominant pole, and the most phase margin that an op amp has is 45 degrees. Phase margins beyond 60 degrees are a waste of op amp bandwidth. When poles and zeros are contained in Z_F and Z_G, they can compensate for the op amp phase shift or add to its instability. In any case, the gain setting components always affect stability. When the closed-loop gain is high, the loop gain is low, and low loop gain circuits are more stable than high loop gain circuits.

Wiring the op amp to a printed circuit board always introduces components formed from stray capacitance and inductance. Stray inductance becomes dominant at very high frequencies, hence, in VFAs, it does not interfere with stability as much as it does with signal handling properties. Stray capacitance causes stability to increase or decrease depending on its location. Stray capacitance from the input or output lead to ground induces instability, while the same stray capacitance in parallel with the feedback resistor increases stability.

The loop gain for a CFA with no input buffer output impedance, R_B, is given in Eq. (10.6). Examining Eq. (10.6), we see that the stability of a CFA depends on two items: the op amp transfer function, Z, and the gain setting component, Z_F. The op amp contains many poles, thus they require external compensation. Fortunately, the external compensation for a CFA is done with Z_F. The factory applications engineer does extensive testing to determine the optimum value of R_F for a given gain. This value should be used in all applications at that gain, but increased stability and less peaking can be obtained by increasing R_F. Essentially this is sacrificing bandwidth for lower-frequency performance, but in applications not requiring the full bandwidth, it is a wise tradeoff.

The CFA stability is not constrained by the closed-loop gain, thus a stable operating point can be found for any gain, and the CFA is not limited by the gain-bandwidth constraint. If the optimum feedback resistor value is not given for a specific gain, one must test to find the optimum feedback resistor value.

Stray capacitance from any node to ground adversely affects the CFA performance. Stray capacitance of just a couple of picofarads from any node to ground causes 3 dB or more of peaking in the frequency response. Stray capacitance across the CFA feedback resistor, quite unlike that across the VFA feedback resistor, always causes some form of instability. CFAs are applied at very high frequencies, so the printed circuit board inductance associated with the trace length and pins adds another variable to the stability equation. Inductance cancels out capacitance at some frequency, but this usually seems to happen in an adverse manner. The wiring of VFAs is critical, but the wiring of CFAs is a science. Stay with the layout recommended by the manufacturer whenever possible.

10.5 Impedance

The input impedance of a VFA and CFA differs dramatically because their circuit configurations are very different. The VFA input circuit is a long-tailed pair, and this configuration gives the advantages that both input impedances match. Also, the input signal looks into an emitter-follower circuit that has high input impedance. The emitter-follower input impedance is $\beta(r_e + R_E)$, where R_E is a discrete emitter resistor. At low input currents, R_E is very high and the input impedance is very high. If a higher input impedance is required, the op amp uses a Darlington circuit that has an input impedance of $\beta^2(r_e + R_E)$.

So far, the implicit assumption is that the VFA is made with a bipolar semiconductor process. Applications requiring very high input impedances often use a FET process. Both BIFET and CMOS processes offer very high input impedance in any long-tailed pair configuration. It is easy to get matched and high input impedances at the amplifier inputs. Do not confuse the matched input impedance at the op amp leads with the overall circuit input impedance. The input impedance looking into the inverting input is R_G, and the impedance looking into the noninverting input is the input impedance of the op amp. While these are two different impedances, they are mismatched because of the circuit not the op amp.

The CFA has a radically different input structure that causes it to have mismatched input impedances. The noninverting input lead of the CFA is the input of a buffer that has very high input impedance. The inverting input lead is the output of a buffer that has very low impedance. There is no possibility that these two input impedances can be matched.

Again, because of the circuit, the inverting circuit input impedance is R_G. Once the circuit gain is fixed, the only way to increase R_G is to increase R_F. But, R_F is determined by a tradeoff between stability and bandwidth. The circuit gain and bandwidth requirements fix R_F, hence there is no room to further adjust R_F to raise the resistance of R_G. If the manufacturer's data sheet says that $R_F = 100\ \Omega$ when the closed-loop gain is 2, then

Table 10.1: Tabulation of Pertinent Voltage-Feedback Amplifier and Current-Feedback Amplifier Equations

Circuit Configuration	Current-Feedback Amplifier	Voltage-Feedback Amplifier
Noninverting		
Forward or direct gain	$$\dfrac{Z\left(1+\dfrac{Z_F}{Z_G}\right)}{Z_F\left(1+\dfrac{Z_B}{Z_F\ \|\ Z_G}\right)}$$	a
Ideal loop gain	$$\dfrac{Z}{Z_F}\left(1+\dfrac{Z_B}{Z_F\ \|\ Z_G}\right)$$	$\dfrac{aZ_F}{(Z_G+Z_F)}$
Actual closed-loop gain	$$\dfrac{Z_F\left(1+\dfrac{Z_B}{Z_G}\right)}{\dfrac{Z_F\left(1+\dfrac{Z_B}{Z_F\ \|\ Z_G}\right)}{1+\dfrac{Z}{Z_F\left(1+\dfrac{\frac{Z_B}{1+Z_B}}{Z_F\ \|\ Z_G}\right)}}}$$	$\dfrac{a}{1+\dfrac{aZ_G}{Z_F\ \|\ Z_G}}$
Closed-loop gain	$1+Z_F/Z_G$	$1+Z_F/Z_G$
Inverting		
Forward or direct gain	$$\dfrac{Z}{Z_G\left(1+\dfrac{Z_B}{Z_F\ \|\ Z_G}\right)}$$	$\dfrac{aZ_F}{(Z_F+Z_G)}$
Ideal loop gain	$$\dfrac{Z}{Z_F}\left(1+\dfrac{Z_B}{Z_F\ \|\ Z_G}\right)$$	$\dfrac{aZ_G}{(Z_G+Z_F)}$
Actual closed-loop gain	$$\dfrac{-Z_G\left(1+\dfrac{Z_B}{Z_F\ \|\ Z_G}\right)}{1+\dfrac{Z}{Z_F\left(1+\dfrac{Z_B}{Z_F\ \|\ Z_G}\right)}}$$	$\dfrac{\dfrac{-aZ_F}{Z_F+Z_G}}{1+\dfrac{aZ_G}{Z_F\ \|\ Z_G}}$
Closed-loop gain	$-Z_F/Z_G$	$-Z_F/Z_G$

$R_G = 100$ or $50\,\Omega$ depending on the circuit configuration. This sets the circuit input impedance at $100\,\Omega$. This analysis is not entirely accurate because R_B adds to the input impedance, but this addition is very small and dependent on IC parameters. CFA op amp circuits are usually limited to noninverting voltage applications, but they serve very well in inverting applications that are current-driven.

The CFA is limited to the bipolar process because that process offers the highest speed. The option of changing process to BIFET or CMOS to gain increased input impedance is not attractive today. Although this seems like a limiting factor, it is not because CFAs are often used in low impedance where the inputs are terminated in 50 or $75\,\Omega$. Also, most very high-speed applications require low impedances.

10.6 Equation Comparison

The pertinent VFA and CFA equations are repeated in Table 10.1. Notice that the ideal closed-loop gain equations for the inverting and noninverting circuits are identical. The ideal equations for the VFA depend on the op amp gain, a, being very large thus making $A\beta$ large compared to one. The CFA needs two assumptions to be valid to obtain the ideal equations. First, the ideal equations for the CFA depend on the op amp transimpedance, Z, being very large, thus making $A\beta$ large compared to one. Second, R_B must be very small compared to $Z_F \| Z_G$.

The ideal gain equations are identical, but the applications are very different because the VFA is best applied to lower-frequency precision jobs, while the CFA applications are in the very high-frequency realm. The transimpedance in a CFA acts much like the gain does in a VFA. In each case, transimpedance or gain, it is the parameter that enables the use of feedback.

CHAPTER 11

Fully Differential Op Amps

11.1 Introduction

The term "fully differential op amp" probably brings chills to the spine of designers. Such thoughts as "oh no—now I have to learn something new." What most designers do not realize is that op amps began as fully differential components over 50 years ago. Techniques about how to use the fully differential versions have been almost lost over the decades. Today's fully differential op amps offer performance advantages unheard of in those first units.

This chapter will just present the facts a designer needs to get started, and some resources for further design assistance. Hopefully, after reading this chapter, a designer can approach a fully differential op amp design with confidence and excitement.

11.2 What Does "Fully Differential" Mean?

Designers should already be familiar with single-ended op amps after reading the other chapters of this book. Briefly, single-ended op amps have two inputs—a positive and negative input—which are understood to be fully differential. They have a single output, which is referenced to system ground (Fig. 11.1).

The op amp also has two power supply inputs, which are connected to bipolar power supplies (equal and opposite positive and negative potentials), or a single potential, with a positive supply and a ground connected to the power supply pins. These power supply pins are often omitted from the schematic symbol, when power supply connections are implied elsewhere on the schematic.

Fully Differential op amps add a second output (Fig. 11.2).

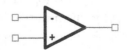

Figure 11.1
Single-ended op amp schematic symbol.

Op Amps for Everyone. http://dx.doi.org/10.1016/B978-0-12-811648-7.00011-X

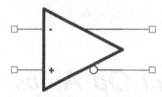

Figure 11.2
Fully differential op amp schematic symbol.

The second output is fully differential—the two outputs are called "positive output" and "negative output"—similar terminology to the two inputs. Like the inputs, they are differential. The output voltages will be equal and opposite in polarity (referenced to the common-mode operating point of the circuit).

11.3 How is the Second Output Used?

An op amp is used as a closed-loop device. Most designers know how to close the loop on a single-ended op amp.

Whether the single-ended op amp is used in an inverting or a noninverting mode, the loop is closed from the output to the inverting input.

11.4 Differential Gain Stages

So how is the loop closed on a fully differential op amp? It stands to reason, if there are two outputs, both of them have to be operated closed loop. Therefore, the equivalent way of closing the loop on a fully differential op amp is:

Two feedback loops are required to close the loops for a fully differential op amp. If the loops are not matched, there can be significant second-order harmonic distortion, but there are special cases where the output pathways can be different, and one will be discussed later in this chapter.

Note that for a fully differential op amp, each feedback loop is an inverting feedback loop. Both polarities of output are available, so terms like "inverting" and "noninverting" are meaningless. Instead, think of the single-ended schematics in Fig. 11.3. In both cases, the loop goes from the (noninverting) output to the inverting input, introducing a 180 degrees phase shift. For the fully differential op amp, the top feedback loop has a 180 degrees phase shift from the noninverting output to the inverting input, and the bottom feedback loop has a 180 degrees phase shift from the inverting output to the noninverting input. Both feedback paths are therefore inverting. There is no "noninverting" fully differential op amp gain circuit.

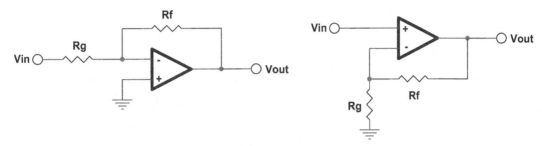

Figure 11.3
Closing the loop on a single-ended op amp.

The gain of the (ideal) fully differential op amp stage is

$$\frac{V_O}{V_I} = \frac{R_F}{R_G} \tag{11.1}$$

Exercise the proper caution in applying fully differential op amps, or the same type of errors introduced in Chapter 7 will ensue. Keep away from that op loop limitation!

11.5 Single-Ended to Differential Conversion

The schematic shown in Fig. 11.4 is a fully differential gain circuit. Fully differential applications, however, are somewhat limited. Very often the fully differential op amp is used to convert a single-ended signal to a differential signal—perhaps to connect to the differential input of an A/D converter.

The two configurations shown in Fig. 11.5 are equivalent. At first glance, they look identical, but they are not. The difference is that in the left configuration, the inverting input is used for signal and the noninverting input for reference. In the right configuration,

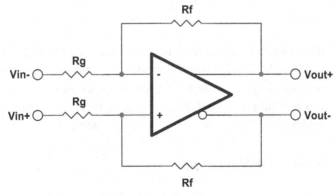

Figure 11.4
Closing the loop on a fully differential op amp.

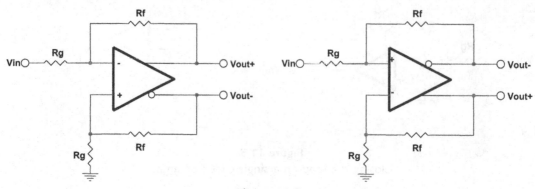

Figure 11.5
Single ended to differential conversion.

the noninverting input is used for signal and the inverting input for reference. They are functionally equivalent, either one will work.

The gain of the single-ended to differential stage is

$$\frac{V_O}{V_I} = \frac{R_F}{R_G} \tag{11.2}$$

The only difference between this configuration and the previous is that one side of the input voltage is referenced to ground.

The dynamics of the gain are sometimes best described pictorially. Fig. 11.6 shows the relationship between V_{IN}, V_{OUT+}, and V_{OUT-} when $R_F = R_G$.

OK, what is going on here? The amplitude of the input, V_{IN}, is twice that of the output? The gain is correct, however, because the value of the differential gain $[(V_{OUT+}) - (V_{OUT-})]$ at any point in Fig. 11.6 is equal to the amplitude of V_{IN}.

11.6 A New Function

Texas Instruments fully differential op amps have an additional pin: V_{ocm}, which stands for "voltage output common mode (level)." The function of this pin can be either and input or an output, because its source is just a voltage divider off of the power supply, but it is seldom used as an output. When it is used as an output, it will correspond to the common-mode voltage about which the V_{OUT+} and V_{OUT-} outputs swing.

11.7 Conceptualizing the V_{ocm} Input

In the mechanical model, which is a more complex version of a child's teeter totter, physically raising the V_{IN-} arm (with a length of R_1) causes the arm to pivot on fulcrum

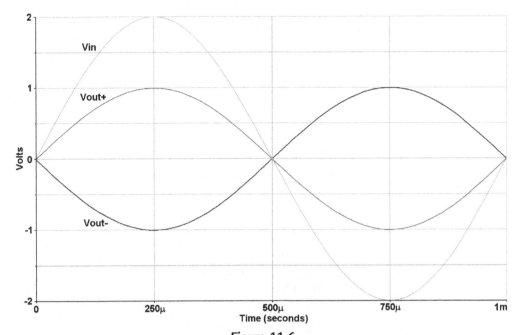

Figure 11.6
Relationship between V_{IN}, V_{OUT+}, and V_{OUT-}.

V_{IN}, and the other side of the arm V_{OUT+}, will rise proportionally to the length of the arm (R_2). A second fulcrum (V_{ocm}), between the two arms, forces the end of the other arm (V_{OUT-}) to go down by the same amount. The second arm also has a fulcrum at Vp, which causes its other side (V_{IN+}) to move the same amount as V_{IN-}, but in the other direction (assuming lengths $R_1 = R_3$, and $R_2 = R_4$). V_{ocm} can be used to raise and lower the average height (offset) of both V_{OUT+} and V_{OUT-} equally. But be careful! It can also exceed the mechanical limits of the model! Too low and the V_{OUT+} and V_{OUT-} ends of the arms will hit "ground." Too high and the V_{IN-} and V_{IN+} ends of the arms can both hit "ground." The base of the teeter totter and its fulcrum represent the "potential" for movement or power supply (Figs. 11.7 and 11.8).

We actually implemented this teeter totter model with a well-known set of child's building blocks, complete with a motor for excitation. If the reader has a set of these toys, you might consider building up the model for your enjoyment, education, and a way to introduce your children to the concepts of differential transmission (Fig. 11.9).

The most common use of the V_{ocm} pin is to set the output common-mode level of the fully differential op amp. This is a very useful function, because it can be used to match the common-mode point of a data converter to which the fully differential amplifier is connected. High-precision/high-speed data converters often employ differential inputs and provide a reference output.

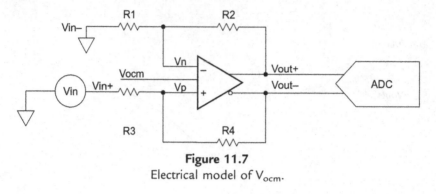

Figure 11.7
Electrical model of V_{ocm}.

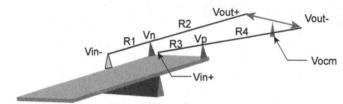

Figure 11.8
Mechanical model of V_{ocm}.

Figure 11.9
A mechanical fully differential amplifier.

The schematic of Fig. 11.10 is simplified, and does not show compensation, termination, or decoupling components for clarity. Nevertheless, it shows the basic concept. This is an important type of interface and will be elaborated on further in a later chapter.

The remainder of this chapter presents the designer with a basic set of applications based on fully differential applications.

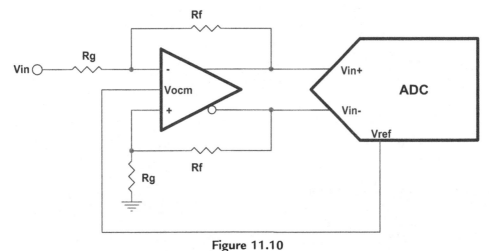

Figure 11.10
Using a fully differential op amp to drive a analog-to-digital converter

11.8 Instrumentation

An instrumentation amplifier can be constructed from two single-ended amplifiers and a fully differential amplifier as shown in Fig. 11.11. Both polarities of the output signal are available, of course, and there is no ground dependence.

11.9 Filter Circuits

Filtering is done to eliminate unwanted content in audio, among other things. Differential filters that do the same job to differential signals as their single-ended cousins do to single-ended signals can be applied.

For differential filter implementations, the components are simply mirror imaged for each feedback loop. The components in the top feedback loop are designated *A*, and those in the bottom feedback loop are designated *B*.

For clarity, decoupling components are not shown in the following schematics. Proper operation of high-speed op amps requires proper decoupling techniques. That does not mean a shotgun approach of using inexpensive 0.1-µF capacitors. Decoupling component selection should be based on the frequencies that need to be rejected and the characteristics of the capacitors used at those frequencies.

11.9.1 Single-Pole Filters

Single-pole filters are the simplest filters to implement with single-ended op amps and the same holds true with fully differential amplifiers.

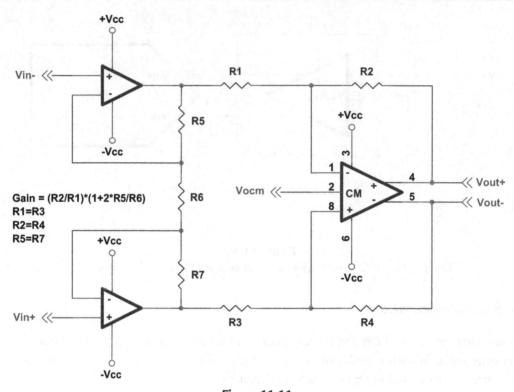

Figure 11.11
Instrumentation amplifier.

A low-pass filter can be formed by placing a capacitor in the feedback loop of a gain stage, in a manner similar to single-ended op amps (Fig. 11.12).

A high-pass filter can be formed by placing a capacitor in series with an inverting gain stage as shown in Fig. 11.13:

11.9.2 Double-Pole Filters

Many double-pole filter topologies incorporate positive and negative feedback, and therefore have no differential implementation. Others employ only negative feedback, but use the noninverting input for signal input and also have no differential implementation. This limits the number of options for designers, because both feedback paths must return to an input.

The good news, however, is that there are topologies available to form differential low-pass, high-pass, band-pass, and notch filters. However, the designer might have to use an unfamiliar topology or more op amps than would have been required for a single-ended circuit.

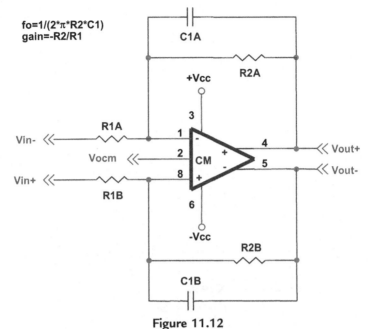

Figure 11.12
Single-pole differential low-pass filter.

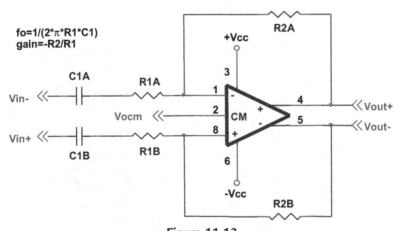

Figure 11.13
Single-pole differential high-pass filter.

11.9.3 Multiple Feedback Filters

Multiple feedback (MFB) filter topology is the simplest topology that will support fully differential filters. Unfortunately, the MFB topology is a bit hard to work with, but component ratios are shown for common unity gain filters (Figs. 11.14 and 11.15).

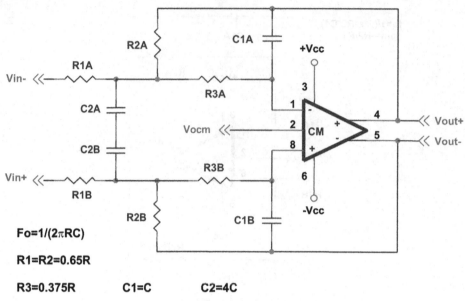

Fo=1/(2πRC)

R1=R2=0.65R

R3=0.375R   C1=C   C2=4C

Figure 11.14
Differential low-pass filter.

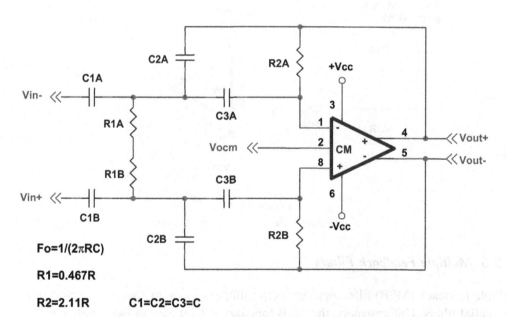

Fo=1/(2πRC)

R1=0.467R

R2=2.11R   C1=C2=C3=C

Figure 11.15
Differential high-pass filter.

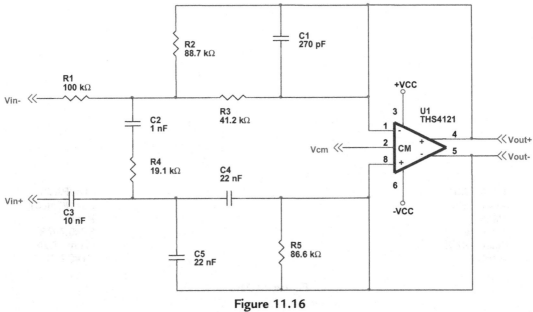

Figure 11.16
Differential speech filter.

There is no reason why the feedback paths have to be identical. A band-pass filter can be formed by using nonsymmetrical feedback pathways (one low pass and one high pass). Fig. 11.16 shows a band-pass filter that passes the range of human speech (300 Hz–3 kHz). Fig. 11.17 shows the response.

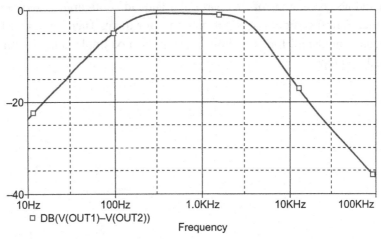

Figure 11.17
Differential speech filter response.

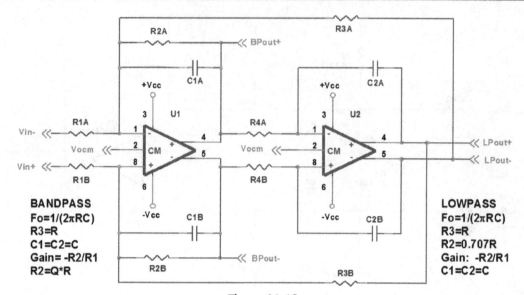

Figure 11.18
Differential biquad filter.

11.9.4 Biquad Filter

Biquad filter topology is a double-pole topology that is available in low pass, high pass, band pass, and notch. The single-ended implementation of this filter topology has three op amps, with the third op amp included only to invert the output of the previous op amp. That inversion is inherent in the fully differential op amp, and therefore eliminates the third op amp, reducing the total number of op amps required to 2 (Fig. 11.18).

The high-pass and notch versions of this particular Biquad configuration require additional op amps, and therefore this topology is not optimum for them. There are other topologies, however, that can generate all four functions and can also be implemented with only two fully differential op amps.

CHAPTER 12

Different Types of Op Amps

12.1 Introduction

The previous chapters have covered voltage- and current-feedback op amps, as well as fully differential op amps. These are the main types of op amps, but certainly not the only types—there are a few other types of op amps, most of which are special cases or applications of voltage-feedback op amps, integrated onto a single piece of silicon for the designer.

12.2 Uncompensated/Undercompensated Voltage-Feedback Op Amps

I have alluded to these before, but reintroduce the topic here because they are still available for designers who know how to use them. Fig. 12.1 (adapted from a real data sheet) shows the open loop plot for an undercompensated op amp.

If you have been diligent in understanding the previous chapters on stability, you can see that at unity gain (0 dB), the op amp above will be very unstable. Not only is a secondary pole clearly visible in the open loop response, but the phase plot, and therefore the phase margin, has taken a nosedive. This amplifier will be unstable at unity gain, in danger of

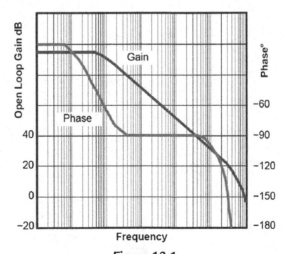

Figure 12.1

Undercompensated op amp.

Op Amps for Everyone. http://dx.doi.org/10.1016/B978-0-12-811648-7.00012-1

sustained oscillation. In this case, the op amp would be stable at a gain of 10 (20 dB), but no lower. In fact, this figure was adapted from the data sheet of an op amp that is advertised as stable at gains of 10 and above. Let the designer beware, and not apply it at lower gains!

There are only a couple of completely uncompensated op amps on the market. Pay close attention to the data sheet, and place external compensation components as recommended by the data sheet, or they will be unstable.

12.3 Instrumentation Amplifier

An instrumentation amplifier is used to amplify a differential signal when both inputs need to be high impedance, usually because the source is high impedance. Fig. 12.2 shows a common application that of a strain gauge. A strain gauge consists of four resistive elements, one or more of which varies with applied mechanical stress. If the traditional differential amplifier of Fig. 2.9 were used, the input impedances of the stage would load the strain gauge down, invalidating the measurement. The only way to combat this loading effect is with the three-amplifier implementation of Fig. 12.2.

In this circuit configuration, both sources connect to the noninverting input of two op amps. This impedance is very high, and if the op amps are identical, both impedances are very nearly equal.

When $R_7 = R_6$, $R_5 = R_2$, $R_1 = R_4$,

$$V_{OUT} = (V_{IN2} - V_{IN1})\left(\frac{2R_1}{R_3} + 1\right)\left(\frac{R_6}{R_2}\right) + V_{REF} \qquad (12.1)$$

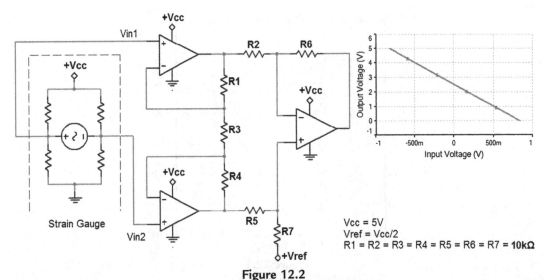

Vcc = 5V
Vref = Vcc/2
R1 = R2 = R3 = R4 = R5 = R6 = R7 = 10kΩ

Figure 12.2
Instrumentation amplifier.

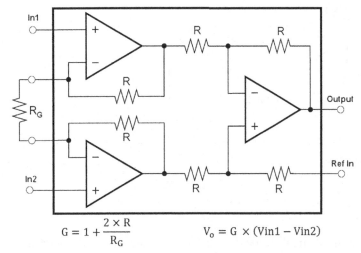

$$G = 1 + \frac{2 \times R}{R_G}$$

$$V_o = G \times (Vin1 - Vin2)$$

Figure 12.3
High-precision differential amplifier.

This differential amplifier has the unique feature that the gain can be changed with only one resistor R_3. Implementing instrumentation amplifiers, however, can become troublesome, especially if board space and power are at a premium. Resistors should be matched with more precision than is expected from the circuit. Resistor mismatching increases distortion due to unequal gains, and it increases the common-mode voltage feed through. Resistors are hard to match, and matched resistor sets/arrays are expensive with long lead times. Many semiconductor manufacturers, therefore, have implemented the three op amp topology directly in silicon, freeing the designer from the need to do so. Fig. 12.3 shows a typical offering from a semiconductor company.

Resistors integrated on silicon can be matched to a high degree of precision, and board space is recovered because many of these IC's are implemented in eight pin packages. Topologies inside the IC may differ from that shown above, so the designer is cautioned to read the data sheet carefully to design a circuit with the correct input impedance and gain characteristics. The output amplifier, for example, may not be a unity gain stage, the ratio of R_F to R_G on the output stage may be 10 or even 100, making extremely high gains possible for very low input signal levels. Also note that the topology above will still work even if R_G is omitted. Stage gain for Fig. 12.3 would revert to 1.

12.4 Difference Amplifier

A variation on the instrumentation amplifier is the difference amplifier. These are used when the input voltage is larger than the supply voltage of the chip, and are, by design, stable attenuators. They can be followed by gain stages so the net effect is to translate

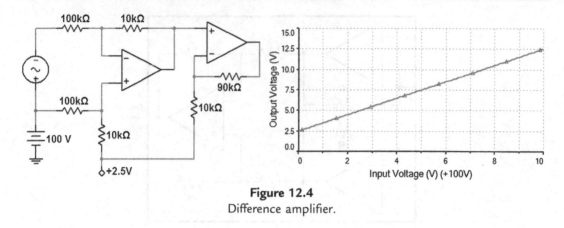

Figure 12.4
Difference amplifier.

voltage levels that are difficult to deal with into a manageable voltage range. These most commonly involve voltage levels with high DC offsets. Fig. 12.4 shows the use of a difference amplifier.

The difference amplifier of Fig. 12.4 has an input of 0−10 V, with a DC offset of 100 V (making the actual input 100−110 V). This voltage is to be monitored at the output of the second op amp. The circuit operates off of a power supply of 0−15 V.

The first stage performs the operation of removing the common-mode 100 V DC offset. Because the input resistors are 100 k and the gain resistors 10 k, they effectively operate as voltage dividers, taking the input voltage down into the range where the op amp can handle the voltage. The input stage is a differential attenuator stage, which divides the input voltage by a factor of 10. Additionally, a 2.5 V offset lifts the output off of ground so the op amp V_{OL} does not clip the response. Therefore, the output of the first op amp is 2.5−3.5 V. The good news is that the 100 V DC offset has been eliminated. The bad news is that the voltage swing on top of the 100 V offset has been attenuated by a factor of 10:1. This does not take good advantage of the available voltage range of 0−15 V.

The second stage corrects this by amplifying the voltage by a factor of 10, while preserving the 2.5 V offset. This produces the output characteristic shown in Fig. 12.4 where a swing of 100−110 V on the input produces an output swing of 2.5−12.5 V, easily within the V_{OH}/V_{OL} range of the output op amp, and ready to be interpreted by a data converter or meter calibrated to eliminate the 2.5 V offset (see Appendix B).

Difference amplifiers are most commonly used as "high side current monitors" for power supplies. Fig. 12.5 illustrates this application:

This figure is actually a minor variation of Fig. 12.5. The schematic has been rearranged a bit to show the components commonly integrated into an IC (the dashed line). The 100 V DC offset is the power supply to be monitored, and the signal source that was above it has been

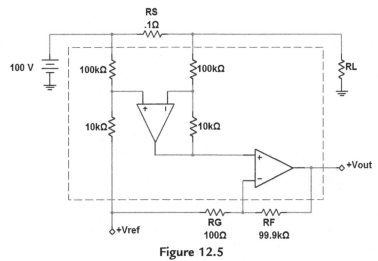

Figure 12.5
High side current monitor.

replaced with a current sense resistor RS. The output of the sense resistor is connected to a load RL which will be assumed to be 99.9 Ω. The astute designer might recognize at this point that RS and RL form a voltage divider. Because RS is such a small value, it does not contribute much voltage drop compared to RL. The designer should also exercise caution—RS will probably need to be a larger wattage resistor if RL is drawing appreciable current. In this case the total load on the 100 V power supply is 100 Ω, and therefore the current through RS and RL is 1 A. The wattage of RS is therefore 0.1 W. The wattage in RL is the responsibility of the designer and is assumed to be distributed among many active components in an application circuit. The voltage drop across RS is limited to 0.1 V, which leaves 99.9 V available to the load, a drop of only 0.1%. $+V_{REF}$ is still +2.5 V. The second amplifier is now operated at a gain of 1000, which will give an output at V_{OUT} of 12.5 V for the fully loaded power supply operating at 1 A, and V_{OUT} will be 2.5 V for an open circuit. R_F could be made a standard value of 100 kΩ with only a 0.1% measurement error.

An example difference amplifier is shown in Fig. 12.6. This particular model is the AD628 or INA146, which are functional but not pin equivalent devices, and happen to be applicable to the design examples above. These devices have made provision for an RC low-pass filter to reduce noise, which should not be confused with a compensation network. Other models in the product lines from semiconductor manufacturers are designed for different levels of attenuation (allowing more or less common mode voltage). Some may omit the output amplifier. The designer should read the data sheet carefully to know what device is best for their application.

So why buy a special difference amplifier IC instead of just implementing with ordinary op amps? Referring back to earlier discussions, it should be clear that op amp inverting

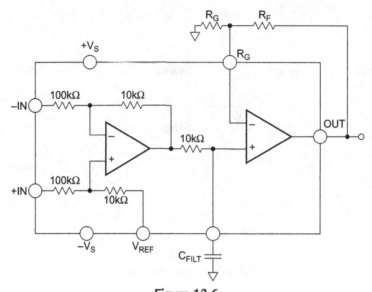

Figure 12.6
Commercial difference amplifier.

attenuators are inherently unstable—this applies to differential stages as well as simple inverting stages. Think of the differential stage in Fig. 12.4 as a simple inverting attenuator with an offset applied to the noninverting input, it is still an inverting attenuator and therefore unstable. The input amplifier in IC difference amplifiers have been designed to be gain of 0.1 stable, while the output amplifier is designed for high gains with low offsets, which are two very different stability and performance criteria. Of course, the designer can always keep this in their arsenal of design tricks if they really want to design an inverting op amp attenuator!

12.5 Buffer Amplifiers

The op amps discussed until this point have had something in common. They are somewhat limited in the amount of power that they can drive with their output. Voltage-feedback amplifiers, in general, can drive a 600 Ω load fairly well, but are not designed for lower impedances. Current-feedback amplifiers, on the other hand, are often designed with very robust output stages. In fact, a whole class of line driver current-feedback amplifiers was designed for digital subscriber line (DSL) applications. Unfortunately, DSL is losing traction in favor of cable and fiber to the home networking. The high output power op amps remain for now, but may become obsolete as time goes by.

Fortunately, another class of amplifiers is available—buffer amplifiers. Buffer amplifiers can be thought of as integrated unity gain buffers, hook up a power supply, bypass it properly, apply an input signal, and connect the output to the load. No resistors required! A very slick way to drive heavy loads such as long cables, audio loads, etc.

Buffer amplifiers, being active devices, will add their own characteristics to the signal. Fortunately, there is a way to partially cancel undesirable buffer amplifier effects—put them inside the loop!

The circuit of Fig. 12.7 assumes a unity gain, noninverting buffer amplifier. The performance of this "hybrid amplifier" circuit can actually be better than just using a single amplifier, because the precision input amplifier is free from supplying load current to V_{OUT}, and therefore will not heat up as much and drift. Fig. 12.7 shows split supply operation, single supply operation is also possible—as long as the designer keeps the V_{OH} and V_{OL} specifications of the buffer in mind—they may also vary with the load. There is no reason why the buffer amplifier cannot be operated off of a different (higher) set of rails, but the designer must be careful that the input amplifier never sees a voltage greater than its normal operating input range. This means there must be an appreciable gain in the circuit so that R_F and R_G form a voltage divider on the output voltage keeping the inverting input of the hybrid stage in the normal operating region.

Power stages have sources of instability that have not been discussed here, associated with their heavy and often times capacitive loads. The designer should carefully follow data sheet instructions for bypassing, maximum inductance and capacitance on inputs and outputs. There also may be recommendations for snubber networks to suppress unwanted high frequency oscillation. Suffice it to say — dealing with heavy loads is a complex task and should be taken seriously.

The buffer amplifiers can be replaced with noninverting op amp circuits. Current feedback amplifiers often have an output stage more robust than voltage feedback amplifiers, and as long as the designer is careful to observe stability recommendations for current feedback amplifiers, they will function well as buffer amplifiers. Power voltage feedback amplifiers

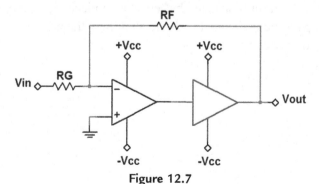

Figure 12.7
A better way to use a buffer amplifier.

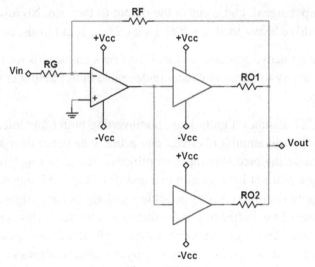

Figure 12.8
Paralleling buffer amplifiers.

are also available, some of which can drive several amps of load current at high voltage levels. The designer must pay special attention to the data sheet characteristics and recommendations for these devices, and they must often have a heatsink to operate at their recommended loads.

If even more output power is required, it is possible to parallel buffer amplifiers (Fig. 12.8). To insure proper current sharing, small values of output resistor (RO1 and RO2), usually between 1 and 5 Ω are placed at each amplifier output. This causes a decrease in output voltage swing, because the series resistors act as a voltage divider with the load. But without the series resistors, the buffers would tend to drive each other into oscillation. Usually a bit of experimentation is required to find the correct value. The designer should also make sure the resistors are of the correct wattage.

There is no theoretical limit on how many buffers can be placed in parallel; however, PC board parasitics and parasitic associated with the amplifiers themselves usually limit the number to a handful of devices.

CHAPTER 13

Troubleshooting—What to Do When Things Go Wrong

13.1 Introduction

You now have a good understanding of the basics of op amp design. Before wandering off into op amp applications, I thought I would take time to offer some guidance to those of you who have a circuit and it does not work or does not work properly. This has been the number one most requested topic for the fifth edition of *Op Amps for Everyone*, and I will strive not to disappoint. I should tell you right up front, there is no such thing as an exhaustive list of fixes. As soon as I come up with a fix for one application, another, different problem will arise, sometimes in the same circuit! After handling hundreds of applications inquiries, and being a designer myself on dozens if not hundreds of circuits, I believe that I have invented quite a few ways of messing things up. We all learn as much from our failures as by our successes (at least those of us who are wise enough to do so). A design failure should be looked at as an opportunity to learn something new—you have learned one more thing that does not work—not as a personal failure or proof of your incompetence. If only management and other team members could always be so enlightened!

13.2 Simple Things First—Check the Power!

Even with close to 40 years of experience, you would be surprised at the number of times I have simply forgotten to hook up the power! Or I have done something that mangles the power. This is easy to do if a CAD program is incorrectly applied. If the power pins are not shown on the op amp, and have to be separately placed—it is easy to forget them! The very first step when the circuit is not operating—or not doing what you expect—is to check the power pins of the op amp. If you are operating split supply, the positive power input should be the positive analog voltage you are supplying to the op amp. And the negative power input should be the negative analog voltage you are supplying to the op amp. If you are operating single supply, the negative voltage terminal of the op amp should be at ground. Verifying this should be the first thing you do because nothing is going to be right without power. There are sneak paths and leakage paths that can allow a circuit to have an output, just not the right one, or not a good one with no power applied.

Op Amps for Everyone. http://dx.doi.org/10.1016/B978-0-12-811648-7.00013-3

It is easy to mistake this for a bad circuit—or failed op amp—when all that is wrong is that power is not properly applied.

13.3 Do Not Forget That Enable Pin

This can be a sneaky one, because not all enables are created equal. In a split supply system, it may be referenced to the negative supply voltage, not to ground. This makes it difficult to interface to microcontrollers or other logic sources. Some enable pins can be left open if not used, others must be tied to a potential. Suffice it to say, if you are using an op amp with an enable pin, read the data sheet carefully to make sure you have utilized it correctly.

13.4 Check the DC Operating Point

Remember in our discussion of the ideal op amp, I gave you all op amp theory in one sentence: "The output will do whatever it has to do to make the two input voltages equal." This is also an excellent troubleshooting technique. With no signal input, apply power and check the output, the inverting input, and the noninverting input. If they are not all at the same level, or very close to the same level, something is wrong. This holds true for split supply circuits or single supply circuits.

- If the output is near the positive supply voltage, the noninverting input is probably higher than the inverting input, and for some reason the op amp cannot equalize the two inputs.
- If the output is near the negative supply voltage, the inverting input is probably higher than the noninverting input, and again for some reason the op amp cannot equalize the two inputs.

This is called "hitting the rail," and obviously the circuit will not operate in this configuration. If it responds to an AC signal at all, part of the output waveform will be clamped to a DC potential, while the other excursion may be present in part. This is oftentimes caused by too much gain amplifying the input offset voltage. Or perhaps, if this is an AC application, one of the DC blocking capacitors has been forgotten. Even in a split supply circuit, it is oftentimes easy to eliminate the DC offset by proper application of DC blocking capacitors.

13.5 The Gain Is Wrong

You should not have this problem, if you have been reading this book, especially Section 7.6. Assuming that you do not have a broken resistor somewhere, you should double check the open loop gain of your op amp. You may have run out of operating room, and need to select another. With hundreds of op amps on the market, chances are good you can find one with more open loop bandwidth and therefore more operating room where you need it. If not, you are going to consider two op amp stages, and divide up your gain between them. This is probably not a disaster, because many op amps come in dual versions, which would be more passive components but still only one IC.

13.6 The Output Is Noisy

All semiconductors generate noise, this is just the nature of the semiconductor physics involved. With that said, there can be a number of reasons why your circuit generates excessive noise. I cover the subject of op amp noise in Appendix C if you want more information. Op amp noise is dependent on frequency, you can always look at a different op amp with different noise specifications. Before spending money on an expensive low noise op amp, you should rule out external noise sources.

For now, I will cover the topic of noise from the standpoint of military specifications—at least in broad, general terms. There are four categories of noise:

- Conducted emissions—noise produced by the circuit conducted to other places over power or signal connections.
- Conducted susceptibility—noise produced elsewhere in the circuit that gets conducted into the op amp stage over power or signal connections.
- Radiated emissions—noise produced by the circuit that radiates wirelessly to other parts of the system.
- Radiated susceptibility—external RF noise that affects the circuit.

Let us look at these separately, and establish protocols to deal with them.

First and foremost, make sure your power pins—including the reference if used—are bypassed. This is your first line of defense. Proper bypassing does not mean slapping a 0.1 µF capacitor on the power pin. I will have a lot more to say about this later, but at least read the data sheet and follow any specific recommendations. That is a starting point, not the finishing point of bypassing.

13.6.1 Conducted Emissions and Radiated Emissions

If you have either of these conditions, you have my sympathy, because you probably have a completely unfunctional circuit. You have probably loaded down the output with capacitance, making the stage unstable, or you have completely ignored and violated stability criteria. A thorough reevaluation of your circuit is the only cure, with special emphasis on stability. Even I occasionally hang too much capacitance on an op amp output and it oscillates. Or I forget that I have selected a gain of 10 stable op amps, and am using it at a gain of one. An oscillating circuit will put noise on the power lines, conducting to other stages, or it will actually broadcast, and you better fix that before encountering the Federal Communications Commission (FCC)!

The most common reason for too much capacitance on an output is the presence of a coax cable. Think of a coaxial cable as long, skinny capacitor. Loading an op amp that is not designed to drive capacitive loads with a capacitor will never do! The simplest method of combating this is to select an op amp with the ability to drive capacitive loads. These will be advertised as such, and the capacitance they can drive will be clearly specified. However, this will severely limit the number of op amps from which you can select.

There are techniques to counter the capacitance of coaxial cable (and other capacitive loads). One of the simplest is to place a simple series resistor on the output of the op amp.

This will convert the capacitive load of the coaxial cable into the capacitor in a simple one pole low-pass filter, so you need to make sure the roll-off associated with the low-pass filter does not adversely affect the response of your system.

A lot of the coaxial cables you will encounter have a characteristic impedance of 50 Ω. For higher frequency applications, you can use this characteristic to your advantage and borrow a technique from RF design. You can create a 50 Ω balanced system. I will not go into all the particulars of such a system, which are evaluated in any number of resources on RF design but suffice it to say that it will isolate your system from many of the adverse parasitic effects of having a long cable in the system. It is much superior to the simple series resistance approach above, but comes at a cost. A 50 Ω transmission system assumes a 50 Ω series resistance at the source and depends on a 50 Ω termination resistor at the destination for impedance matching. Remembering the voltage divider rule, this means that your output voltage amplitude is divided by two. This can be countered in our simple schematic above by converting the noninverting buffer in Fig. 13.1 above to a noninverting gain of two by adding two resistors as shown in Fig. 13.2:

If R_F and R_G are the same value, follow the stage gain. The noninverting gain on V_{IN} will be 2, so V_{OUT} from the op amp stage is $V_{IN} \times 2$. The voltage divider consisting of the series and termination resistors form a voltage divider of ½, so V_{OUT} ultimately equals V_{IN}.

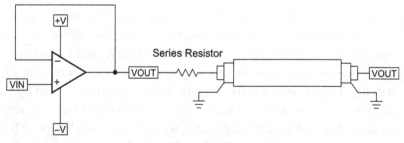

Figure 13.1
Simple method to isolate coaxial cable capacitance.

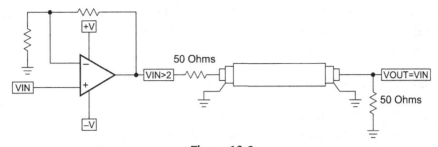

Figure 13.2
50 Ω transmission method.

Fifty Ohm transmission requires an op amp that can drive a 100 Ω load, and this also limits the number of op amps that are available for use. Generally, they are going to be high-speed op amps. A number of voltage-feedback amplifiers are available with 100 Ω drive, but generally 50 Ω transmission is better done with current-feedback op amps. This is because the requirement to use a gain of two to compensate for the voltage divider effect would otherwise limit the bandwidth of a voltage-feedback op amp.

The schematic above shows a split supply, and the astute reader may already have uncovered a flaw. The op amp will have a DC offset due to its input offset voltage. This offset will hopefully be small, but if you use a current-feedback amplifier, they are not known for low offset voltage specifications. This means a DC voltage will appear at the output, and this is not a good thing because the DC voltage will also appear across the 100 Ω load. The best-case scenario is that this will waste power supply current; the worst-case scenarios include clipping on one part of the waveform or another. Be careful with the power rating of the 50 Ω matching resistors if a steady DC voltage is designed into the system!

It is far better to isolate the 50 Ω termination system with a DC blocking capacitor as shown in Fig. 13.3.

If you consult textbooks on RF design, you will probably find this DC blocking capacitor, because many of those textbooks discuss transistor stages, which have DC bias voltages on

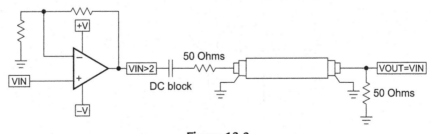

Figure 13.3
50 Ω transmission method, with DC blocking capacitor.

each stage that must also be blocked. The capacitor can be selected to be large enough to have minimal effect on the signal. It should be a good quality dielectric, especially if temperature of the circuit is subject to changes.

13.6.2 Radiated Susceptibility

The best piece of advice I will give you in this chapter—get that cell phone out of your shirt pocket! If you are like me, you forget this all the time while you are working in the lab. The GSM tones periodically coming out of some cell phones are powerful and can affect your circuit.

Noise, especially "oscillation," may result from op amp instability, or from external sources being amplified by your circuit. The very first thing you should do is to terminate the input of your op amp circuit, and check the output. If the oscillation goes away, it was not oscillation at all. It was noise amplified from an external source. You may even have to put your circuit inside a Faraday cage to isolate it from external sources of noise. Before you complain about buying an expensive Faraday cage, you should know that I use a tin candy box with a ground strap soldered to the box, lid, and ground of the board. I close the lid over the power connections and scope probe lead. This makes an inexpensive and effective Faraday cage that does a good job of isolating the board inside from outside sources of radio frequency interference. If you still see sustained oscillation, your circuit is unstable and needs to be redesigned.

If your circuit must work in the lab in the presence of strong RF signals, there are some things you can do to reject RF, assuming they are much higher in frequency compared to the bandwidth you are interested in for your application.

The first simplest thing you can do is simply install a low-pass filter at the input of your circuit, designed to attenuate high frequencies while not affecting the frequencies you are interested in as shown in Fig. 13.4

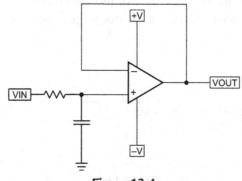

Figure 13.4
Simple method to reject RF

I have shown the low-pass filter on a simple noninverting buffer, your application circuit can be anything you want. I have used this technique successfully to design precision transducer circuits only two miles from an antenna farm with 20 full power FM stations and a dozen full power TV stations. Without the low-pass filter, the output was completely saturated with pickup from broadcast stations, with the low-pass filter, the output was completely clean DC level. If your application frequency is closer in frequency to major broadcast facilities, this technique may not work. Also remember that cell phone. Two watts is small compared to a 5 million watt UHF TV station, but that 2 W in your pocket is also a lot closer to your application circuit!

The next thing you can do to combat RF is to slow down your op amp. Modern op amps are implemented on a single piece of silicon, and except for the very rare uncompensated amplifier and amplifiers with offset null pins, you have no access to their internal circuitry to accomplish slowing. There is a way, however, to slow down an op amp. It involves slowing down the slew rate. Fig. 13.5 shows the first stage of a voltage-feedback op amp.

V_{INs} are the input pins of the op amp. V_{OUTs} are connected to the next internal stage of the op amp. This is not intended to be tutorial on the exact schematic of the input stage of every op amp. It is one possible variation, there are many others. C slew rate is not a component intentionally added to the emitters of the input transistors (which have been given the name "long-tail pair" by op amp designers). C slew rate is a representation of distributed parasitic capacitance that appears in the stage, including between the bases of the transistors (or gates in the case of an FET input op amp). Power supply designers are

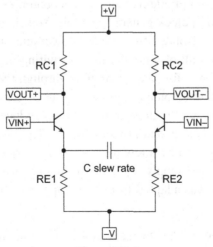

Figure 13.5
Typical op amp input stage.

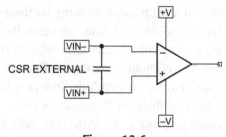

Figure 13.6
External slew rate reduction capacitor.

very familiar with parasitic gate capacitance! The point here is that parasitic capacitance will happen, and it will act to limit the slew rate of the op amp, and therefore slow it down. You do not have access to the emitters of the transistors, but you do have access to the bases! They are, after all, the inputs. So if you want to intentionally slow an op amp down, all you have to do is to connect a capacitor between the two inputs as shown in Fig. 13.6.

The slew rate reduction capacitor shown above is in addition to the components of your application, of course. It is no way negates the requirement to close the loop of the op amp, I merely omitted all other components to show the addition of one part. It is also good to keep this capacitor to a few picofarads or tens of picofarads, because it can introduce noise.

13.6.3 Conducted Susceptibility

This is one of the most common problems you will encounter, especially on boards that contain a microcontroller with a clock generator circuit. You should always attempt to have analog ground and power isolated from digital power and ground. Making a continuous ground plane under all of your circuitry is inviting high-speed digital signals to capacitively couple to your low-noise analog circuitry through the ground. This is especially the case for boards that contain high-power bus drivers such as CAN transceivers. The energy associated with logic transitions produce ringing on the digital edges, these can affect the ground potential momentarily. The logic-level transitions can also change the power supply load by a few milliamperes, producing power supply ripple. Unfortunately for you, the power supply rejection of op amps becomes worse at high frequencies. Fig. 13.7, which was adapted from a real op amp data sheet, shows the problem.

The horizontal axis has been hidden to make it general, but the scale is log frequency. Op Amp parameters such as power supply rejection ratio are covered in

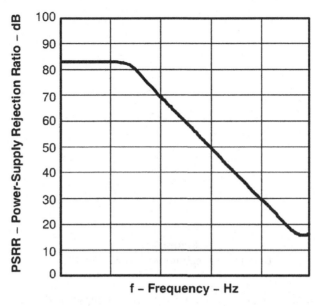

Figure 13.7
Power supply rejection ratio.

Appendix B if you want more information. Two things you should take away from this figure:

• The graph looks a whole lot like the open loop response graph of an op amp. It does not have anything to do with the open loop response, however. Do not confuse the two!
• The important thing is the power supply rejection gets worse at high frequencies, and these are the frequencies that are likely to get conducted into an op amp from high-speed digital circuitry!

Fortunately, there is a very simple thing you can do to combat this problem. You can place a small series resistor in line with the power supply inputs, and put the decoupling caps after the series resistor as shown in Fig. 13.8.

You have, in effect, made a single-pole low-pass filter on the power input, which will reject the power supply with the same degree of rejection that the PSRR is deteriorating! Before you get too excited, I need to give you some cautions:

• Op Amp supply current, particularly for high-speed op amps, can be above 10 mA. And if you are using an IC with more than one op amp, that current is multiplied by the number of op amps. This will limit your choice of series resistors to just a few Ohms at most. A current-starved op amp will also produce many strange symptoms! Reduce the series resistance if you encounter unusual effects.

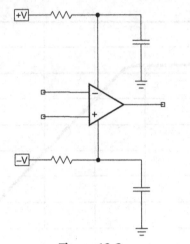

Figure 13.8
Decoupling technique to reduce PSRR.

- If you place too much series resistance on the power input of the op amp, the supply voltage on its power pins will be reduced. This will reduce the output voltage swing of the op amp. Remember that you can always measure the voltage, and if it is too low, you need to reduce the value of series resistance.
- The decoupling capacitors are now serving double duty as the frequency element in a low-pass filter. If there is more than one capacitor, you add the capacitance. It is seldom necessary to exactly counter the op amp power supply rejection ratio; you only need to achieve a flat response at your bandwidth of interest that is sufficiently high to isolate your circuit from the effects of high frequencies on the board.
- Because you are now effectively designing a single-pole low-pass filter on the power supply, and the capacitor will form the low-pass breakpoint with the series resistor and likely be a relative large value, use a good quality capacitor. If your circuit encounters thermal effects, be aware of the effect of temperature on capacitance value.
- Nothing in this scheme of reducing PSRR negates data sheet requirements for decoupling capacitor (see Appendix B). Your low-pass filter capacitor should be at least the value(s) recommended in the data sheet, and the capacitor should still be located close to the power supply pins of the op amp.

13.7 The Output Has an Offset

This error can appear in both DC and AC circuits. A DC circuit is more difficult to deal with and usually involves selecting a different op amp, or carefully canceling the offset by using an op amp with an "offset null" function. I have not discussed the offset null function of op amps, because it is a feature that is seldom used or necessary. As op amps get better and better, the offset nulling function is being increasingly omitted; however, some older op amps persist in the marketplace due to their popularity, particularly in military applications.

The offset null pins are usually pins 1 and 5 of the standard 8 pin single op amp package, although pin 8 is sometimes used (Fig. 13.9).

The usual method of offset nulling involves a potentiometer between the two pins, and the center adjustment connected to V^-, although some op amps connect to V^+ instead. Follow the manufacturer's recommendation (Fig. 13.10).

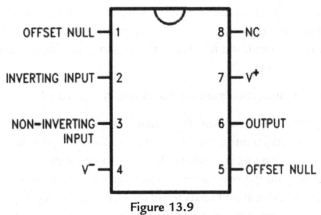

Figure 13.9
Op amp package with offset null pins.

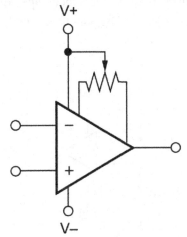

Figure 13.10
One possible offset null correction.

Going back to the previous section—if you need to slow down the slew rate of an op amp with offset null pins—they are direct connections to the long-tail pair emitters! So added a slew rate reduction capacitor to an op amp with offset null function is extremely easy.

A lot of people get into trouble applying inverting op amp circuits that must include DC gain when they fail to take input bias current into account. All op amp inputs require input bias current, which must be supplied by the user. For the common mode rejection of an op amp to work properly (and reject the DC offset), the input bias currents of the inverting and noninverting inputs must be balanced. The easiest way to do this is with a resistor inserted between the noninverting input and ground (or virtual ground). The value of this resistor is $R_F \| R_G$. The reason for this value is intuitive. Replacing the input voltage V_{IN} with ground, the input gain resistor R_G is connected to ground. But the output of the op amp in this condition is also at ground potential, so R_F is effectively also connected to ground. Therefore, the input bias current of the inverting input is $R_F \| R_G$. So if you want to balance the input bias current, the resistor at the noninverting input should also be $R_F \| R_G$ as shown in Fig. 13.11.

Consider the case of a noninverting gain circuit shown in Fig. 13.12.

You almost certainly selected the noninverting gain configuration because of its high input impedance, but $R_F \| R_G$ works against that high input impedance, particularly in the case of high gain, high-speed op amp circuits, where R_G tends to be small. So most designers choose to omit it and rely on the resistance of V_{IN} to provide input bias. However, that unbalances the input bias current and leads to offset. Fortunately, high-speed application circuits are seldom concerned with DC accuracy. If this is a concern, then you can easily get rid of DC at the output by adding one capacitor as shown in Fig. 13.13.

The DC gain of this circuit is unity, because the added capacitor blocks DC, converting the stage to a unity gain buffer for DC. $R_F \| R_G$ can be omitted, provided that the source of

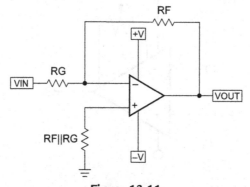

Figure 13.11
Method to balance input bias current.

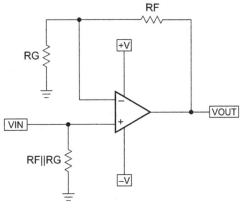

Figure 13.12
Noninverting gain circuit with input bias current balancing.

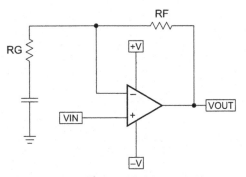

Figure 13.13
Noninverting gain circuit with low DC offset.

V_{IN} provides a path for input bias current. If not, it can be made large enough not to load the input, because you no longer have to worry about amplifying DC errors.

13.8 Conclusion

This chapter has given you a basic set of tools to combat some of the common problems that arise in applying op amps. It is by no means comprehensive, I cannot anticipate everything that might go wrong, but I believe this should allow you to rule out the most common mistakes and troublesome annoyances. You might also want to consult Chapter 25—which is a compilation of actual problems reported to me by customers when I was an applications engineer—and their solutions.

CHAPTER 14

Interfacing a Transducer to an Analog to Digital Converter

14.1 Introduction

One of the most common questions asked by engineers is "which op amp should I use with a (substitute a part number) ADC?"

The question of which data converter to use with any given data converter is one that makes marketing people cringe, because the answer touches on different technical areas that they are probably not conversant in. It is also an uncomfortable question for applications support engineers, because the customer may not have a clear idea of all the issues and tradeoffs involved.

The process of selecting an op amp to drive a data converter is—as much as anything—an exercise in weeding out those op amps that will clearly not work. The subset of op amps left at the end of an elimination process is much more manageable.

Op amp manufacturers have simultaneously made things easier—and harder for you. Some manufacturer websites have website sections targeted to major product groups, and op amps appropriate for that product group will pop to the top of the list. If that is not the case, the number of op amp listings is sometimes vast, and it may not be evident which op amp has been optimized for a product group. But each new op amp manufactured is the result of a product group meeting, where a group of IC designers, managers, and marketing people have decided that optimizing four or five parameters will allow them to sell an op amp to manufacturers of a particular type of product. Your job is to discover that op amp, and this procedure will lead you in that direction.

Make no mistake—the list of questions below is daunting. I know that—but these are important questions that must be answered for the most part, before doing the design. You have probably already answered most of them without realizing it. Or the answer is so self-evident that the question was not asked. The answer to one question can automatically eliminate many other questions. Hopefully this will lead to an organized, methodical approach to the question that will guide you to a good choice.

Op Amps for Everyone. http://dx.doi.org/10.1016/B978-0-12-811648-7.00014-5

I have divided the questions into broad categories:

- The system as a whole—understanding what the product is supposed to do can lead you in the right direction.
- The power supply: power supply voltages in electronics, especially portable devices, have been trending downwards. Customers want lower and lower battery voltage, but with no compromise in performance. The first regency transistor radio, for example, used a 22.5 V battery. My childhood transistor radio used a 9 V battery. My current portable radio runs off of a single 1.5 V battery, yet offers much better performance than the older radios.
- Input signal characteristics: the type of signal being input to the stage can greatly influence the choice of op amp. Signals that are audio or RF can be AC coupled, and DC specifications of the op amp are not important. Some sensors, on the other hand, are almost entirely DC in nature, so AC performance of the op amp is not important.
- Analog to digital converter (ADC) characteristics: a single-ended data converter will use single-ended op amps, whereas a fully differential data converter will require fully differential drive.
- Op amp characteristics: sometimes packaging, temperature range, or other considerations influence which op amp is used.

And finally, I will offer some hints about how to properly drive a fully differential ADC.

Since there are so many questions, I suggest that they be discussed in a "kick-off" meeting environment, where all designers associated with the system can freely input ideas. Each one of the questions is not intended to be a "sticking point" or closed door, merely a suggested consideration that might affect system design. I know how meetings go, how one point can grow into a half hour debate, but that is not the intention. Each of these questions should be simple; the answer self-evident when asked of the right person, so system level discussions should go very fast.

14.2 System Information

The overall characteristics of the system often times yield valuable information. A clear understanding of the product and its function is imperative to design success and can start the process of weeding out unsuitable parts.

- Exactly what is the end-equipment and its application? Different systems have different requirements.
  Examples:
  - A transducer interface design requires DC accuracy and lead to DC accurate op amps.
  - Wireless communication systems will require high speed op amps with good RF specs.

- In general terms, what is the function of this signal-acquisition chain in the system? Where does the input signal come from and what happens to it once it is digitized?
  Examples:
  - A slowly changing DC signal can utilize slow op amps optimized for DC accuracy.
  - An RF system will be AC coupled and must run at least as fast as they Nyquist frequency.
  - Audio systems require op amps with low levels of total harmonic distortion.
- How many signal-acquisition chains are used in the product? Channel density can influence system design in numerous ways, including space constraints, thermal requirements, and amplifier channel density per package.
  Example:
  - A medical ultrasound device can have 100 or more channels, leading to challenges of component count, board size, power consumption, and heating—so you will want quad op amps, low in power consumption, in small packages, probably operating off of a low voltage.
- Where will the system be used? What temperature conditions will the system operate in?
  Examples:
  - Military, space, downhole, and geothermal are all applications where the signal chain will be subjected to extremes of temperature. They will require high reliability components that are probably already on an approved list.
  - Consumer electronics will probably be subjected to nothing more severe than a hot car dashboard or overnight freeze; therefore, the op amps will not have to be nearly as temperature tolerant as the categories listed above.

14.3 Power Supply Information

Power supply rails can quickly rule out op amps. This is similar to clothing shopping, the style may be desirable, but if the size does not fit, the style is useless. So a wise shopper finds the options in the size first, before becoming attached to a style. Similarly, power supply information is collected first, because it will narrow op amp choices (Fig. 14.1).

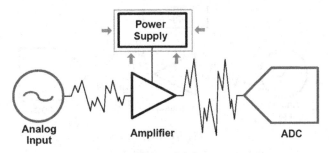

Figure 14.1

Focusing on the power supply characteristics.

- What power supply voltages are present in the system? Are ±15 V supplies available? ±5 V? No negative supplies at all? Only low voltage, +3 V from batteries or even +1.5 V? An op amp with fantastic specifications at ±15 V may not operate at all from +3 V—remember the V_{OL} and V_{OH} specifications of the op amp. Therefore, if +3 V is all that is available in the system, you can rule out all but single supply, rail-to-rail devices and must design within their limitations.

  Examples:
  - Legacy analog systems almost always have ±15 V rails. Upgrading them with newer generation op amps requires that the new op amps can operate at these voltages.
  - Many high end data acquisition systems have standardized on ±5 V rails. Many devices are offered, perhaps the majority of op amps in this voltage rating. They may not be able to operate off of ±15 V rails at all.
  - Portable equipment (battery operated) tends to operate off batteries that will provide multiples of 1.5 V. Even systems that use button cells have multiples of 3 V. Lithium ion battery systems will have multiples of 4 V. Of course, the ultimate in light weight, small, portable electronics will operate off of a single 1.5 or 3 V battery. However, because the cell voltage droops as the battery is depleted, these systems must operate off of a voltage lower than 1.5 V, sometimes as little as 0.8 V, creating an extreme limitation on the number of suitable op amps.

- Is a precision reference available in the system? In single-supply systems, it is important to supply a virtual ground to the op amp circuitry.

  Example:
  - Higher end data acquisition systems tend to use ADC's with built in references. You should use this reference if at all possible.

14.4 Input Signal Characteristics

Understanding the input source is a key to properly designing the interface circuitry between the source and the ADC (Fig. 14.2).

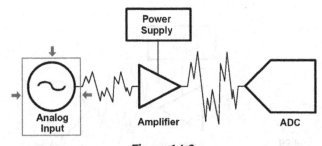

Figure 14.2
Focusing on the input signal.

- What is the output amplitude range of the source? This information is used in conjunction with the cases in Chapter 4. The source determines V_{IN} minimum and V_{IN} maximum.
- Does the source produce a current output? This requires a different topology.
  Example:
  - Some temperature sensors
- Is the signal source output single-ended or differential? A differential input signal may lend itself to an instrumentation amplifier rather than a single-ended op amp.
  Examples:
  - 600 Ω balanced audio
  - Pressure transducers (strain gauges)
- What is the output impedance of the signal source? Very high impedance sources require even higher impedance amplifiers. This will definitely dictate a noninverting op amp topology, but even that may not suffice. It may require extremely high impedance J-FET op amps or instrumentation amplifiers.
  Examples:
  - Photomultiplier tubes
  - Pressure transducers (strain gauges)

14.5 Analog to Digital Converter Characteristics

Now that the power supply and input signal have been defined, it is time to focus on the device that the op amp will drive—the ADC (Fig. 14.3).

- What is the full-scale input range of the data converter? The ADC input low and high voltages, along with values from the input signal section above, determine the "case" of Chapter 4.
- Will the data converter be used with single-ended or differential inputs? Typically, most high-performance data converters have differential inputs and require their use for optimum performance. You will, however, have to convert a single-ended signal to a fully differential signal to get maximum performance from the ADC.

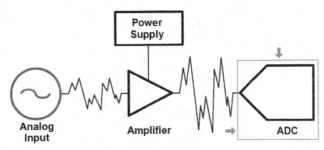

Figure 14.3
Focusing on the analog to digital converter.

- What is the desired resolution and effective number of bits? A 14 bit converter will not effectively yield 14 bits. The true resolution will probably be closer to 12 or 13. If 14 bit performance is really desired, ask if a 16 bit converter can be substituted. Often times there are a family of similar data converters, and a higher resolution converter may be pin for pin compatible.
- What is the desired sampling rate? Often, people assume that a data converter is going to be used at its maximum sampling rate, but sometimes this comes at the cost of accuracy. For example, an 80 mega-sample per second converter might be given a sampling frequency of 60 MSps to get more accuracy.
- Are there any compensation requirements for the input of the data converter? Normally, a small RC filter is required at the input of the data converter to compensate for its capacitive input. These components are specified in the converter data sheet and should be included as part of the interface. Otherwise, the op amp interface circuit may exhibit instability.

14.6 Interface Characteristics

By now you have narrowed down the potential choices for an op amp by its supply voltages. You know what "case" your interface requires, which will give you a rough idea of the schematic of the interface. You also know whether you need a single-ended to fully differential conversion stage. But there are some other pieces of information you need to flesh out the complete input signal to ADC input interface (Fig. 14.4).

- Is the signal DC accurate, high speed, audio, or RF? You can consult Appendix B for a reference on which op amp parameters are important for different types of applications. **Examples**:
  - Pressure and temperature transducer circuits are almost exclusively DC accurate. You need an op amp with good DC specifications.
  - High speed systems may go hand-in-hand with RF, or just require fast op amps because of filter constraints—consult Chapter 18 for high speed filter considerations; you may be surprised just how fast an op amp needs to be to operate in an active filter circuit.

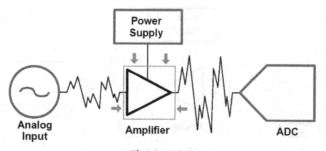

Figure 14.4
Focusing on the operational amplifiers.

- Audio requires op amps with low noise in the audio bandwidth. Some low noise high speed op amps may have bad noise specifications at low frequency.
- RF applications require a completely different set of specifications. Consult Chapter 19.
- Do you need to filter the signal?
  Examples:
  - Reduce high frequency noise (low pass filter)
  - Reduce low frequency noise (high pass filter)
  - Detect a single frequency (band pass filter)
  - Reject an interfering frequency (notch filter)

These filters will be discussed in Chapter 16.

- Are there specific requirements for the package of the amplifiers?
  Examples:
  - Small, surface mount? Or does package size not matter?
  - Does it have to be a high reliability ceramic package?

Now, you have a really good idea of what op amps you can not use, and hopefully a small list of the ones that you can. You should take that small list and evaluate them based on their suitability. Perhaps make use of one of a free simulation program, or even order them and try them in your prototype circuit. Do not be afraid to experiment. Options are sometimes good, so if half a dozen parts meet your requirements, be glad because you can put them as alternates on your bill of materials!

14.7 Architectural Decisions

Even with your op amp selection(s) made, the job is still not quite done.

- As I mentioned above, you may need a compensation network. This network will also form a low pass filter, but fortunately the low pass characteristic is above the operating frequency of the ADC.
- I would put gain stages first; then filter stages. In the case of a lot of gain, I would use two or more gain stages.
- Of course, if you have a fully differential ADC and a single-ended signal source, you need to do a conversion. I will spend the rest of this section on this interface.

Fig. 14.5 shows a typical single-ended to fully differential interface circuit. The input signal is referenced to ground, while the common mode operating point of the op amp interface is set by the data converter. The op amp interface can be run off of a single supply. DC blocking capacitors C_1 and C_2 prevent the common mode point from being

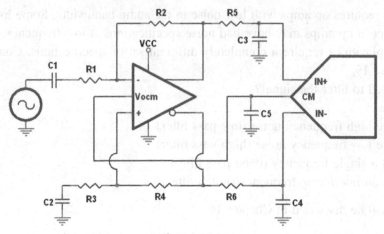

Figure 14.5
Single-ended to fully differential AC coupled interface.

affected by the input signal. Response down to DC, of course, has been sacrificed, but this may be acceptable in most applications (Fig. 14.5).

R_5, C_3, R_6, C_4 form the compensation networks defined by the ADC data sheet. If this circuit requires only gain and no filtering, it may be possible to use the fully differential

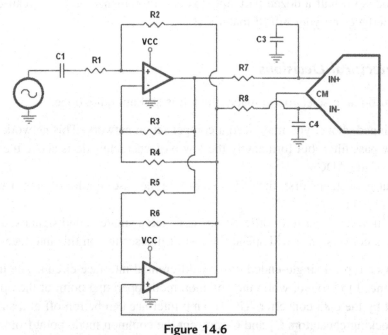

Figure 14.6
Preferred single-ended to fully differential AC coupled interface.

op amp as the entire interface. If filtering is required, it can be done single-ended and input to C_1.

It is not absolutely necessary to use a fully differential amplifier to drive a differential A/D converter. Fig. 14.6 shows the preferred way to convert single-ended signals to differential without a transformer (Fig. 14.6).

While this circuit looks a bit unusual, the strategy is to equalize the delay for IN+ and IN− by forcing each phase of the signal to go through two op amps before being applied to the inputs. This may not be intuitive at first glance! Each amplifier, though, is in the feedback loop for the other. Think of this as an inverting op amp circuit—gain is adjusted by changing R_1 (corresponding to R_G), R_2 through R_6 are equal values (corresponding to R_F). This circuit is referenced to the ADC reference in the bottom op amp. Unfortunately, nothing comes for free. You have doubled the noise of the circuit because noise is uncorrelated and now is cascaded through two op amps instead of one.

14.8 Conclusions

Selecting the right op amp(s) for an interface between a sensor and a data converter can be a daunting challenge. Selecting the op amps is more of an exercise in weeding out unsuitable devices, at least in the early stages. You can use a systematic approach to approach the problem, and tackle the issues and questions one section of circuitry at a time. You can narrow down the op amp choices by:

- analyzing the nature of the system,
- knowing the power supply voltages in the system,
- knowing the input signal characteristics, and
- knowing the ADC characteristics.

And finally

- knowing something about the op amps themselves and
- knowing how to interface with the ADC.

These questions may be formidable, but they are essential if you want to develop a working interface. Take the time to ask them and understand the implications of the answers.

CHAPTER 15

Interfacing D/A Converters to Loads

15.1 Introduction

A digital-to-analog converter, or D/A, is a component that takes a digital word and converts it to a corresponding analog voltage. It has the opposite function of an A/D converter. The D/A is only capable of producing a quantized representation of an analog voltage, not an infinite range of output voltages.

The application will almost always dictate the selection of the D/A converter, leaving the designer the task of interfacing that converter with the output load.

A D/A converter interfaces with a buffer op amp. Most D/A converters are manufactured with a process that is incompatible with op amps. Therefore, the op amp cannot be manufactured on the same IC. It must be external, and its characteristics are an integral part of the conversion process. In most cases, the data sheet will make a recommendation for the selection of a buffer op amp. Follow the recommendation, unless there is a compelling reason not to do so. Performance can be improved only if you know exactly what op amp specifications need to be optimized.

Signal conditioning—low-pass filtering, DC offsets, and power stages—should all be placed after the recommended op amp buffer. Do not attempt to combine these functions with the buffer unless you are an experienced designer with a good grasp of all of the implications.

15.2 Load Characteristics

There are two main types of loads that a D/A may have to drive—AC and DC. Each has different characteristics and will require different interface circuitry.

15.2.1 DC Loads

These include linear actuators such as those used on 3D printers, positioning tables, motors, programmable power supplies, outdoor displays, and lighting systems. Large load currents and/or high voltages characterize some of these loads. DC accuracy is important, because it is related to a series of desired mechanical positions or intensities in the load device.

Op Amps for Everyone. http://dx.doi.org/10.1016/B978-0-12-811648-7.00015-7

15.2.2 AC Loads

These include things such as audio chains, frequency generators, IF outputs—any load that does not have a DC component.

15.3 Understanding the D/A Converter and Its Specifications

It is important to understand the D/A converter and its specifications before discussing interfaces.

15.3.1 Types of D/A Converters—Understanding the Trade-offs

D/A converters are available in several types, the most common of which is the resistor ladder type. There are several variations on the resistor ladder technique, with the R/2R configuration being the most common.

15.3.2 The Resistor Ladder D/A Converter

In this type of converter, a precision voltage reference is divided into 2^{N-1} parts in an internal voltage divider, where N is the number of bits specified for the converter. One switch at a time turns on, corresponding to the correct DC level (Fig. 15.1).

Unfortunately, the number of resistors and switches doubles for each additional bit of resolution. This means that an 8-bit D/A converter would have 255 resistors and 256 switches, and a 16-bit D/A converter would have 65,535 resistors and 65,536 switches. For this reason, this architecture is almost never used for higher-resolution D/A converters.

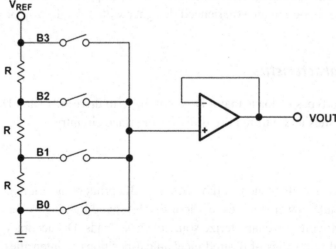

Figure 15.1
Resistor ladder D/A converter.

15.3.3 The Weighted Resistor D/A Converter

This type of converter is very similar to the resistor ladder D/A converter. In this case, however, each resistor in the string is given a value proportional to the binary value of the bit it represents. Currents are then summed from each active bit to achieve the output (Fig. 15.2).

The number of resistors and switches reduced to one per bit, but the range of the resistors is extremely wide for high-resolution converters, making it hard to fabricate all of them on the IC. The resistor used for B0 in Fig. 15.2 is the limiting factor for power dissipation from V_{REF} to ground.

This converter architecture is often used to make logarithmic converters. In this case, the R, 2R, 4R, 8R ... resistors are replaced with logarithmically weighted resistors.

This type of converter, and the R/2R converter described in the next paragraph, uses a feedback resistor fabricated on the D/A IC itself. This feedback resistor is not an optional convenience for the designer—it is crucial to the accuracy of the D/A. It is fabricated on the same silicon as the resistor ladder. Therefore, it experiences the same thermal drift as the resistor ladder. The gain of the buffer amplifier is fixed, with a full-scale output voltage limited to V_{REF}. If a different full-scale D/A output voltage is needed, change V_{REF}. If the full-scale V_{OUT} must exceed the maximum rating of the D/A reference voltage, use a gain stage after the buffer op amp (see Section 15.7.2).

The op amp must be selected carefully, because it will be operated in much less than unity gain mode for some combinations of bits. This is probably one of the main reasons why this architecture is not popular, as well as the requirement for a wide range of resistor values for high-precision converters.

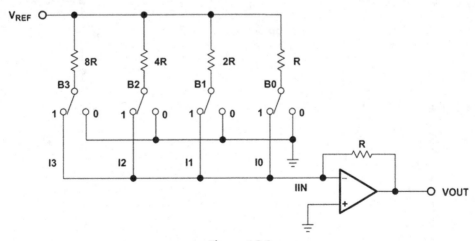

Figure 15.2
Binary weighted D/A converter.

15.3.4 The R/2R D/A Converter

An R/R$_2$ network can be used to make a D/A converter that has none of the disadvantages of the types mentioned above (Fig. 15.3).

For a given reference voltage V$_{REF}$, a current I flows through resistor R. If two resistors, each the same value (2R), are connected from V$_{REF}$ to ground, a current I/2 flows through each leg of the circuit. But the same current will flow if one leg is made up of two resistors, each with the value of R. If two resistors in parallel whose value is 2R replace the bottom resistor, the parallel combination is still R. I/4 flows through both legs, adding to I/2. Extending the network for 4 bits as shown on the right, the total current on the bottom leg is I/4 plus I/8 plus I/16 plus I/16 in the resistor to ground. Kirchhoff's current law is satisfied, and convenient tap points have been established to construct a D/A converter (Fig. 15.4):

This converter architecture has advantages over the types previously mentioned. The number of resistors has doubled from the number required for the current-summing type, but there are only two values. Usually, the 2R resistors are composed of two resistors in series, each with a value of R. The feedback resistor for the buffer amplifier is again fabricated on the converter itself for maximum accuracy. Although the op amp is still not

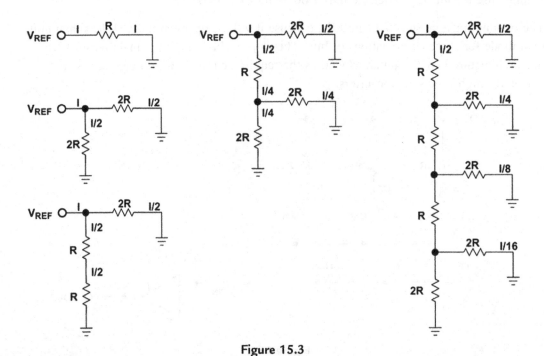

Figure 15.3
R/2R resistor array.

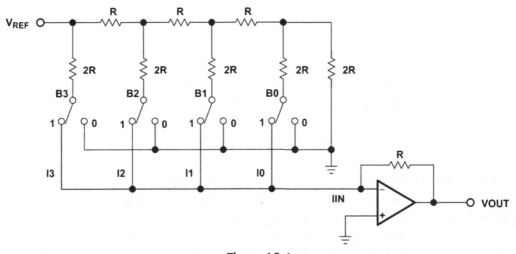

Figure 15.4
R/2R D/A converter.

operated in unity gain mode for all combinations of bits, it is much closer to unity gain with this architecture.

The important op amp parameters for all resistor ladder D/As are as follows:

- Input offset voltage—the lower the better. It adds to the converter offset error.
- Input bias current—the lower the better. The product of the bias current and the feedback resistance creates an output offset error.
- Output voltage swing—it must meet or preferably exceed zero to full-scale swing from the D/A.
- Settling time and slew rate—must be fast enough to allow the op amp to settle before the next digital bit combination is presented to the D/A input register.

15.3.5 The Sigma Delta D/A Converter

The sigma delta D/A converter takes advantage of the speed of advanced IC processes to do a conversion as a series of approximations summed together. A phase-locked loop (PLL) derived sample clock operates at many times the overall conversion frequency—in the case shown in Fig. 15.5, it is 128×. The PLL is used to drive an interpolation filter, a digital modulator, and a 1-bit D/A converter. The conversion is done by using the density ratio of the voltage out of the 1-bit D/A as the analog signal. As the pattern of 1s and 0s is presented to the 1-bit converter, their time average at the sample frequency recreates the analog waveform.

Sigma delta converters are popular for audio frequencies, particularly CD players. The primary limiting factor is the sample clock. The original CD players operated at a sample

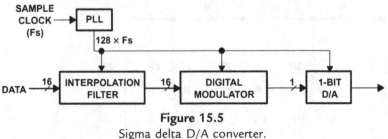

Figure 15.5
Sigma delta D/A converter.

rate of 44.1 kHz, which means that according to Nyquist sampling theory, the maximum audio frequency that can be reproduced is 22.05 kHz. If an audio frequency of 23.05 kHz is present in the recorded material, it will alias back into the audio output at 1 kHz—producing an annoying whistle. This places a tremendous constraint on the low-pass filter following the D/A in a CD player. It must reject all audio frequencies above 22.05 kHz while passing those up to 20 kHz, the commonly accepted upper limit of human hearing. While this can be done in conventional filter topologies, they are extremely complex (nine or more poles). Inevitably, phase shift and amplitude roll-off or ripple will start far below 20 kHz. The original CD players often sounded a bit "harsh" or "dull" because of this.

The solution was to overclock the sample clock. To keep things simple, designers made it a binary multiple of the original sampling frequency. Today, 8× or even higher oversampling is standard in CD players. Little do the audio enthusiasts know that the primary reason why this was done was not to improve the sound, it was done to substantially reduce the cost of the CD player! A faster sample clock is very cheap. Nine-pole audio filters are not. At 8× oversampling, the CD player only needs to achieve maximum roll-off at 352.8 kHz—a very easy requirement. Instead of the filter having to roll off in a mere 2 kHz of bandwidth, now it has 332 kHz of bandwidth to accomplish the roll-off. The sound of an oversampled CD player really is better, but it comes at the cost of increased radiated radio frequency interference, coming from the sample clock.

Sigma delta converters introduce a great deal of noise onto the power rails, because the internal digital circuitry is continually switching to the power supply rails at the sample clock frequency F_S.

15.4 D/A Converter Error Budget

The system designer must do an error budget to know how many bits are actually needed to meet the system requirements—how much "graininess" or what step size is acceptable in the output signal. A personal pet peeve of mine is the D/A converter in radios with digital displays. The D/A voltage driving the tuning voltage for varactor diodes is not accurate enough, particularly on AM. So they widen the IF bandwidth so that it is sloppy enough to still tune stations, even though the tuning voltage is not accurate. If you try to improve the radio by putting in narrower ceramic filters, some stations are impossible to tune because the tuning voltage combined with narrow IF response will miss some stations in part of the band!

15.4.1 Accuracy Versus Resolution

It is important for the designer to understand the difference between converter accuracy and converter resolution. The number of bits determines resolution of a converter. Insufficient resolution is not an error—it is a design characteristic of the D/A. If a given converter's resolution is insufficient, use a converter with better resolution (more bits).

Accuracy is the error in the analog output from the theoretical value for a given digital input. Errors are described in the next paragraph. A very common method of compensating for D/A error is to use a converter that has one or two bits more resolution than the application requires. With the cost of converters coming down, and more advanced models being introduced every day, this may be cost-effective.

15.4.2 DC Application Error Budget

DC applications will depend on the value of DC voltage coming out of the converter. Total harmonic distortion (THD) and signal-to-noise (S/N) will not be important because the frequency coming out of the converter is almost DC.

The resolution of a converter is $\pm 1/2$ LSB, where an LSB is defined as:

$$1 \text{ LSB} = \frac{V_{FS}}{2^N - 1} \tag{15.1}$$

where, V_{FS}, full-scale output voltage; N, number of converter bits.

The number of bits in a DC system determines the DC step size that corresponds to a bit. Table 15.1 shows the number of bits and the corresponding voltage step size for three popular voltages.

The bit step size can get critical, especially for portable equipment. There is a requirement to operate off of low voltage, to minimize the number of batteries. The buffer amplifier, if it includes gain, will use large resistor values, lowering its noise immunity. Fortunately, the vast majority of DC applications are not portable; they are in an industrial environment.

For example, a converter is used to position a drill on a table used to drill printed circuit board holes. The positions of the holes are specified as 0.001 in., ± 0.0003 in. The

Table 15.1: DC Step Size for D/A Converters

Bits	States	3 V	5 V	10 V
4	16	0.1875	0.3125	0.625
8	256	0.011719	0.019531	0.039063
10	1,024	0.00293	0.004883	0.009766
12	4,096	0.000732	0.001221	0.002441
14	16,384	0.000183	0.000305	0.00061
16	65,536	4.58E−05	7.63E−05	0.000153
18	262,144	1.14E−05	1.91E−05	3.81E−05
20	1,048,576	2.86E−06	4.77E−06	9.54E−06
22	4,194,304	7.15E−07	1.19E−06	2.38E−06
24	16,777,216	1.79E−07	2.98E−07	5.96E−07

actuators are centered on the table at 0 V, with full negative position of −12 in. occurring at −5 V, and full positive position of +12 in. occurring at +5 V. There are two actuators, one for vertical and one for horizontal.

This example has several aspects. The first is that the positioning voltage has to swing both positive and negative. In the real world, it may be necessary to add (or subtract in this case) a fixed offset to the D/A output. The output voltage has to swing over a 10 V range, which may mean that the output of the D/A has to be amplified. The actuators themselves probably operate off of higher current than the D/A is designed to provide. Section 15.7 covers some methods for meeting these requirements.

Assume, for now, that the D/A has the necessary offset and gain. A ±12 in. position is 24 in. total, which corresponds to ±5 V from the D/A circuitry. The 24 in. range must be divided into equal 0.0003 in. steps to meet the resolution requirement, which is 80,000 steps. From Table 15.1, an 18-bit D/A converter is required. The actual system will be able to position with a step size of 0.0000916 in. Two-independent conversion systems are needed, one for horizontal and one for vertical.

15.4.3 AC Application Error Budget

The error budget for an AC application will most likely be specified as THD, dynamic range, or S/N ratio. Assuming no internal noise, and no noise in the buffer op amp circuitry, the inverse of the dynamic range is the S/N ratio of the converter D/A. Of course, noise is always present and is measured with all input data set to zero. Noise will make the S/N ratio decrease.

The number of converter bits, however, is the overwhelming factor determining these parameters. Technically, they are not "errors," because the design of the converter sets them. If the designer cannot live with these design limits, the only choice is to specify a converter with better resolution (more bits).

15.4.3.1 Total Harmonic Distortion

The THD of an ideal D/A converter is the quantization noise due to the converter resolution. The number of bits of the converter determines the lowest possible THD. The greater the number of bits, the lower the amplitude of the harmonics, as shown in Fig. 15.6.

Assuming ideal D/A conversion, there is a direct relationship between the number of bits and the THD caused by the resolution:

$$\text{THD}(\%) = \frac{1}{2^N} \times 100 \tag{15.2}$$

where N is the number of converter bits. Of course, this is the limit for ideal conversion.

15.4.3.2 Dynamic Range

There is also a direct relationship between the number of bits (n) and the maximum dynamic range of the D/A (Eq. 15.2 and Table 15.2):

$$\text{Dynamic range} = 6.20 \times n + 1.76 \tag{15.3}$$

Notice that there is approximately a 6 dB improvement in dynamic range per bit. This is an easy way to figure out what improvement can be realized by increasing the number of bits from one value to another.

For example, if the designers of a CD player want to have a 90 dB S/N ratio, they would pick a 16-bit converter from Table 15.2. The THD is 0.0015% minimum.

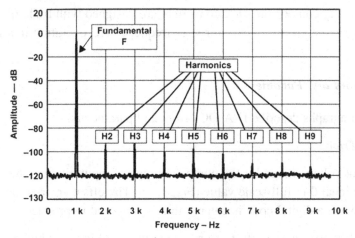

Figure 15.6
Total harmonic distortion.

Table 15.2: Converter Bits, Total Harmonic Distortion (THD), and Dynamic Range

Bits	States	THD (%)	Dynamic Range
4	16	6.25	25.8
8	256	0.390625	49.9
10	1,024	0.097656	62.0
12	4,096	0.024414	74.0
14	16,384	0.006104	86.0
16	65,536	0.001526	98.1
18	262,144	0.000381	110.1
20	1,048,576	0.000095	122.2
22	4,194,304	0.000024	134.2
24	16,777,216	0.000006	146.2

15.4.4 RF Application Error Budget

RF applications are a high-frequency subset of AC applications. RF applications may be concerned with the position and relative amplitude of various harmonics. Minimizing one harmonic at the expense of another may be acceptable if the overall RF spectrum is within specified limits.

15.5 D/A Converter Errors and Parameters

The D/A errors described in this section will add to the errors caused by the resolution of the converter.

This section is divided into DC and AC sections, but many of the DC errors masquerade as AC errors. A given D/A may or may not include either DC or AC error specifications. This should give the designer a clue that the device is optimized for DC or AC applications. Like any component, D/A converters are designed with trade-offs. It is possible to misapply a converter meant for high-frequency AC operation in a DC application, etc.

15.5.1 DC Errors and Parameters

The following paragraphs describe D/A DC errors and parameters.

15.5.1.1 Offset Error

The analog output voltage range for the complete range of input bits may be shifted linearly from the ideal 0 to full-scale value (Fig. 15.7). The offset error is the $\pm\Delta$ V from 0 V that results when a digital code is entered that is supposed to produce 0.

Related to the offset error is the offset error temperature coefficient, which is the change in offset over temperature. This is usually specified in ppm/°C.

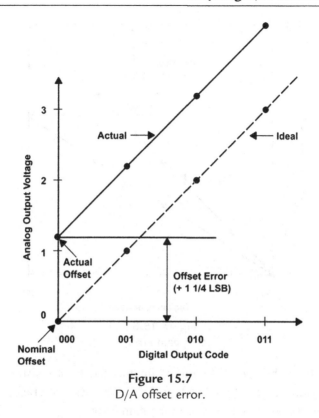

Figure 15.7
D/A offset error.

Offset error is critical in DC applications. For this reason, a buffer op amp must be selected that does not contribute to the problem—its own offset voltages should be much less than that of the converter. In AC applications, the offset error is not important and can be ignored. The buffer op amp can be selected for low THD, high slew rate, or whatever other parameters are important for the application.

15.5.1.2 Gain Error

The gain of the D/A converter may be greater than or less than the gain needed to produce the desired full-scale analog voltage (Fig. 15.8). The gain error is the difference in slope between the ideal D/A output gain and the actual gain.

Related to the gain error is the gain error temperature coefficient, which is the change in gain over temperature.

Gain error can be critical in both AC and DC applications. For example,

- An RF predriver must not cause the output stage to exceed FCC license requirements.
- A mechanical positioner must not stop short of or go past its intended position.

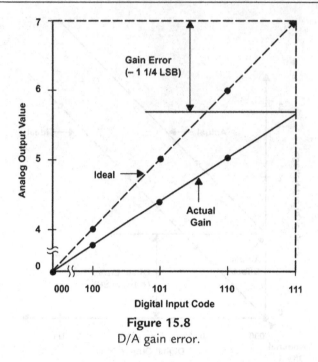

Figure 15.8
D/A gain error.

The op amp buffer should be operated with the internal feedback resistor. If possible, full-scale amplitude adjustments should be made to V_{REF}. This way, tolerances and thermal drift in external resistors do not contribute to the gain error.

15.5.1.3 Differential Nonlinearity Error

When the increase in output voltage (ΔV) is not the same for every combination of bits, the converter has a differential nonlinearity (DNL) error. If the DNL exceeds 1 LSB, the converter is nonmonotonic. This can cause a problem for some servo control loops. A nonmonotonic D/A would appear in Fig. 15.9 as a momentary dip in the analog output characteristic.

15.5.1.4 Integral Nonlinearity Error

The integral nonlinearity (INL) error is similar to the DNL error, except it is a first-order effect that stretches across the entire range from 0 to full-scale output voltage (Fig. 15.10).

Both the INL and DNL errors affect AC applications as distortion and spectral harmonics (spurs). In DC applications, they will result in an error in the DC output voltage. The mechanical steps of a positioning table, for instance, may not be exact increments.

15.5.1.5 Power Supply Rejection Ratio

The power supply rejection ratio is sometimes called the *power supply sensitivity*. It is the ability of the converter to reject ripple and noise on its power inputs. DC

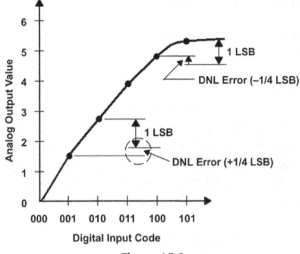

Figure 15.9
Differential nonlinearity error.

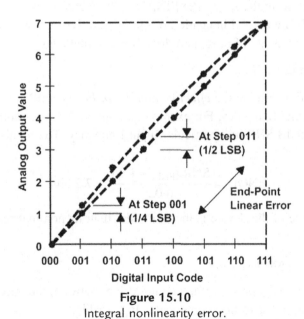

Figure 15.10
Integral nonlinearity error.

applications may not be adversely affected. Poor power supply rejection can cause spurs and harmonic distortion in AC applications, as external frequency components leak into the output and modulate with it. The designer must decouple the D/A and buffer op amp carefully to combat these problems.

15.5.2 AC Application Errors and Parameters

The following paragraphs describe D/A AC errors and parameters.

15.5.2.1 THD + N

There will always be some noise that is generated internally in the converter and buffer amp. A useful specification for audio and communication system designers is the THD + N (total harmonic distortion plus noise). The distortion plus noise (THD + N) is the ratio of the sum of the harmonic distortion and noise to the root-mean-square (RMS) power of the input signal. As was the case with op amp parameters (Appendix B), the noise sources add according to the RMS law. The distortion and noise are measured separately and then added together to form the ratio. The noise voltage relates to the measured bandwidth.

15.5.2.2 Signal-to-Noise and Distortion

The signal-to-noise and distortion (SINAD) is the ratio of the input signal to the sum of the harmonic distortion and noise. The distortion and noise are measured separately and than added together to form the ratio. The SINAD is the reciprocal to the THD + N. The SINAD and THD + N are a good indication of the overall dynamic performance of the ADC, because all components of noise and distortion are included.

15.5.2.3 Effective Number of Bits

The SINAD is used to determine the *effective number of bits* (ENOB) of accuracy the converter displays at that frequency. For example, a nominal 8-bit resolution D/A may be specified as having 45 dB SNR at a particular input frequency. The number of effective bits is defined as

$$\text{ENOB} = \frac{\text{SNR}_{\text{REAL}} - 1.76}{6.02} = 7.2 \text{ bits} \qquad (15.4)$$

The actual performance of the device is therefore less than its nominal resolution at this frequency.

15.5.2.4 Spurious-Free Dynamic Range

Spurious-free dynamic range is the difference in decibel between the maximum signal component and the largest distortion component (Fig. 15.11):

It is an important specification in RF applications, where FCC regulations specify the magnitude of spurs.

Improper decoupling may cause spurs. A notch filter can be used to eliminate a spur, but many RF applications are RF agile—changing the frequency of the spur as well. The notch must catch all spur frequencies, or it is useless.

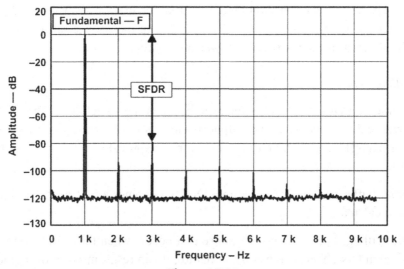

Figure 15.11
Spurious-free dynamic range.

15.5.2.5 *Intermodulation Distortion*

The DNL and INL errors described previously appear in a high-frequency AC application as intermodulation distortion (Fig. 15.12).

The best method of combating intermodulation distortion is to make the buffer amplifier system as linear as possible (beware of rail-to-rail op amps that may not be linear near the

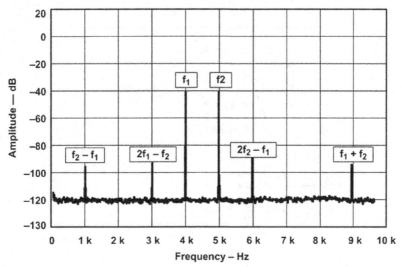

Figure 15.12
Intermodulation distortion.

voltage rails). Try to limit current through the internal feedback resistor in the digital-to-analog converter (DAC). See Section 15.7.1 on increasing the voltage rail for suggestions about reducing internal feedback resistor power dissipation.

15.5.2.6 Settling Time

The settling time of a D/A converter is the time between the switching of the digital inputs of the converter and the time when the output reaches its final value and remains within a specified error band (Fig. 15.13). Settling time is the reciprocal of the maximum D/A conversion rate.

When an output buffer op amp is used with a D/A, it becomes a part of the settling time/conversion rate calculations.

Related to the settling time is a glitch response that occurs when the digital code changes state. Even though this effect is transitory in nature, it can result in noise or harmonics when used in fast AC applications. The best way of reducing the glitch is to properly decouple the D/A and op amp buffer (see Chapter 13). In extreme cases, a deglitching circuit may be needed (Fig. 15.14).

This technique relies on the software designer to balance the timing of the control signal so it activates the hold function right before the D/A input code changes, then releases the hold right after the code has changed. The selection of C_{HOLD} is critical—it must hold the buffer output without droop and without compromising system bandwidth.

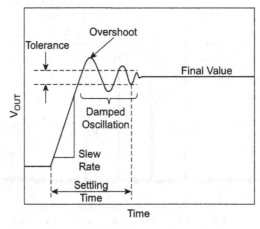

Settling Time

Figure 15.13
D/A settling time.

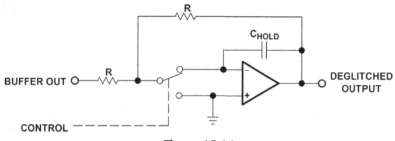

Figure 15.14
D/A deglitch circuit.

15.6 Compensating for DAC Capacitance

D/A converters are constructed of either bipolar or CMOS technology, with CMOS being the more common. CMOS transistors, however, have a lot of capacitance. This capacitance will add in D/A converters, depending on the number of resistors switched on or off. Capacitance at an inverting op amp input is a good way to cause it to oscillate, especially since some buffer amplifiers will be operated at less than unity gain. The converter capacitance C_O must be compensated for externally (Fig. 15.15).

The normal technique for compensating the buffer amplifier for output capacitance is to add a feedback capacitor C_F. C_F is calculated by the following:

$$C_F = 2 \times \sqrt{\frac{C_O}{2\pi R_F} \times \frac{1}{G_{BW}}} \qquad (15.5)$$

where, C_O, the output capacitance from the D/A data sheet; R_F, the feedback resistance from the D/A data sheet; G_{BW}, the small signal unity gain bandwidth product of the output amplifier.

Unfortunately, the feedback capacitors C_F and the internal D/A capacitance C_O will both limit the conversion speed of the D/A. If faster conversion is needed, a D/A with a lower

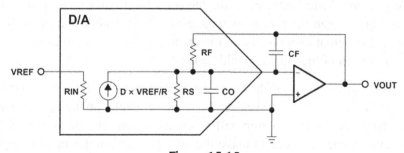

Figure 15.15
Compensating for CMOS DAC output capacitance.

output capacitance, and therefore a lower feedback compensation capacitor will be needed. The overall settling time with the external capacitance is:

$$T_S \approx \sqrt{\frac{R_F(C_O + C_F)}{2G_{BW}}} \qquad (15.6)$$

where, C_O, the D/A internal capacitance; R_F, the feedback resistor; C_F, the compensation capacitance; G_{BW}, the small signal unity gain bandwidth product of the output amplifier.

15.7 Increasing Op Amp Buffer Amplifier Current and Voltage

Process limitations of op amps limit the power that can be dissipated at the output. Unfortunately, there are applications that will require the DAC to interface to loads that dissipate considerable power. These include actuators, position solenoids, stepper motors, loudspeakers, vibration tables, positioning tables—the possibilities are endless.

While several "power op amps" are available that can drive heavy loads, they usually compromise several other specifications to achieve the high-power operation. Input voltage offset, input current, and input capacitance can be decades higher than the designer is accustomed to, and make these power op amps unsuitable for direct interface with a DAC as a replacement for the buffer op amp.

The power booster stage can be designed discretely, or a prepackaged amplifier of some sort, depending on what is needed for the application. Sometimes high current is required for driving loads such as actuators and stepper motors. Audio applications can require a lot of wattage to drive loudspeakers. This implies a higher-voltage rail than op amps commonly operate at. This and other high-voltage applications can operate off of, and generate lethal voltages. The designer needs to be extremely careful not to create an unsafe product, or be electrocuted while developing it.

The power stage is most often included in the feedback loop of the op amp circuit, so that the closed loop can compensate for power stage errors. This is not always possible if the voltage swing of the output exceeds that of the op amp voltage rails. In these cases, a voltage-divided version of the output should be used.

There are three broad categories of booster: the current booster, the voltage booster, and boosters that do both. All of them work on the same principle: anything that is put inside the feedback loop of the op amp will be compensated for—the output voltage will swing to whatever voltage it needs to make the voltage at the buffer op amp inputs equal.

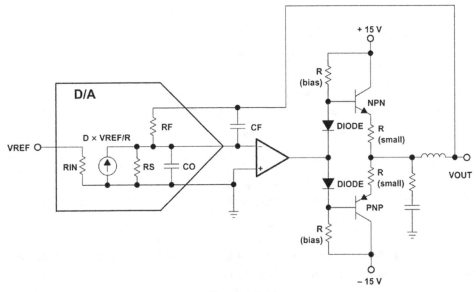

Figure 15.16
D/A output current booster.

15.7.1 Current Boosters

These usually use some variation of the class-B push—pull amplifier topology (Fig. 15.16).

The circuit in Fig. 15.16 has been employed for decades—many resources are available that can be used to design exact component values. It boosts current because the output impedance of the op amp has been bypassed and used as the driver for the base of the NPN and PNP power transistors. The two diodes compensate for the VBE drop in the transistors, whose bases are biased by two resistors off of the supplies. The output of the booster stage is fed back to the feedback resistor in the D/A to complete the feedback loop. The output impedance of the stage is only limited by the characteristics of the output transistors and small emitter resistors. Modern power transistors have such high-frequency response that this circuit may oscillate. The RC snubber network and a small inductor in series with the load can be used to damp the oscillation—or be omitted if oscillation is not a problem. Beware of varying transistor betas, however.

15.7.2 Voltage Boosters

If even more current is needed, or the output voltage swing must be more than ±15 V, the booster stage can be operated at voltages higher than the buffer amplifier potentials. A designer might be tempted to try the circuit of Fig. 15.17.

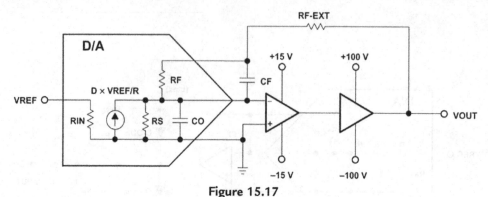

Figure 15.17
Incorrect method of increasing voltage swing of D/A converters.

Anytime there are higher-voltage rails on the output section, there are potential hazards. The circuit above illustrates a common misapplication.

- The whole reason for using the booster amp is to allow the V_{OUT} to swing to a ± 100 V rail. If this circuit was operated in the unity gain mode (external $R_F = 0$), the V_{OUT} will only swing ± 15 V, maximum. There would be no need for the ± 100 V rail. That voltage rail is there to allow voltage gain.
- If the circuit is operated with a gain (external R larger than 0), the external R_F adds to the internal R_F to create the gain:

$$\text{Gain} = \frac{R_{F-EXT} + R_{F-INT}}{R_S} \tag{15.7}$$

The problem with this is that the wattage of the resistors increases as the external voltage rail increases. The designer has control over the wattage of the external R_F, but has no control whatsoever over internal R_F or R_S. Because these resistors are fabricated on the IC, their wattage is limited. Even if the wattage rating of the internal resistors is meticulously observed, they may have undesirable thermal coefficients if allowed to dissipate that wattage. Resistor self-heating will change the resistance according to its rated temperature coefficient (maximum). The external resistor is sure to have a different thermal coefficient from the internal resistors, causing a gain error. The designer may never have encountered the effects of resistor self-heating before, because through-hole and surface mount devices have enough bulk to minimize the effect of self-heating. At the geometries present on IC D/As, resistor self-heating is a much more pronounced effect. It will produce a nonlinearity error in the D/A output.

This effect is most pronounced in high-resolution converters, where the geometry is the smallest. The designer, therefore, must limit the current in the feedback resistor if at all

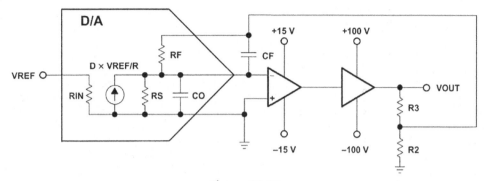

Figure 15.18
Correct method of increasing voltage range.

possible. Fig. 15.18 shows a method of achieving gain control while keeping the high-current path out of the internal feedback resistor.

- R_3 and R_2 are selected to ensure that the feedback voltage to the D/A internal R_F can never exceed the D/A rated limits.
- R_3 and R_2, of course, have to be the correct power rating. R_2, in particular, has to be carefully selected. If it burns out, the feedback loop will present hazardous voltages to the D/A. R_3, which drops the bulk of the voltage, will have to dissipate considerable wattage.

If the combination of voltage swing and power ratings cannot be balanced to achieve a working design, the only choice left to the designer will be to break the feedback loop and live with the loss of accuracy. For AC applications, this may be acceptable.

15.7.3 Power Boosters

The two types of boosters above can, of course, be combined to produce more power. In audio applications, for example, a ±15 V power supply limits the output power to 112.5 W, absolute maximum, into an 8 Ω load. To increase the power, the voltage rails must also be increased, with all of the cautions of the previous paragraph observed.

15.7.4 Single-Supply Operation and DC Offsets

A D/A power circuit is not the right place to try to apply single-supply design techniques. In audio applications, a single-supply design would force a large coupling capacitor, which would distort and limit low-frequency response. In DC applications, a DC offset will continually drive a load, which will have to dissipate the excess voltage through its internal resistance as heat.

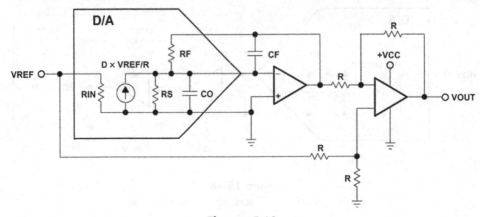

Figure 15.19
Single-supply DAC operation.

Nevertheless, there may be applications that require a DC offset. The designer is fortunate in that there is already a precision reference available in the circuit. The reference drives the resistor network in the D/A and may be external or internal to it. In most cases, an internal reference is brought out to a pin on the device. It is important for the designer not to excessively load the reference, as that would directly affect D/A accuracy (Fig. 15.19).

In the circuit in Fig. 15.19, the output of the buffer amplifier is shifted up in DC level by $1/2 \ V_{REF}$ (not $1/2 \ V_{CC}$). V_{REF} was selected because it is much more stable and accurate than V_{CC}. The four resistors in the level shifter circuit must be highly accurate and matched, or this circuit will contribute to gain and offset errors. Thermal errors, however, cannot be compensated for, because the external resistors are probably going to have a different thermal drift than those on the IC. This technique is limited to applications that will see only a small change in ambient temperature.

CHAPTER 16

Active Filter Design Techniques

16.1 Introduction

What is a filter?

A filter is a device that passes electric signals at certain frequencies or frequency ranges while preventing the passage of others.

Webster

Filter circuits are used in a wide variety of applications. In the field of telecommunication, band-pass filters are used in the audio frequency range (0—20 kHz) for modems and speech processing. High-frequency band-pass filters (several hundred MHz) are used for channel selection in telephone central offices. Data acquisition systems usually require antialiasing low-pass filters as well as low-pass noise filters in their preceding signal-conditioning stages. System power supplies often use band-rejection filters to suppress the 60 Hz line frequency and high-frequency transients.

In addition, there are filters that do not filter any frequencies of a complex input signal, but just add a linear phase shift to each frequency component, thus contributing to a constant time delay. These are called all-pass filters.

At high frequencies (>1 MHz), all of these filters usually consist of passive components such as inductors (L), resistors (R), and capacitors (C). They are then called LRC filters.

In the lower frequency range (1 Hz—1 MHz), however, the inductor value becomes very large and the inductor itself gets quite bulky, making economical production difficult.

In these cases, active filters become important. Active filters are circuits that use an operational amplifier (op amp) as the active device in combination with some resistors and capacitors to provide an LRC-like filter performance at low frequencies (Fig. 16.1).

This chapter covers active filters. It introduces the three main filter optimizations (Butterworth, Tschebyscheff, and Bessel), followed by five sections describing the most common active filter applications: low-pass, high-pass, band-pass, band-rejection, and all-pass filters. Rather than resembling just another filter book, the individual filter sections are written in a cookbook style, thus avoiding tedious mathematical derivations that are covered in books devoted to the subject of active filter design. Each section starts with the general transfer function of a filter,

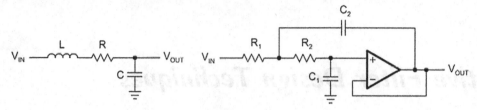

Figure 16.1
Second-order passive low-pass and second-order active low-pass.

followed by the design equations to calculate the individual circuit components. The chapter closes with a section on practical design hints for single-supply filter designs.

If you find this material to be complex, the following chapter introduces techniques that will allow you to very quickly design working filters with a minimum of mathematical calculation. What is lost, though with the rapid techniques in the following chapter is the ability to design every combination of gain, frequency, and quality factor—it makes certain decisions up front that simplify the process. If you need a specialized filter, stick with the techniques in this chapter.

16.2 Fundamentals of Low-Pass Filters

The most simple low-pass filter is the passive RC low-pass network shown in Fig. 16.2.

Its transfer function is

$$A(s) = \frac{\dfrac{1}{RC}}{s + \dfrac{1}{RC}} = \frac{1}{1 + sRC}$$

where the complex frequency variable, $s = j\omega + \sigma$, allows for any time variable signals. For pure sine waves, the damping constant, σ, becomes zero and $s = j\omega$.

For a normalized presentation of the transfer function, s is referred to the filter's corner frequency, or -3 dB frequency, ω_C, and has these relationships:

$$s = \frac{s}{\omega_C} = \frac{j\omega}{\omega_C} = j\frac{f}{f_C} = j\Omega$$

With the corner frequency of the low-pass in Fig. 16.2 being $f_C = 1/2\pi RC$, s becomes $s = sRC$ and the transfer function A(s) results in:

$$A(s) = \frac{1}{1 + s}$$

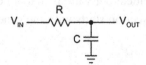

Figure 16.2
First-order passive RC low-pass.

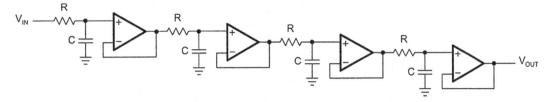

Figure 16.3
Fourth-order passive RC low-pass with decoupling amplifiers.

The magnitude of the gain response is

$$|A| = \frac{1}{\sqrt{1+\Omega^2}}$$

For frequencies $\Omega \gg 1$, the rolloff is 20 dB/decade. For a steeper rolloff, n filter stages can be connected in series as shown in Fig. 16.3. To avoid loading effects, op amps, operating as impedance converters, separate the individual filter stages.

The resulting transfer function is

$$A(s) = \frac{1}{(1+\alpha_1 s)(1+\alpha_2 s)...(1+\alpha_n s)}$$

In the case that all filters have the same cut-off frequency, f_C, the coefficients become $\alpha_1 = \alpha_2 = ...\alpha_n = \alpha = \sqrt{\sqrt[n]{2}-1}$, and f_C of each partial filter is $1/\alpha$ times higher than f_C of the overall filter.

Fig. 16.4 shows the results of a fourth-order RC low-pass filter. The rolloff of each partial filter (Curve 1) is -20 dB/decade, increasing the rolloff of the overall filter (Curve 2) to 80 dB/decade.

The corner frequency of the overall filter is reduced by a factor of $\alpha \approx 2.3$ times versus the -3 dB frequency of partial filter stages.

In addition, Fig. 16.4 shows the transfer function of an ideal fourth-order low-pass function (Curve 3).

In comparison to the ideal low-pass, the RC low-pass lacks in the following characteristics:

- The passband gain varies long before the corner frequency, f_C, thus amplifying the upper passband frequencies less than the lower passband.
- The transition from the passband into the stopband is not sharp, but happens gradually, moving the actual 80-dB rolloff by 1.5 octaves above f_C.
- The phase response is not linear, thus increasing the amount of signal distortion significantly.

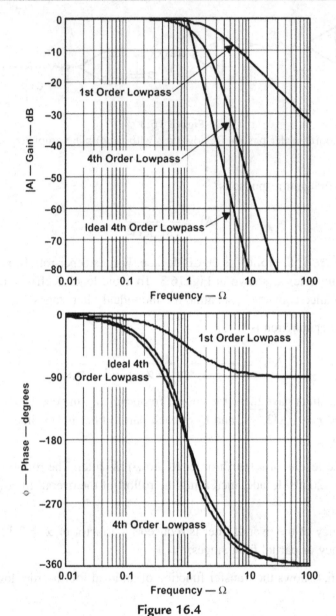

Figure 16.4
Frequency and phase responses of a fourth-order passive RC low-pass filter.
Note: Filter response graphs plot gain versus the normalized frequency axis $\Omega(\Omega = f/f_C)$. Curve 1: first-order partial low-pass filter, Curve 2: fourth-order overall low-pass filter, Curve 3: ideal fourth-order low-pass filter.

The gain and phase response of a low-pass filter can be optimized to satisfy *one* of the following three criteria:

1. a maximum passband flatness,
2. an immediate passband-to-stopband transition, and
3. a linear phase response.

For that purpose, the transfer function must allow for complex poles and needs to be of the following type:

$$A(s) = \frac{A_0}{(1 + a_1 s + b_1 s^2)(1 + a_2 s + b_2 s^2) \ldots (1 + a_n s + b_n s^2)} = \frac{A_0}{\prod_i (1 + a_i s + b_i s^2)}$$

where A_0 is the passband gain at DC, and a_i and b_i are the filter coefficients.

Since the denominator is a product of quadratic terms, the transfer function represents a series of cascaded second-order low-pass stages, with a_i and b_i being positive real coefficients. These coefficients define the complex pole locations for each second-order filter stage, thus determining the behavior of its transfer function.

The following three types of predetermined filter coefficients are available and listed in table format in Section 16.9:

- The Butterworth coefficients, optimizing the passband for maximum flatness.
- The Tschebyscheff coefficients, sharpening the transition from passband into the stopband.
- The Bessel coefficients, linearizing the phase response up to f_C.

The transfer function of a passive RC filter does not allow further optimization, due to the lack of complex poles. The only possibility to produce conjugate complex poles using passive components is the application of LRC filters. However, these filters are mainly used at high frequencies. In the lower frequency range (<10 MHz) the inductor values become very large and the filter becomes uneconomical to manufacture. In these cases active filters are used.

Active filters are RC networks that include an active device, such as an operational amplifier (op amp).

Section 16.3 shows that the products of the RC values and the corner frequency must yield the predetermined filter coefficients a_i and b_i, to generate the desired transfer function.

The following paragraphs introduce the most commonly used filter optimizations.

16.2.1 Butterworth Low-Pass Filters

The Butterworth low-pass filter provides maximum passband flatness. Therefore, a Butterworth low-pass is often used as antialiasing filter in data converter applications where precise signal levels are required across the entire passband.

Fig. 16.5 plots the gain response of different orders of Butterworth low-pass filters versus the normalized frequency axis, $\Omega(\Omega = f/f_C)$; the higher the filter order, the longer the passband flatness.

16.2.2 Tschebyscheff Low-Pass Filters

The Tschebyscheff low-pass filters provide an even higher gain rolloff above f_C. However, as Fig. 16.6 shows, the passband gain is not monotone, but contains ripples of constant magnitude instead. For a given filter order, the higher the passband ripples, the higher the filter's rolloff.

With increasing filter order, the influence of the ripple magnitude on the filter rolloff diminishes.

Each ripple accounts for one second-order filter stage. Filters with even order numbers generate ripples above the 0-dB line, whereas filters with odd order numbers create ripples below 0 dB.

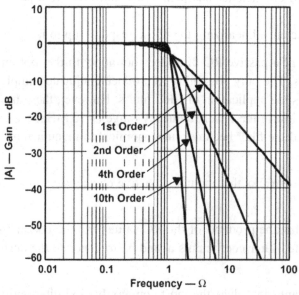

Figure 16.5
Amplitude responses of Butterworth low-pass filters.

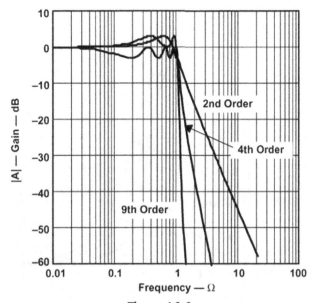

Figure 16.6
Gain responses of Tschebyscheff low-pass filters.

Tschebyscheff filters are often used in filter banks, where the frequency content of a signal is of more importance than a constant amplification.

16.2.3 Bessel Low-Pass Filters

The Bessel low-pass filters have a linear phase response (Fig. 16.7) over a wide frequency range, which results in a constant group delay (Fig. 16.8) in that frequency range. Bessel low-pass filters, therefore, provide an optimum square-wave transmission behavior. However, the passband gain of a Bessel low-pass filter is not as flat as that of the Butterworth low-pass, and the transition from passband to stopband is by far not as sharp as that of a Tschebyscheff low-pass filter (Fig. 16.9).

16.2.4 Quality Factor Q

The quality factor Q is an equivalent design parameter to the filter order n. Instead of designing an nth order Tschebyscheff low-pass, the problem can be expressed as designing a Tschebyscheff low-pass filter with a certain Q.

For band-pass filters, Q is defined as the ratio of the mid-frequency, f_m, to the bandwidth at the two -3 dB points:

$$Q = \frac{f_m}{(f_2 - f_1)}$$

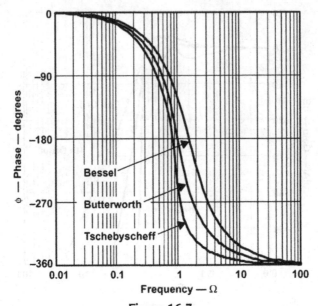

Figure 16.7
Comparison of phase responses of fourth-order low-pass filters.

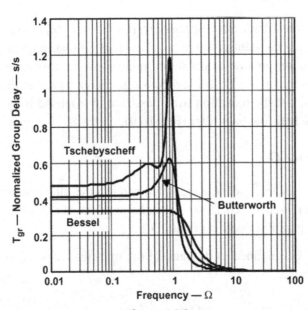

Figure 16.8
Comparison of normalized group delay (t_{gr}) of fourth-order low-pass filters.

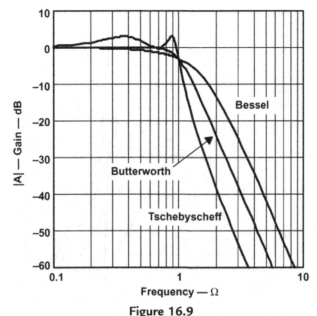

Figure 16.9
Comparison of gain responses of fourth-order low-pass filters.

For low-pass and high-pass filters, Q represents the pole quality and is defined as:

$$Q = \frac{\sqrt{b_i}}{a_i}$$

High Qs can be graphically presented as the distance between the 0 dB line and the peak point of the filter's gain response. An example is given in Fig. 16.10, which shows a 10th-order Tschebyscheff low-pass filter and its five partial filters with their individual Qs.

The gain response of the fifth filter stage peaks at 31 dB, which is the logarithmic value of Q_5:

$$Q_5[dB] = 20\,logQ_5$$

Solving for the numerical value of Q_5 yields

$$Q_5 = 10\,\frac{31}{20} = 35.48$$

which is within 1% of the theoretical value of Q = 35.85 given in Section 16.9, Table 16.11, last row.

The graphical approximation is good for Q > 3. For lower Qs, the graphical values differ from the theoretical value significantly. However, only higher Qs are of concern, since the higher the Q is, the more a filter inclines to instability.

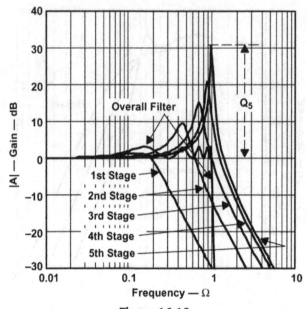

Figure 16.10
Graphical presentation of quality factor Q on a 10th-order Tschebyscheff low-pass filter.

16.2.5 Summary

The general transfer function of a low-pass filter is:

$$A(s) = \frac{A_0}{\prod_i (1 + a_i s + b_i s)} \tag{16.1}$$

The filter coefficients a_i and b_i distinguish between Butterworth, Tschebyscheff, and Bessel filters. The coefficients for all three types of filters are tabulated down to the 10th-order in Section 16.9, Tables 16.6 through 16.12.

The multiplication of the denominator terms with each other yields an nth order polynomial of s, with n being the filter order.

While n determines the gain rolloff above f_C with $n \cdot 20$ dB/decade, a_i and b_i determine the gain behavior in the passband.

In addition, the ratio $\frac{\sqrt{b_i}}{a_i} = Q$ is defined as the pole quality. The higher the Q value, the more a filter inclines to instability.

16.3 Low-Pass Filter Design

Eq. (16.1) represents a cascade of second-order low-pass filters. The transfer function of a single stage is

$$A_i(s) = \frac{A_0}{(1 + a_i s + b_i s^2)} \tag{16.2}$$

For a first-order filter, the coefficient b is always zero ($b_1 = 0$), thus yielding:

$$A(s) = \frac{A_0}{1 + a_1 s} \tag{16.3}$$

The first-order and second-order filter stages are the building blocks for higher-order filters.

Often the filters operate at unity gain ($A_0 = 1$) to lessen the stringent demands on the op amp's open-loop gain.

Fig. 16.11 shows the cascading of filter stages up to the sixth order. A filter with an even order number consists of second-order stages only, while filters with an odd order number include an additional first-order stage at the beginning.

Fig. 16.10 demonstrated that the higher the corner frequency of a partial filter, the higher its Q. Therefore, to avoid the saturation of the individual stages, the filters need to be placed in the order of rising Q values. The Q values for each filter order are listed (in rising order) in Section 16.9, Tables 16.6 through 16.12.

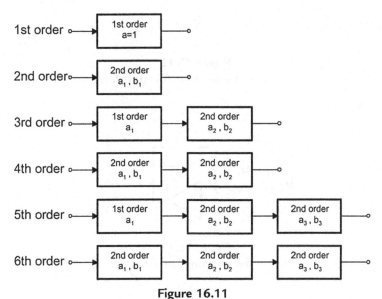

Figure 16.11
Cascading filter stages for higher-order filters.

16.3.1 First-Order Low-Pass Filter

Figs. 16.12 and 16.13 show a first-order low-pass filter in the inverting and in the noninverting configuration.

The transfer functions of the circuits are given as:

$$A(s) = \frac{1 + \dfrac{R_2}{R_3}}{1 + \omega_c R_1 C_1 s} \quad \text{and} \quad A(s) = \frac{-\dfrac{R_2}{R_1}}{1 + \omega_c R_2 C_1 s}$$

The negative sign indicates that the inverting amplifier generates a 180 degrees phase shift from the filter input to the output.

The coefficient comparison between the two transfer functions and Eq. (16.3) yields

$$A_0 = 1 + \frac{R_2}{R_3} \quad \text{and} \quad A_0 = -\frac{R_2}{R_1}$$

$$a_1 = \omega_c R_1 C_1 \quad \text{and} \quad a_1 = \omega_c R_2 C_1$$

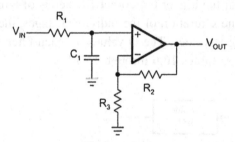

Figure 16.12
First-order noninverting low-pass filter.

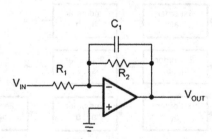

Figure 16.13
First-order inverting low-pass filter.

To dimension the circuit, specify the corner frequency (f_C), the DC gain (A_0), and capacitor C_1, and then solve for resistors R_1 and R_2:

$$R_1 = \frac{a_1}{2\pi f_c C_1} \quad \text{and} \quad R_2 = \frac{a_1}{2\pi f_c C_1}$$

$$R_2 = R_3(A_0 - 1) \quad \text{and} \quad R_1 = -\frac{R_2}{A_0}$$

The coefficient a_1 is taken from one of the coefficient tables, Tables 16.6 through 16.12 in Section 16.9.

Note, that all filter types are identical in their first order and $a_1 = 1$. For higher filter orders, however, $a_1 \neq 1$ because the corner frequency of the first-order stage is different from the corner frequency of the overall filter.

Example 16.1. First-Order Unity-Gain Low-Pass Filter.

For a first-order unity-gain low-pass filter with $f_C = 1$ kHz and $C_1 = 47$ nF, R_1 calculates to:

$$R_1 = \frac{a_1}{2\pi f_c C_1} = \frac{1}{2\pi \cdot 10^3 \text{ Hz} \cdot 47 \cdot 10^{-9} \text{ F}} = 3.38 \text{ k}\Omega$$

However, to design the first stage of a third-order unity-gain Bessel low-pass filter, assuming the same values for f_C and C_1, requires a different value for R_1. In this case, obtain a_1 for a third-order Bessel filter from Table 16.6 in Section 16.9 (Bessel coefficients) to calculate R_1:

$$R_1 = \frac{a_1}{2\pi f_c C_1} = \frac{0.756}{2\pi \cdot 10^3 \text{ Hz} \cdot 47 \cdot 10^{-9} \text{ F}} = 2.56 \text{ k}\Omega$$

When operating at unity gain, the noninverting amplifier reduces to a voltage follower (Fig. 16.14), thus inherently providing a superior gain accuracy. In the case of the inverting amplifier, the accuracy of the unity gain depends on the tolerance of the two resistors, R_1 and R_2.

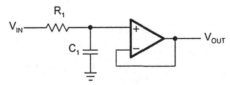

Figure 16.14
First-order noninverting low-pass filter with unity gain.

16.3.2 Second-Order Low-Pass Filter

There are two topologies for a second-order low-pass filter, the Sallen–Key and the multiple feedback (MFB) topology.

16.3.2.1 Sallen–Key Topology

The general Sallen–Key topology in Fig. 16.15 allows for separate gain setting via $A_0 = 1 + R_4/R_3$. However, the unity-gain topology in Fig. 16.16 is usually applied in filter designs with high-gain accuracy, unity gain, and low Qs (Q < 3).

The transfer function of the circuit in Fig. 16.15 is given as:

$$A(s) = \frac{A_0}{1 + \omega_c[C_1(R_1 + R_2) + (1 - A_0)R_1C_2]s + \omega_c^2 R_1 R_2 C_1 C_2 s^2}$$

For the unity-gain circuit in Fig. 16.16 ($A_0 = 1$), the transfer function simplifies to:

$$A(s) = \frac{1}{1 + \omega_c C_1(R_1 + R_2)s + \omega_c^2 R_1 R_2 C_1 C_2 s^2}$$

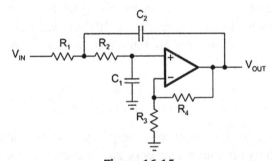

Figure 16.15
General Sallen–Key low-pass filter.

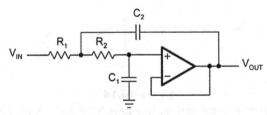

Figure 16.16
Unity-gain Sallen–Key low-pass filter.

The coefficient comparison between this transfer function and Eq. (16.2) yields

$$A_0 = 1$$
$$a_1 = \omega_c C_1(R_1 + R_2)$$
$$b_1 = \omega_c^2 R_1 R_2 C_1 C_2$$

Given C_1 and C_2, the resistor values for R_1 and R_2 are calculated through:

$$R_{1,2} = \frac{a_1 C_2 \mp \sqrt{a_1^2 C_2^2 - 4 b_1 C_1 C_2}}{4 \pi f_c C_1 C_2}$$

To obtain real values under the square root, C_2 must satisfy the following condition:

$$C_2 \geq C_1 \frac{4 b_1}{a_1^2}$$

Example 16.2. Second-Order Unity-Gain Tschebyscheff Low-Pass Filter.

The task is to design a second-order unity-gain Tschebyscheff low-pass filter with a corner frequency of $f_C = 3$ kHz and a 3-dB passband ripple.

From Table 16.11 (the Tschebyscheff coefficients for 3-dB ripple), obtain the coefficients a_1 and b_1 for a second-order filter with $a_1 = 1.0650$ and $b_1 = 1.9305$.

Specifying C_1 as 22 nF yields in a C_2 of:

$$C_2 \geq C_1 \frac{4 b_1}{a_1^2} = 22 . 10^{-9} \text{ nF} \cdot \frac{4 \cdot 1.9305}{1.065^2} \cong 150 \text{ nF}$$

Inserting a_1 and b_1 into the resistor equation for $R_{1,2}$ results in:

$$R_1 = \frac{1.065 \cdot 150 \cdot 10^{-9} - \sqrt{(1.065 \cdot 150 \cdot 10^{-9})^2 - 4 \cdot 1.9305 \cdot 22 \cdot 10^{-9} \cdot 150 \cdot 10^{-9}}}{4 \pi \cdot 3 \cdot 10^3 \cdot 22 \cdot 10^{-9} \cdot 150 \cdot 10^{-9}} = 1.26 \text{ k}\Omega$$

and

$$R_2 = \frac{1.065 \cdot 150 \cdot 10^{-9} - \sqrt{(1.065 \cdot 150 \cdot 10^{-9})^2 - 4 \cdot 1.9305 \cdot 22 \cdot 10^{-9} \cdot 150 \cdot 10^{-9}}}{4 \pi \cdot 3 \cdot 10^3 \cdot 22 \cdot 10^{-9} \cdot 150 \cdot 10^{-9}} = 1.30 \text{ k}\Omega$$

with the final circuit shown in Fig. 16.17.

A special case of the general Sallen–Key topology is the application of equal resistor values and equal capacitor values: $R_1 = R_2 = R$ and $C_1 = C_2 = C$.

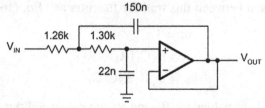

Figure 16.17
Second-order unity-gain Tschebyscheff low-pass with 3 dB ripple.

The general transfer function changes to:

$$A(s) = \frac{A_0}{1 + \omega_c RC(3 - A_0)s + (\omega_c RC)^2 s^2} \quad \text{with} \quad A_0 = 1 + \frac{R_4}{R_3}$$

The coefficient comparison with Eq. (16.2) yields the following:

$$a_1 = \omega RC(3 - A_0)$$
$$b_1 = (\omega RC)^2$$

Given C and solving for R and A_0 results in:

$$R = \frac{\sqrt{b_1}}{2\pi f_c C} \quad \text{and} \quad A_0 = 3 - \frac{a_1}{\sqrt{b_1}} = 3 - \frac{1}{Q}$$

Thus, A_0 depends solely on the pole quality Q and vice versa; Q, and with it the filter type, is determined by the gain setting of A_0:

$$Q = \frac{1}{3 - A_0}$$

The circuit in Fig. 16.18 allows the filter type to be changed through the various resistor ratios R_4/R_3.

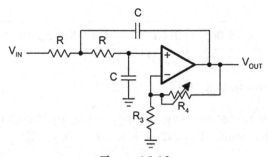

Figure 16.18
Adjustable second-order low-pass filter.

Table 16.1: Second-Order Filter Coefficients

Second-Order	Bessel	Butterworth	3 dB Tschebyscheff
a_1	1.3617	1.4142	1.065
b_1	0.618	1	1.9305
Q	0.58	0.71	1.3
R_4/R_3	0.268	0.568	0.234

Table 16.1 lists the coefficients of a second-order filter for each filter type and gives the resistor ratios that adjust the Q.

16.3.2.2 Multiple Feedback Low Pass Filter Topology

The multiple feedback (MFB) topology is commonly used in filters that have high Qs and require a high gain.

The transfer function of the circuit in Fig. 16.19 is given as follows:

$$A(s) = - \frac{\dfrac{R_2}{R_1}}{1 + \omega_c C_1 \left(R_2 + R_3 + \dfrac{R_2 R_3}{R_1} \right) s + \omega_c^2 C_1 C_2 R_2 R_3 s^2}$$

Through coefficient comparison with Eq. (16.2) one obtains the relation:

$$A_0 = - \frac{R_2}{R_1}$$

$$a_1 = \omega_c C_1 \left(R_2 + R_3 + \frac{R_2 R_3}{R_1} \right)$$

$$b_1 = \omega_c^2 C_1 C_2 R_2 R_3$$

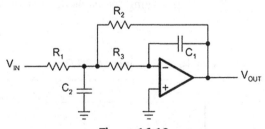

Figure 16.19
Second-order multiple feedback low-pass filter.

Given C_1 and C_2, and solving for the resistors R_1–R_3:

$$R_2 = \frac{a_1 C_2 - \sqrt{a_1^2 C_2^2 - 4b_1 C_1 C_2 (1 - A_0)}}{4\pi f_c C_1 C_2}$$

$$R_1 = \frac{R_2}{-A_0}$$

$$R_3 = \frac{b_1}{4\pi^2 f_c^2 C_1 C_2 R_2}$$

To obtain real values for R_2, C_2 must satisfy the following condition:

$$C_2 \geq C_1 \frac{4b_1 (1 - A_0)}{a_1^2}$$

16.3.3 Higher-Order Low-Pass Filters

Higher-order low-pass filters are required to sharpen a desired filter characteristic. For that purpose, first-order and second-order filter stages are connected in series, so that the product of the individual frequency responses results in the optimized frequency response of the overall filter.

To simplify the design of the partial filters, the coefficients a_i and b_i for each filter type are listed in the coefficient tables (Tables 16.6 through 16.12 in Section 16.9), with each table providing sets of coefficients for the first 10 filter orders.

Example 16.3. Fifth Order Filter (Table 16.2).

The task is to design a fifth-order unity-gain Butterworth low-pass filter with the corner frequency $f_C = 50$ kHz.

First the coefficients for a fifth-order Butterworth filter are obtained from Table 16.7, Section 16.9.

Then dimension each partial filter by specifying the capacitor values and calculating the required resistor values.

Table 16.2: Example 16.3 Filters

	a_i	b_i
Filter 1	$A_1 = 1$	$b_1 = 0$
Filter 2	$a_2 = 1.6180$	$b_2 = 1$
Filter 3	$a_3 = 0.6180$	$b_3 = 1$

16.3.3.1 First Filter

With $C_1 = 1$ nF,

$$R_1 = \frac{a_1}{2\pi f_c C_1} = \frac{1}{2\pi \cdot 50 \cdot 10^3 \text{ Hz} \cdot 1 \cdot 10^{-9} \text{ F}} = 3.18 \text{ k}\Omega$$

The closest 1% value is 3.16 kΩ (See Fig 16.20).

16.3.3.2 Second Filter

With $C_1 = 820$ pF,

$$C_2 \geq C_1 \frac{4b_2}{a_2^2} = 820 \cdot 10^{-12} \text{ F} \cdot \frac{4 \cdot 1}{1.618^2} = 1.26 \text{ nF}$$

The closest 5% value is 1.5 nF (See Fig 16.21).

With $C_1 = 820$ pF and $C_2 = 1.5$ nF, calculate the values for R_1 and R_2 through:

$$R_1 = \frac{a_2 C_2 - \sqrt{a_2^2 C_2^2 - 4b_2 C_1 C_2}}{4\pi f_c C_1 C_2} \quad \text{and} \quad R_1 = \frac{a_2 C_2 + \sqrt{a_2^2 C_2^2 - 4b_2 C_1 C_2}}{4\pi f_c C_1 C_2}$$

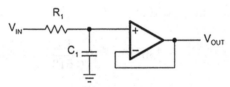

Figure 16.20
First-order unity-gain low-pass.

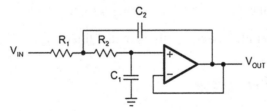

Figure 16.21
Second-order unity-gain Sallen–Key low-pass filter.

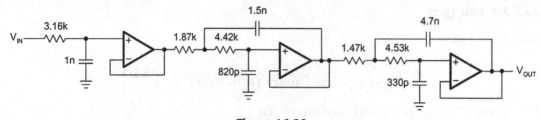

Figure 16.22
Fifth-order unity-gain Butterworth low-pass filter.

and obtain

$$R_1 = \frac{1.618 \cdot 1.5 \cdot 10^{-9} - \sqrt{\left(1.618 \cdot 1.5 \cdot 10^{-9}\right)^2 - 4 \cdot 1 \cdot 820 \cdot 10^{-12} \cdot 1.5 \cdot 10^{-9}}}{4\pi \cdot 50 \cdot 10^3 \cdot 820 \cdot 10^{-12} \cdot 1.5 \cdot 10^{-9}} = 1.87 \text{ k}\Omega$$

$$R_2 = \frac{1.618 \cdot 1.5 \cdot 10^{-9} - \sqrt{\left(1.618 \cdot 1.5 \cdot 10^{-9}\right)^2 - 4 \cdot 1 \cdot 820 \cdot 10^{-12} \cdot 1.5 \cdot 10^{-9}}}{4\pi \cdot 50 \cdot 10^3 \cdot 820 \cdot 10^{-12} \cdot 1.5 \cdot 10^{-9}} = 4.42 \text{ k}\Omega$$

R_1 and R_2 are available 1% resistors.

16.3.3.3 Third Filter

The calculation of the third filter is identical to the calculation of the second filter, except that a_2 and b_2 are replaced by a_3 and b_3, thus resulting in different capacitor and resistor values.

Specify C_1 as 330 pF, and obtain C_2 with:

$$C_2 \geq C_1 \frac{4b_3}{a_3^2} = 330 \cdot 10^{-12} \text{ F} \cdot \frac{4.1}{0.618^2} = 3.46 \text{ nF}$$

The closest 10% value is 4.7 nF.

With $C_1 = 330$ pF and $C_2 = 4.7$ nF, the values for R_1 and R_2 are as follows:

- $R_1 = 1.45$ kΩ, with the closest 1% value being 1.47 kΩ, and
- $R_2 = 4.51$ kΩ, with the closest 1% value being 4.53 kΩ.

Fig. 16.22 shows the final filter circuit with its partial filter stages.

16.4 High-Pass Filter Design

By replacing the resistors of a low-pass filter with capacitors, and its capacitors with resistors, a high-pass filter is created (See Fig. 16.23).

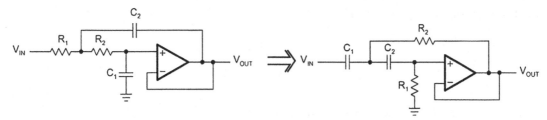

Figure 16.23
Low-pass to high-pass transition through components exchange.

To plot the gain response of a high-pass filter, mirror the gain response of a low-pass filter at the corner frequency, $\Omega = 1$, thus replacing Ω with $1/\Omega$ and S with $1/S$ in Eq. (16.1) (See Fig. 16.24).

The general transfer function of a high-pass filter is then:

$$A(s) = \frac{A_\infty}{\prod\limits_i \left(1 + \dfrac{a_i}{s} + \dfrac{b_i}{s^2}\right)} \tag{16.4}$$

with A_∞ being the passband gain.

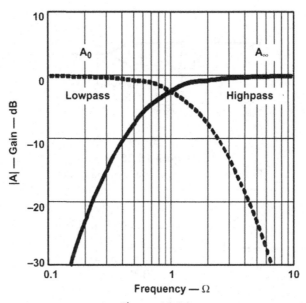

Figure 16.24
Developing the gain response of a high-pass filter.

Since Eq. (16.4) represents a cascade of second-order high-pass filters, the transfer function of a single stage is given as:

$$A_i(s) = \frac{A_\infty}{\left(1 + \dfrac{a_i}{s} + \dfrac{b_i}{s^2}\right)} \tag{16.5}$$

With $b = 0$ for all first-order filters, the transfer function of a first-order filter simplifies to:

$$A(s) = \frac{A_0}{1 + \dfrac{a_i}{s}} \tag{16.6}$$

16.4.1 First-Order High-Pass Filter

Figs. 16.25 and 16.26 show a first-order high-pass filter in the noninverting and the inverting configuration.

The transfer functions of the circuits are as follows:

$$A(s) = \frac{1 + \dfrac{R_2}{R_3}}{1 + \dfrac{1}{\omega_c R_1 C_1} \cdot \dfrac{1}{s}} \quad \text{and} \quad A(s) = -\frac{\dfrac{R_2}{R_1}}{1 + \dfrac{1}{\omega_c R_1 C_1} \cdot \dfrac{1}{s}}$$

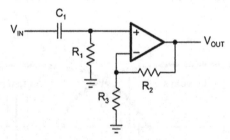

Figure 16.25
First-order noninverting high-pass filter.

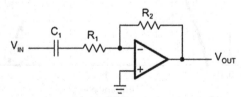

Figure 16.26
First-order inverting high-pass filter.

The negative sign indicates that the inverting amplifier generates a 180°phase shift (from the filter input to the output).

The coefficient comparison between the two transfer functions and Eq. (16.6) provides two different passband gain factors:

$$A_\infty = 1 + \frac{R_2}{R_3} \quad \text{and} \quad A_\infty = -\frac{R_2}{R_1}$$

while the term for the coefficient a_1 is the same for both circuits:

$$a_1 = \frac{1}{\omega_c R_1 C_1}$$

To dimension the circuit, specify the corner frequency (f_C), the DC gain (A_∞), and capacitor (C_1), and then solve for R_1 and R_2:

$$R_1 = \frac{1}{2\pi f_c a_1 C_1}$$
$$R_2 = R_3(A_\infty - 1) \quad \text{and} \quad R_2 = -R_1 A_\infty$$

16.4.2 Second-Order High-Pass Filter

High-pass filters use the same two topologies as the low-pass filters: Sallen–Key and MFB. The only difference is that the positions of the resistors and the capacitors have changed.

16.4.2.1 Sallen–Key Topology

The general Sallen–Key topology in Fig. 16.27 allows for separate gain setting via $A_0 = 1 + R_4/R_3$.

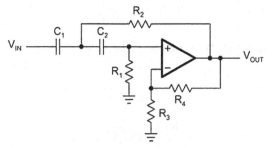

Figure 16.27
General Sallen–Key high-pass filter.

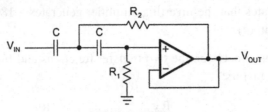

Figure 16.28
Unity-gain Sallen—Key high-pass filter.

The transfer function of the circuit in Fig. 16.27 is given as:

$$A(s) = \frac{\alpha}{1 + \dfrac{R_2(C_1 + C_2) + R_1 C_2(1 - \alpha)}{\omega_c R_1 R_2 C_1 C_2} \cdot \dfrac{1}{s} + \dfrac{1}{\omega_c^2 R_1 R_2 C_1 C_2} \cdot \dfrac{1}{s^2}} \qquad \text{with} \quad \alpha = 1 + \frac{R_4}{R_3}$$

The unity-gain topology in Fig. 16.28 is usually applied in low-Q filters with high gain accuracy.

To simplify the circuit design, it is common to choose unity-gain ($\alpha = 1$) and $C_1 = C_2 = C$. The transfer function of the circuit in Fig. 16.28 then simplifies to:

$$A(s) = \frac{1}{1 + \dfrac{2}{\omega_c R_1 C} \cdot \dfrac{1}{s} + \dfrac{2}{\omega_c^2 R_1 R_2 C^2} \cdot \dfrac{1}{s^2}}$$

The coefficient comparison between this transfer function and Eq. (16.5) yields the following:

$$A_\infty = 1$$

$$a_1 = \frac{2}{\omega_c R_1 C}$$

$$b_1 = \frac{1}{\omega_c^2 R_1 R_2 C^2}$$

Given C, the resistor values for R_1 and R_2 are calculated through:

$$R_1 = \frac{1}{\pi f_c C a_1}$$

$$R_2 = \frac{a_1}{4\pi f_c C b_1}$$

16.4.2.2 Multiple Feedback High Pass Filter Topology

To simplify the computation of the circuit, capacitors C_1 and C_3 assume the same value $(C_1 = C_3 = C)$ as shown in Fig. 16.29.

The transfer function of the circuit in Fig. 16.29 is

$$A(s) = \frac{-\dfrac{C}{C_2}}{1 + \dfrac{2C_2 + C}{\omega_c R_1 C_2 C} \cdot \dfrac{1}{s} + \dfrac{2C_2 + C}{\omega_c^2 R_2 R_1 C_2 C} \cdot \dfrac{1}{s^2}}$$

Through coefficient comparison with Eq. (16.5), obtain the following relations:

$$A_\infty = \frac{C}{C_2}$$

$$a_1 = \frac{2C + C_2}{\omega_c R_1 C C_2}$$

$$b_1 = \frac{2C + C_2}{\omega_c R_1 C C_2}$$

Given capacitors C and C_2, and solving for resistors R_1 and R_2:

$$R_1 = \frac{1 - 2A_\infty}{2\pi f_c \cdot C \cdot a_1}$$

$$R_2 = \frac{a_1}{2\pi f_c \cdot b_1 C_2 (1 - 2A_\infty)}$$

The passband gain (A_∞) of an MFB high-pass filter can vary significantly due to the wide tolerances of the two capacitors C and C_2. To keep the gain variation at a minimum, it is necessary to use capacitors with tight tolerance values.

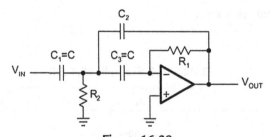

Figure 16.29
Second-order multiple feedback high-pass filter.

16.4.3 Higher-Order High-Pass Filter

Likewise, as with the low-pass filters, higher-order high-pass filters are designed by cascading first-order and second-order filter stages. The filter coefficients are the same ones used for the low-pass filter design and are listed in the coefficient tables (Tables 16.6 through 16.12 in Section 16.9).

Example 16.4. Third Order High-Pass Filter with $f_C = 1$ kHz (Table 16.3).

The task is to design a third-order unity-gain Bessel high-pass filter with the corner frequency $f_C = 1$ kHz. Obtain the coefficients for a third-order Bessel filter from Table 16.6, Section 16.9 and compute each partial filter by specifying the capacitor values and calculating the required resistor values.

16.4.3.1 First Filter

With $C_1 = 100$ nF,

$$R_1 = \frac{1}{2\pi f_c a_1 C_1} = \frac{1}{2\pi \cdot 10^3 \text{ Hz} \cdot 0.756 \cdot 100 \cdot 10^{-9} \text{ F}} = 2.105 \text{ k}\Omega$$

Closest 1% value is 2.1 kΩ.

16.4.3.2 Second Filter

With $C = 100$ nF,

$$R_1 = \frac{1}{\pi f_c C a_1} = \frac{1}{\pi \cdot 10^3 \cdot 100 \cdot 10^{-9} \cdot 0.756} = 3.18 \text{ k}\Omega$$

Closest 1% value is 3.16 kΩ.

$$R_2 = \frac{a_1}{4\pi f_c C b_1} = \frac{0.9996}{4\pi \cdot 10^3 \cdot 100 \cdot 10^{-9} \cdot 0.4772} = 1.67 \text{ k}\Omega$$

Closest 1% value is 1.65 kΩ.

Fig. 16.30 shows the final filter circuit.

Table 16.3 Example 16.4 Filters

	a_i	b_i
Filter 1	$a_1 = 0.756$	$b_1 = 0$
Filter 2	$a_2 = 0.9996$	$b_2 = 0.4772$

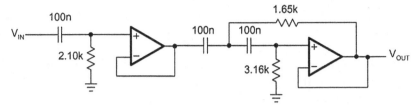

Figure 16.30
Third-order unity-gain Bessel high-pass.

16.5 Band-Pass Filter Design

In Section 16.4, a high-pass response was generated by replacing the term S in the low-pass transfer function with the transformation $1/S$. Likewise, a band-pass characteristic is generated by replacing the S term with the transformation:

$$\frac{1}{\Delta\Omega}\left(s + \frac{1}{s}\right) \tag{16.7}$$

In this case, the passband characteristic of a low-pass filter is transformed into the upper passband half of a band-pass filter. The upper passband is then mirrored at the mid frequency, f_m ($\Omega = 1$), into the lower passband half (See Fig. 16.31).

The corner frequency of the low-pass filter transforms to the lower and upper -3 dB frequencies of the band-pass, Ω_1 and Ω_2. The difference between both frequencies is defined as the normalized bandwidth $\Delta\Omega$:

$$\Delta\Omega = \Omega_2 - \Omega_1$$

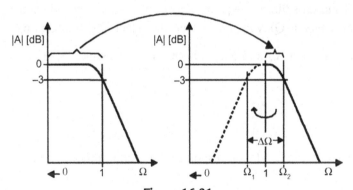

Figure 16.31
Low-pass to band-pass transition.

The normalized mid frequency, where $Q = 1$, is

$$\Omega_m = 1 = \Omega_2 \cdot \Omega_1$$

In analogy to the resonant circuits, the quality factor Q is defined as the ratio of the mid frequency (f_m) to the bandwidth (B):

$$Q = \frac{f_m}{B} = \frac{f_m}{f_2 - f_1} = \frac{1}{\Omega_2 - \Omega_1} = \frac{1}{\Delta\Omega} \tag{16.8}$$

The simplest design of a band-pass filter is the connection of a high-pass filter and a low-pass filter in series, which is commonly done in wide-band filter applications. Thus, a first-order high-pass and a first-order low-pass provide a second-order band-pass, whereas a second-order high-pass and a second-order low-pass result in a fourth-order band-pass response.

In comparison to wide-band filters, narrow-band filters of higher order consist of cascaded second-order band-pass filters that use the Sallen–Key or the MFB topology.

16.5.1 Second-Order Band-Pass Filter

To develop the frequency response of a second-order band-pass filter, apply the transformation in Eq. (16.7) to a first-order low-pass transfer function:

$$A(s) = \frac{A_0}{1 + s}$$

Replacing s with $\frac{1}{\Delta\Omega}\left(s + \frac{1}{s}\right)$ yields the general transfer function for a second-order band-pass filter:

$$A(s) = \frac{A_0 \cdot \Delta\Omega \cdot s}{1 + \Delta\Omega \cdot s + s^2} \tag{16.9}$$

When designing band-pass filters, the parameters of interest are the gain at the mid frequency (A_m) and the quality factor (Q), which represents the selectivity of a band-pass filter.

Therefore, replace A_0 with A_m and $\Delta\Omega$ with $1/Q$ (Eq. 16.7) and obtain:

$$A(s) = \frac{\dfrac{A_m}{Q} \cdot s}{1 + \dfrac{1}{Q} \cdot s + s^2} \tag{16.10}$$

Fig. 16.32 shows the normalized gain response of a second-order band-pass filter for different Qs.

The graph shows that the frequency response of second-order band-pass filters gets steeper with rising Q, thus making the filter more selective.

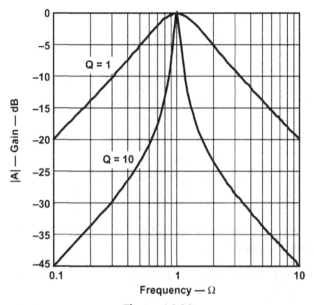

Figure 16.32
Gain response of a second-order band-pass filter.

16.5.1.1 Sallen–Key Topology

The Sallen–Key band-pass circuit in Fig. 16.33 has the following transfer function:

$$A(s) = \frac{G \cdot RC\omega_m \cdot s}{1 + RC\omega_m(3 - G) \cdot s + R^2C^2\omega_m^2 \cdot s^2}$$

Through coefficient comparison with Eq. (16.10), obtain the following equations:

$$\text{mid} - \text{frequency:} \quad f_m = \frac{1}{2\pi RC}$$

$$\text{inner gain:} \quad G = 1 + \frac{R_2}{R_1}$$

$$\text{gain at } f_m: \quad A_m = \frac{G}{3 - G}$$

$$\text{filter quality:} \quad Q = \frac{1}{3 - G}$$

The Sallen–Key circuit has the advantage that the quality factor (Q) can be varied via the inner gain (G) without modifying the mid frequency (f_m). A drawback is, however, that Q and A_m cannot be adjusted independently.

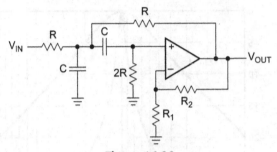

Figure 16.33
Sallen—Key band-pass.

Care must be taken when G approaches the value of 3, because then A_m becomes infinite and causes the circuit to oscillate.

To set the mid frequency of the band-pass, specify f_m and C and then solve for R:

$$R = \frac{1}{2\pi f_m C}$$

Because of the dependency between Q and A_m, there are two options to solve for R_2; either to set the gain at mid frequency:

$$R_2 = \frac{2A_m - 1}{1 + A_m}$$

or to design for a specified Q:

$$R_2 = \frac{2Q - 1}{Q}$$

16.5.1.2 Multiple Feedback Band Pass Filter Topology

The MFB band-pass circuit in Fig. 16.34 has the following transfer function:

$$A(s) = \frac{-\dfrac{R_2 R_3}{R_1 + R_3} C \omega_m \cdot s}{1 + \dfrac{2R_1 R_3}{R_1 + R_3} C \omega \cdot s + \dfrac{R_1 R_2 R_3}{R_1 + R_3} C^2 \cdot \omega_m^2 \cdot s^2}$$

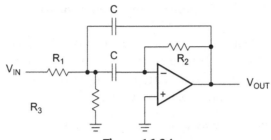

Figure 16.34
Multiple feedback band-pass.

The coefficient comparison with Eq. (16.9), yields the following equations:

$$\text{mid} - \text{frequency}: \quad f_m = \frac{1}{1\pi C} \sqrt{\frac{R_1 + R_3}{R_1 R_2 R_3}}$$

$$\text{gain at } f_m: \quad -A_m = \frac{R_2}{2R_1}$$

$$\text{filter quality}: \quad Q = \pi f_m R_2 C$$

$$\text{bandwidth}: \quad B = \frac{1}{\pi R_2 C}$$

The MFB band-pass allows to adjust Q, A_m, and fm independently. Bandwidth and gain factor do not depend on R_3. Therefore, R_3 can be used to modify the mid frequency without affecting bandwidth, B, or gain, A_m. For low values of Q, the filter can work without R_3; however, Q then depends on A_m via:

$$-A_m = 2Q^2$$

Example 16.5. Second-Order MFB Band-Pass Filter with $f_m = 1$ kHz.

To design a second-order MFB band-pass filter with a mid frequency of $f_m = 1$ kHz, a quality factor of $Q = 10$, and a gain of $A_m = -2$, assume a capacitor value of $C = 100$ nF, and solve the previous equations for R_1 through R_3 in the following sequence:

$$R_2 = \frac{Q}{\pi f_m C} = \frac{10}{\pi \cdot 1 \text{ kHz} \cdot 100 \text{ nF}} = 31.8 \text{ k}\Omega$$

$$R_1 = \frac{R_2}{-2A_m} = \frac{31.8 \text{ k}\Omega}{4} = 7.96 \text{ k}\Omega$$

$$R_3 = \frac{-A_m R_1}{2Q^2 + A_m} = \frac{2 \cdot 7.96 \text{ k}\Omega}{200 - 2} = 80.4 \ \Omega$$

16.5.2 Fourth-Order Band-Pass Filter (Staggered Tuning)

Fig. 16.32 shows that the frequency response of second-order band-pass filters gets steeper with rising Q. However, there are band-pass applications that require a flat gain response close to the mid frequency as well as a sharp passband-to-stopband transition. These tasks can be accomplished by higher-order band-pass filters.

Of particular interest is the application of the low-pass to band-pass transformation onto a second-order low-pass filter, since it leads to a fourth-order band-pass filter.

Replacing the S term in Eq. (16.2) with Eq. (16.7) gives the general transfer function of a fourth-order band-pass:

$$A(s) = \frac{\dfrac{s^2 \cdot A_0 (\Delta\Omega)^2}{b_1}}{1 + \dfrac{a_1}{b_1}\Delta\Omega \cdot s + \left[2 + \dfrac{(\Delta\Omega)^2}{b_1}\right] \cdot s^2 + \dfrac{a_1}{b_1}\Delta\Omega \cdot s^3 + s^4} \tag{16.11}$$

Similar to the low-pass filters, the fourth-order transfer function is split into two second-order band-pass terms. Further mathematical modifications yield:

$$A(s) = \frac{\dfrac{A_{mi}}{Q_i} \cdot \alpha s}{\left[1 + \dfrac{\alpha s}{Q_1} + (\alpha s)^2\right]} \cdot \frac{\dfrac{A_{mi}}{Q_i} \cdot \dfrac{s}{\alpha}}{\left[1 + \dfrac{1}{Q_i}\left(\dfrac{s}{\alpha}\right) + \left(\dfrac{s}{\alpha}\right)^2\right]} \tag{16.12}$$

Eq. (16.12) represents the connection of two second-order band-pass filters in series, where

- A_{mi} is the gain at the mid frequency, f_{mi}, of each partial filter,
- Q_i is the pole quality of each filter, and
  - α and $1/\alpha$ are the factors by which the mid frequencies of the individual filters, f_{m1} and f_{m2}, derive from the mid frequency, f_m, of the overall bandpass.

In a fourth-order band-pass filter with high Q, the mid frequencies of the two partial filters differ only slightly from the overall mid frequency. This method is called staggered tuning.

Factor α needs to be determined through successive approximation, using Eqs. (16)–(13):

$$\alpha^2 + \left[\frac{\alpha \cdot \Delta\Omega \cdot a_1}{b_1(1 + \alpha^2)}\right]^2 + \frac{1}{\alpha^2} - 2 - \frac{(\Delta\Omega)^2}{b_1} = 0 \tag{16.13}$$

with a_1 and b_1 being the second-order low-pass coefficients of the desired filter type.

To simplify the filter design, Table 16.4 lists those coefficients and provides the α values for three different quality factors, Q = 1, Q = 10, and Q = 100.

Table 16.4: Values of α for Different Filter Types

Bessel				Butterworth				Tschebyscheff			
a_1		1.3617		a_1		1.4142		a_1		1.0650	
b_1		0.6180		b_1		1.0000		b_1		1.9305	
Q	100	10	1	Q	100	10	1	Q	100	10	1
$\Delta\Omega$	0.01	0.1	1	$\Delta\Omega$	0.01	0.1	1	$\Delta\Omega$	0.01	0.1	1
α	1.0032	1.0324	1.438	α	1.0035	1.036	1.4426	α	1.0033	1.0338	1.39

After α has been determined, all quantities of the partial filters can be calculated using the following equations:

The mid frequency of filter 1 is

$$f_{m1} = \frac{f_m}{\alpha} \tag{16.14}$$

the mid frequency of filter 2 is

$$f_{m2} = f_m \cdot \alpha \tag{16.15}$$

with f_m being the mid frequency of the overall forth-order band-pass filter.

The individual pole quality, Q_i, is the same for both filters:

$$Q_i = Q \cdot \frac{(1 + \alpha^2)b_1}{\alpha \cdot a_1} \tag{16.16}$$

with Q being the quality factor of the overall filter.

The individual gain (A_{mi}) at the partial mid frequencies, f_{m1} and f_{m2}, is the same for both filters:

$$A_{mi} = \frac{Q_i}{Q} \cdot \sqrt{\frac{A_m}{B_1}} \tag{16.17}$$

with A_m being the gain at mid frequency, f_m, of the overall filter.

Example 16.6. Fourth-Order Butterworth Band-Pass Filter.

The task is to design a fourth-order Butterworth band-pass with the following parameters:

- mid frequency, $f_m = 10$ kHz,
- bandwidth, B = 1000 Hz, and
- and gain, $A_m = 1$.

From Table 16.4 the following values are obtained:

- $a_1 = 1.4142$,
- $b_1 = 1$, and
- $\alpha = 1.036$.

In accordance with Eqs. (16.14) and (16.15), the mid frequencies for the partial filters are

$$f_{mi} = \frac{10 \text{ kHz}}{1.036} = 9.653 \text{ kHz} \quad \text{and} \quad f_{m2} = 10 \text{ kHz} \cdot 1.036 = 10.36 \text{ kHz}$$

The overall Q is defined as $Q = f_m/B$, and for this example results in $Q = 10$.

Using Eq. (16.16), the Q_i of both filters is

$$Q_i = 10 \cdot \frac{(1 + 1.036^2) \cdot 1}{1.036 \cdot 1.4142} = 14.15$$

With Eq. (16.17), the passband gain of the partial filters at f_{m1} and f_{m2} calculates to:

$$A_{mi} = \frac{14.15}{10} \cdot \sqrt{\frac{1}{1}} = 1.415$$

Eqs. (16.16) and (16.17) show that Q_i and A_{mi} of the partial filters need to be independently adjusted. The only circuit that accomplishes this task is the MFB band-pass filter in Section 16.5.1.2.

To design the individual second-order band-pass filters, specify $C = 10$ nF, and insert the previously determined quantities for the partial filters into the resistor equations of the MFB band-pass filter. The resistor values for both partial filters are calculated below.

Filter 1 :

$$R_{21} = \frac{Q_i}{\pi f_{m1} C} = \frac{14.15}{\pi \cdot 9.653 \text{ kHz} \cdot 10 \text{ nF}} = 46.7 \text{ k}\Omega$$

$$R_{11} = \frac{R_{21}}{-2A_{mi}} = \frac{46.7 \text{ k}\Omega}{-2 \cdot - 1.415} = 16.5 \text{ k}\Omega$$

$$R_{31} = \frac{-A_{mi}R_{11}}{2Q_i^2 + A_{mi}} = \frac{1.415 \cdot 16.5 \text{ k}\Omega}{2 \cdot 14.15^2 + 1.415} = 58.1 \text{ }\Omega$$

Filter 2 :

$$R_{22} = \frac{Q_i}{\pi f_{m2} C} = \frac{14.15}{\pi \cdot 10.36 \text{ kHz} \cdot 10 \text{ nF}} = 43.5 \text{ k}\Omega$$

$$R_{12} = \frac{R_{22}}{-2A_{mi}} = \frac{43.5}{-2 \cdot - 1.415} = 15.4 \text{ k}\Omega$$

$$R_{32} = \frac{A_{mi}R_{12}}{2Q_i^2 + A_{mi}} = \frac{1.415 \cdot 15.4 \text{ k}\Omega}{2 \cdot 14.15^2 + 1.415} = 54.2 \text{ }\Omega$$

Fig. 16.35 compares the gain response of a fourth-order Butterworth band-pass filter with $Q = 1$ and its partial filters to the fourth-order gain of Example 16.4 with $Q = 10$.

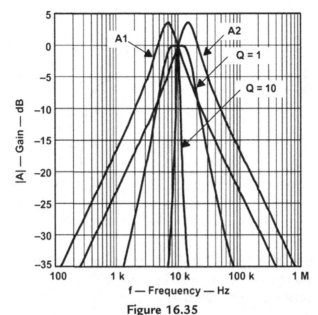

Figure 16.35
Gain responses of a fourth-order Butterworth band-pass and its partial filters.

16.6 Band-Rejection Filter Design

A band-rejection filter is used to suppress a certain frequency rather than a range of frequencies.

Two of the most popular band-rejection filters are the active twin-T and the active Wien-Robinson circuit, both of which are second-order filters.

To generate the transfer function of a second-order band-rejection filter, replace the s term of a first-order low-pass response with the transformation in Eq. (16.18):

$$\frac{\Delta\Omega}{s + \dfrac{1}{s}} \tag{16.18}$$

which gives

$$A(s) = \frac{A_0\left(1 + s^2\right)}{1 + \Delta\Omega \cdot s + s_2} \tag{16.19}$$

Thus the passband characteristic of the low-pass filter is transformed into the lower passband of the band-rejection filter. The lower passband is then mirrored at the mid frequency, f_m ($\Omega = 1$), into the upper passband half (Fig. 16.36).

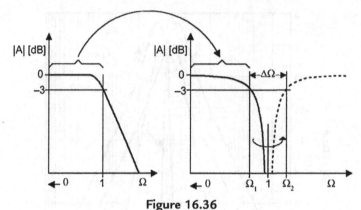

Figure 16.36
Low-pass to band-rejection transition.

The corner frequency of the low-pass transforms to the lower and upper -3-dB frequencies of the band-rejection filter Ω_1 and Ω_2. The difference between both frequencies is the normalized bandwidth $\Delta\Omega$:

$$\Delta\Omega = \Omega_{max} - \Omega_{min}$$

Identical to the selectivity of a band-pass filter, the quality of the filter rejection is defined as:

$$Q = \frac{f_m}{B} = \frac{1}{\Delta\Omega}$$

Therefore, replacing $\Delta\Omega$ in Eq. (16.19) with $1/Q$ yields:

$$A(s) = \frac{A_0\left(1 + s^2\right)}{1 + \dfrac{1}{Q}\cdot s + s^2} \tag{16.20}$$

16.6.1 Active Twin-T Filter

The original twin-T filter, shown in Fig. 16.37, is a passive RC-network with a quality factor of $Q = 0.25$. To increase Q, the passive filter is implemented into the feedback loop of an amplifier, thus turning into an active band-rejection filter, shown in Fig. 16.38.

The transfer function of the active twin-T filter is

$$A(s) = \frac{k\left(1 + s^2\right)}{1 + 2(2 - k)\cdot s + s^2} \tag{16.21}$$

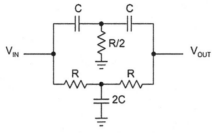

Figure 16.37
Passive Twin-T filter.

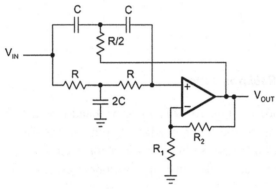

Figure 16.38
Active Twin-T filter.

Comparing the variables of Eq. (16.21) with Eq. (16.20) provides the equations that determine the filter parameters:

$$\text{mid} - \text{frequency:} \quad f_m = \frac{1}{2\pi RC}$$

$$\text{inner gain:} \quad G = 1 + \frac{R_2}{R_1}$$

$$\text{passband gain:} \quad A_0 = G$$

$$\text{rejection quality:} \quad Q = \frac{1}{2(2 - G)}$$

The twin-T circuit has the advantage that the quality factor (Q) can be varied via the inner gain (G) without modifying the mid frequency (f_m). However, Q and Am cannot be adjusted independently.

To set the mid frequency of the band-pass, specify f_m and C, and then solve for R:

$$R = \frac{1}{2\pi f_m C}$$

Because of the dependency between Q and A_m, there are two options to solve for R_2; either to set the gain at mid frequency:

$$R_2 = (A_0 - 1)R_1$$

or to design for a specific Q:

$$R_2 = R_1 \left(1 - \frac{1}{2Q}\right)$$

16.6.2 Active Wien-Robinson Filter

The Wien-Robinson bridge in Fig. 16.39 is a passive band-rejection filter with differential output. The output voltage is the difference between the potential of a constant voltage divider and the output of a band-pass filter. Its Q-factor is close to that of the twin-T circuit. To achieve higher values of Q, the filter is connected into the feedback loop of an amplifier.

The active Wien-Robinson filter in Fig. 16.40 has the transfer function:

$$A(s) = -\frac{\dfrac{\beta}{1+\alpha}\left(1 + s^2\right)}{1 + \dfrac{3}{1+\alpha}\cdot s + s^2} \tag{16.22}$$

with $\alpha = \dfrac{R_2}{R_3}$ and $\beta = \dfrac{R_2}{R_4}$

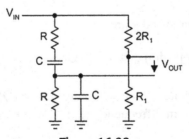

Figure 16.39
Passive Wien-Robinson bridge.

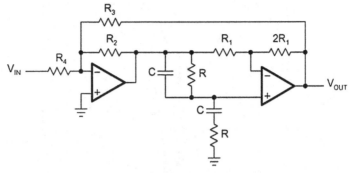

Figure 16.40
Active Wien-Robinson filter.

Comparing the variables of Eq. (16.22) with Eq. (16.20) provides the equations that determine the filter parameters:

$$\text{mid} - \text{frequency}: \quad f_m = \frac{1}{2\pi RC}$$

$$\text{passband gain}: \quad A_0 = -\frac{\beta}{1+\alpha}$$

$$\text{rejection quality}: \quad Q = \frac{1+\alpha}{3}$$

To calculate the individual component values, establish the following design procedure:

1. Define f_m and C and calculate R with:

$$R = \frac{1}{2\pi f_m C}$$

2. Specify Q and determine α via:

$$\alpha = 3Q - 1$$

3. Specify A_0 and determine β via:

$$\beta = -A_0 \cdot 3Q$$

4. Define R_2 and calculate R_3 and R_4 with:

$$R_3 = \frac{R_2}{\alpha}$$

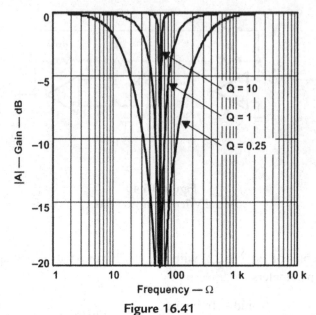

Figure 16.41
Comparison of Q between passive and active band-rejection filters.

and

$$R_4 = \frac{R_2}{\beta}$$

In comparison to the twin-T circuit, the Wien-Robinson filter allows modification of the passband gain, A_0, without affecting the quality factor, Q.

If fm is not completely suppressed due to component tolerances of R and C, a fine-tuning of the resistor $2R_2$ is required.

Fig. 16.41 shows a comparison between the filter response of a passive band-rejection filter with $Q = 0.25$ and an active second-order filter with $Q = 1$ and $Q = 10$.

16.7 All-Pass Filter Design

In comparison to the previously discussed filters, an all-pass filter has a constant gain across the entire frequency range, and a phase response that changes linearly with frequency.

Because of these properties, all-pass filters are used in phase compensation and signal delay circuits.

Similar to the low-pass filters, all-pass circuits of higher order consist of cascaded first-order and second-order all-pass stages. To develop the all-pass transfer function from a low-pass response, replace A_0 with the conjugate complex denominator.

The general transfer function of an all-pass is

$$A(s) = \frac{\prod\limits_{i}(1 - a_i s + b_i s^2)}{\prod\limits_{i}(1 + a_i s + b_i s^2)} \qquad (16.23)$$

with a_i and b_i being the coefficients of a partial filter. The all-pass coefficients are listed in Table 16.12 of Section 16.9.

Expressing Eq. (16.23) in magnitude and phase yields

$$A(s) = \frac{\prod\limits_{i} \sqrt{\left(1 - b_i \Omega^2\right)^2 + a_i^2 \Omega^2} \cdot e^{-ja}}{\prod\limits_{i} \sqrt{\left(1 - b_i \Omega^2\right)^2 + a_i^2 \Omega^2} \cdot e^{+ja}} \qquad (16.24)$$

This gives a constant gain of 1, and a phase shift, φ, of:

$$\varphi = -2\alpha = -2 \sum_i \arctan \frac{a_i \Omega}{1 - b_i \Omega^2} \qquad (16.25)$$

To transmit a signal with minimum phase distortion, the all-pass filter must have a constant group delay across the specified frequency band. The group delay is the time by which the all-pass filter delays each frequency within that band.

The frequency at which the group delay drops to $1/\sqrt{2}$ times its initial value is the corner frequency, f_C.

The group delay is defined through:

$$t_{gr} = -\frac{d\varphi}{d\omega} \qquad (16.26)$$

To present the group delay in normalized form, refer t_{gr} to the period of the corner frequency, T_C, of the all-pass circuit:

$$T_{gr} = \frac{t_{gr}}{T_c} = t_{gr} \cdot f_c = t_{gr} \cdot \frac{\omega_c}{2\pi} \qquad (16.27)$$

Substituting t_{gr} through Eq. (16.26) gives

$$T_{gr} = -\frac{1}{2\pi} \cdot \frac{d\varphi}{d\Omega} \qquad (16.28)$$

Inserting the φ term in Eq. (16.25) into Eq. (16.28) and completing the derivation, results in:

$$T_{gr} = \frac{1}{\pi} \sum_i \frac{a_i\left(1 + b_i\Omega^2\right)}{1 + \left(a_i^2 - 2b_1\right)\cdot\Omega^2 + b_i^2\Omega^4} \qquad (16.29)$$

Setting $\Omega = 0$ in Eq. (16.29) gives the group delay for the low frequencies, $0 < \Omega < 1$, which is

$$T_{gr0} = \frac{1}{\pi} \sum_i a_i \qquad (16.30)$$

The values for T_{gr0} are listed in Table 16.12, Section 16.9, from the first to the tenth order.

In addition, Fig. 16.42 shows the group delay response versus the frequency for the first ten orders of all-pass filters.

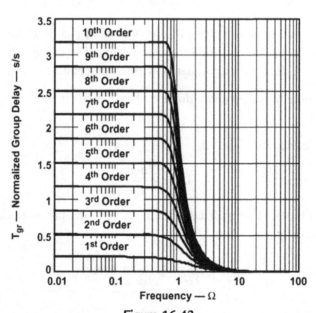

Figure 16.42
Frequency response of the group delay for the first 10 filter orders.

16.7.1 First-Order All-Pass Filter

Fig. 16.43 shows a first-order all-pass filter with a gain of $+1$ at low frequencies and a gain of -1 at high frequencies. Therefore, the magnitude of the gain is 1, while the phase changes from 0 degrees to -180 degrees.

The transfer function of the circuit above is

$$A(s) = \frac{1 - RC\omega_c \cdot s}{1 + RC\omega_c \cdot s}$$

The coefficient comparison with Eq. (16.23) ($b1 = 1$) results in:

$$a_i = RC \cdot 2\pi f_c \qquad (16.31)$$

To design a first-order all-pass, specify f_C and C and then solve for R:

$$R = \frac{a_i}{2\pi f_c \cdot C} \qquad (16.32)$$

Inserting Eq. (16.31) into Eq. (16.30) and substituting ω_C with Eq. (16.27) provides the maximum group delay of a first-order all-pass filter:

$$t_{gr0} = 2RC \qquad (16.33)$$

16.7.2 Second-Order All-Pass Filter

Fig. 16.44 shows that one possible design for a second-order all-pass filter is to subtract the output voltage of a second-order band-pass filter from its input voltage.

The transfer function of the circuit in Fig. 16.44 is

$$A(s) = \frac{1 + (2R_1 - \alpha R_2)C\omega_c \times s + R_1 R_2 C^2 \omega_c^2 \times s^2}{1 + 2R_1 C\omega_c \times s + R_1 R_2 C^2 \omega_c^2 \times s^2}$$

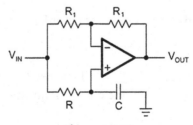

Figure 16.43
First-order all-pass.

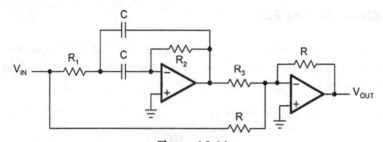

Figure 16.44
Second-order all-pass filter.

The coefficient comparison with Eq. (16.23) yields

$$a_1 = 4\pi f_c R_1 C \qquad (16.34)$$

$$b_1 = a_1 \pi f_c R_2 C \qquad (16.35)$$

$$\alpha = \frac{a_1^2}{b_1} = \frac{R}{R_3} \qquad (16.36)$$

To design the circuit, specify f_C, C, and R, and then solve for the resistor values:

$$R_1 = \frac{a_1}{4\pi f_c C} \qquad (16.37)$$

$$R_2 = \frac{b_1}{a_1 \pi f_c C} \qquad (16.38)$$

$$R_3 = \frac{R}{\alpha} \qquad (16.39)$$

Inserting Eq. (16.34) into Eq. (16.30) and substituting ω_C with Eq. (16.27) gives the maximum group delay of a second-order all-pass filter:

$$t_{gr0} = 4R_1 C \qquad (16.40)$$

16.7.3 Higher-Order All-Pass Filter

Higher-order all-pass filters consist of cascaded first-order and second-order filter stages.

Example 16.7. 2-ms Delay All-Pass Filter.

A signal with the frequency spectrum, $0 < f < 1$ kHz, needs to be delayed by 2 ms. To keep the phase distortions at a minimum, the corner frequency of the all-pass filter must be $f_C \geq 1$ kHz.

Eq. (16.27) determines the normalized group delay for frequencies below 1 kHz:

$$T_{gr0} = \frac{t_{gr0}}{T_c} = 2 \text{ ms} \cdot 1 \text{ kHz} = 2.0$$

Fig. 16.42 confirms that a seventh-order all-pass is needed to accomplish the desired delay. The exact value, however, is $T_{gr0} = 2.1737$. To set the group delay to precisely 2 ms, solve Eq. (16.27) for f_C and obtain the corner frequency:

$$f_C = \frac{T_{gr0}}{t_{gr0}} = 1.087 \text{ kHz}$$

To complete the design, look up the filter coefficients for a seventh-order all-pass filter, specify C, and calculate the resistor values for each partial filter.

Cascading the first-order all-pass with the three second-order stages results in the desired seventh-order all-pass filter (See Fig. 16.45).

16.8 Practical Design Hints

This section introduces DC-biasing techniques for filter designs in single-supply applications, which are usually not required when operating with dual supplies. It also provides recommendations on selecting the type and value range of capacitors and resistors as well as the decision criteria for choosing the correct op amp.

16.8.1 Filter Circuit Biasing

The filter diagrams in this chapter are drawn for dual supply applications. The op amp operates from a positive and a negative supply, while the input and the output voltage are referenced to ground (Fig. 16.46).

For the single supply circuit in Fig. 16.47, the lowest supply voltage is ground. For a symmetrical output signal, the potential of the noninverting input is level-shifted to midrail.

The coupling capacitor, C_{IN} in Fig. 16.47, AC-couples the filter, blocking any unknown DC level in the signal source. The voltage divider, consisting of the two equal-bias resistors R_B, divides the supply voltage to V_{MID} and applies it to the inverting op amp input.

For simple filter input structures, passive RC networks often provide a low-cost biasing solution. In the case of more complex input structures, such as the input of a second-order low-pass filter, the RC network can affect the filter characteristic. Then it is necessary to

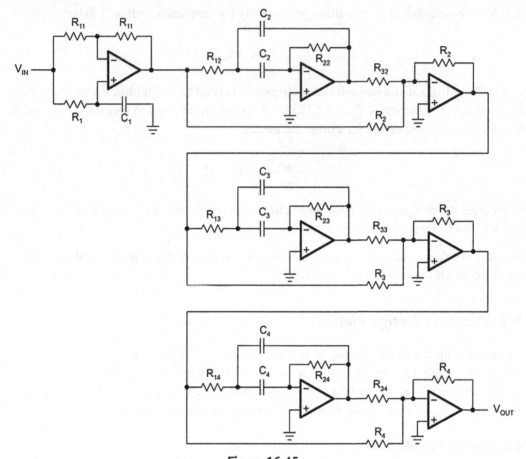

Figure 16.45
Seventh-order all-pass filter.

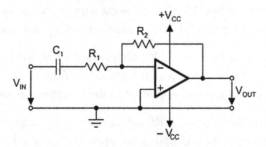

Figure 16.46
Dual-supply filter circuit.

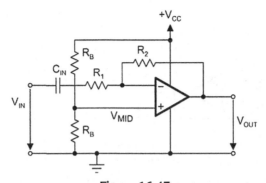

Figure 16.47
Single-supply filter circuit.

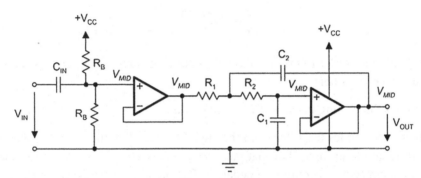

Figure 16.48
Biasing a Sallen–Key low-pass.

either include the biasing network into the filter calculations or to insert an input buffer between biasing network and the actual filter circuit, as shown in Fig. 16.48.

C_{IN} AC-couples the filter, blocking any DC level in the signal source. V_{MID} is derived from V_{CC} via the voltage divider. The op amp operates as a voltage follower and as an impedance converter. V_{MID} is applied via the DC path, R_1 and R_2, to the noninverting input of the filter amplifier.

Note that the parallel circuit of the resistors, R_B, together with C_{IN} creates a high-pass filter. To avoid any effect on the low-pass characteristic, the corner frequency of the input high-pass must be low versus the corner frequency of the actual low-pass.

The use of an input buffer causes no loading effects on the low-pass filter, thus keeping the filter calculation simple.

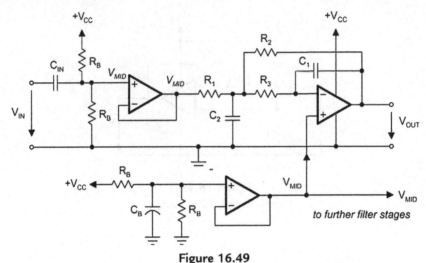

Figure 16.49

Biasing a second-order multiple feedback low-pass filter.

In the case of a higher-order filter, all following filter stages receive their bias level from the preceding filter amplifier.

Fig. 16.49 shows the biasing of an MFB low-pass filter.

The input buffer decouples the filter from the signal source. The filter itself is biased via the noninverting amplifier input. For that purpose, the bias voltage is taken from the output of a V_{MID} generator with low output impedance. The op amp operates as a difference amplifier and subtracts the bias voltage of the input buffer from the bias voltage of the V_{MID} generator, thus yielding a DC potential of V_{MID} at zero input signal.

A low-cost alternative is to remove the op amp and to use a passive biasing network instead. However, to keep loading effects at a minimum, the values for R_B must be significantly higher than without the op amp.

The biasing of a Sallen–Key and an MFB high-pass filter is shown in Fig. 16.50.

The input capacitors of high-pass filters already provide the AC-coupling between filter and signal source. Both circuits use the V_{MID} generator from Fig. 16.50 for biasing. While the MFB circuit is biased at the noninverting amplifier input, the Sallen–Key high-pass is biased via the only DC path available, which is R_1. In the AC circuit, the input signals travel via the low output impedance of the op amp to ground.

16.8.2 Capacitor Selection

The tolerance of the selected capacitors and resistors depends on the filter sensitivity and on the filter performance.

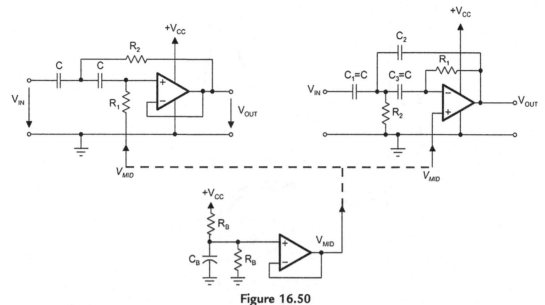

Figure 16.50
Biasing a Sallen–Key and a multiple feedback high-pass filter.

Sensitivity is the measure of the vulnerability of a filter's performance to changes in component values. The important filter parameters to consider are the corner frequency, f_C, and Q.

For example, when Q changes by $\pm2\%$ due to a $\pm5\%$ change in the capacitance value, then the sensitivity of Q to capacity changes is expressed as: $s\dfrac{Q}{C} = \dfrac{2\%}{5\%} = 0.4\dfrac{\%}{\%}$. The following sensitivity approximations apply to second-order Sallen–Key and MFB filters:

$$s\frac{Q}{C} \approx s\frac{Q}{C} \approx s\frac{f_c}{C} \approx \pm\, 0.5\frac{\%}{\%}$$

Although 0.5%/% is a small difference from the ideal parameter, in the case of higher-order filters, the combination of small Q and f_C differences in each partial filter can significantly modify the overall filter response from its intended characteristic.

Figs. 16.51 and 16.52 show how an intended eighth-order Butterworth low-pass can turn into a low-pass with Tschebyscheff characteristic mainly due to capacitance changes from the partial filters.

Fig. 16.51 shows the differences between the ideal and the actual frequency responses of the four partial filters. The overall filter responses are shown in Fig. 16.52. The difference between ideal and real response peaks with 0.35 dB at approximately 30 kHz, which is equivalent to an enormous 4.1% gain error can be seen.

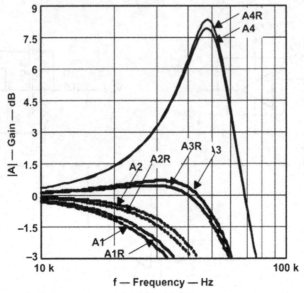

Figure 16.51

Differences in Q and f_C in the partial filters of an eighth-order Butterworth low-pass filter.

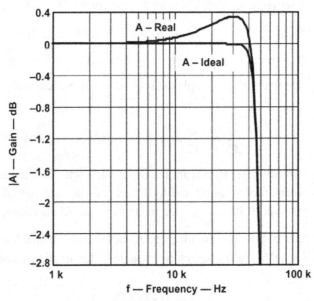

Figure 16.52

Modification of the intended Butterworth response to a Tschebyscheff-type characteristic.

If this filter is intended for a data acquisition application, it could be used at best in a 4-bit system. In comparison, if the maximum full-scale error of a 12-bit system is given with ½ LSB, then maximum pass-band deviation would be −0.001 dB, or 0.012%.

To minimize the variations of f_C and Q, NPO (COG) ceramic capacitors are recommended for high-performance filters. These capacitors hold their nominal value over a wide temperature and voltage range. The various temperature characteristics of ceramic capacitors are identified by a three-symbol code such as: COG, X7R, Z5U, and Y5V.

COG-type ceramic capacitors are the most precise. Their nominal values range from 0.5 pF to approximately 47 nF with initial tolerances from ±0.25 pF for smaller values and up to ±1% for higher values. Their capacitance drift over temperature is typically 30 ppm/°C.

X7R-type ceramic capacitors range from 100 pF to 2.2 µF with an initial tolerance of +1% and a capacitance drift over temperature of ±15%.

For higher values, tantalum electrolytic capacitors should be used.

Other precision capacitors are silver mica, metallized polycarbonate, and for high temperatures, polypropylene or polystyrene.

Since capacitor values are not as finely subdivided as resistor values, the capacitor values should be defined prior to selecting resistors. If precision capacitors are not available to provide an accurate filter response, then it is necessary to measure the individual capacitor values and to calculate the resistors accordingly.

For high performance filters, 0.1% resistors are recommended.

16.8.3 Component Values

Resistor values should stay within the range of 1−100 kΩ. The lower limit avoids excessive current draw from the op amp output, which is particularly important for single-supply op amps in power-sensitive applications. Those amplifiers have typical output currents of between 1 mA and 5 mA. At a supply voltage of 5 V, this current translates to a minimum of 1 kΩ.

The upper limit of 100 kΩ is to avoid excessive resistor noise.

Capacitor values can range from 1 nF to several µF. The lower limit avoids coming too close to parasitic capacitances. If the common-mode input capacitance of the op amp, used in a Sallen−Key filter section, is close to 0.25% of C_1, ($C_1/400$), it must be considered for accurate filter response. The MFB topology, in comparison, does not require input-capacitance compensation.

16.8.4 Op Amp Selection

The most important op amp parameter for proper filter functionality is the unity-gain bandwidth. In general, the open-loop gain (AOL) should be 100 times (40 dB above) the peak gain (Q) of a filter section to allow a maximum gain error of 1% (See Fig. 16.53).

The following equations are good rules of thumb to determine the necessary unity-gain bandwidth of an op amp for an individual filter section.

1. First-order filter:

$$f_T = 100 \cdot \text{Gain} \cdot f_c$$

2. Second-order filter ($Q < 1$):

$$f_T = 100 \cdot \text{Gain} \cdot f_c \cdot k_i \quad \text{with} \quad k_i = \frac{f_{ci}}{f_c}$$

3. Second-order filter ($Q > 1$):

$$f_T = 100 \cdot \text{Gain} \cdot \frac{f_C}{a_i} \sqrt{\frac{Q_i^2 - 0.5}{Q_i^2 - 0.25}}$$

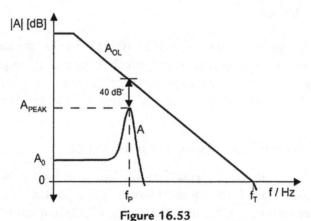

Figure 16.53
Open-loop gain (A$_{OL}$) and filter response (A).

For example, a fifth-order, 10-kHz, Tschebyscheff low-pass filter with 3-dB passband ripple and a DC gain of $A_0 = 2$ has its worst case Q in the third filter section. With $Q_3 = 8.82$ and $a_3 = 0.1172$, the op amp needs to have a unity-gain bandwidth of:

$$f_T = 100 \cdot 2 \cdot \frac{10 \text{ kHz}}{0.1172} \sqrt{\frac{8.82^2 - 0.5}{8.82^2 - 0.25}} = 17 \text{ MHz}$$

In comparison, a fifth-order unity-gain, 10-kHz, Butterworth low-pass filter has a worst case Q of $Q_3 = 1.62$; $a_3 = 0.618$. Owing to the lower Q value, f_T is also lower and calculates to only:

$$f_T = 100 \cdot \frac{10 \text{ kHz}}{0.618} \sqrt{\frac{1.62^2 - 0.5}{1.62^2 - 0.25}} = 1.5 \text{ MHz}$$

Besides good DC performance, low noise, and low signal distortion, another important parameter that determines the speed of an op amp is the slew rate (SR). For adequate full-power response, the SR must be greater than:

$$SR = \pi \cdot V_{pp} \cdot f_C$$

For example, a single-supply, 100-kHz filter with 5 V_{PP} output requires an SR of at least:

$$SR = \pi \cdot 5 \text{ V} \cdot 100 \text{ kHz} = 1.57 \frac{V}{\mu s}$$

Texas Instruments offers a wide range of op amps for high-performance filters in single-supply applications. Table 16.5 provides a selection of single-supply amplifiers sorted in order of rising SR.

Table 16.5: Single-Supply Op Amp Selection Guide ($T_A = 25°C$, $V_{CC} = 5$ V)

Op Amp	Bandwidth (MHz)	Power Bandwidth (kHz)	Slew rate (V/μs)	V_{IO} (mV)	Noise (nV/Hz)
TLV2721	0.51	11	0.18	0.6	20
TLC2201A	1.8	159	2.5	0.6	8
TLV2771A	4.8	572	9	1.9	21
TLC071	10	1000	16	1.5	7
TLE2141	5.9	2800	45	0.5	10.5
THS4001	270	127 MHz (1V_{PP})	400	6	7.5

16.9 Filter Coefficient Tables

The following tables contain the coefficients for the three filter types; Bessel, Butterworth, and Tschebyscheff. The Tschebyscheff tables (Table 16.11) are split into categories for the following passband ripples: 0.5, 1, 2, and 3 dB.

The table headers consist of the following quantities:

n is the filter order,
i is the number of the partial filter,
a_i and b_i are the filter coefficients,

Table 16.6: Bessel Coefficients

n	i	a_i	b_i	$k_i = f_{Ci}/f_C$	Q_i
1	1	1.0000	0.0000	1.000	—
2	1	1.3617	0.6180	1.000	0.58
3	1	0.7560	0.0000	1.323	—
	2	0.9996	0.4772	1.414	0.69
4	1	1.3397	0.4889	0.978	0.52
	2	0.7743	0.3890	1.797	0.81
5	1	0.6656	0.0000	1.502	—
	2	1.1402	0.4128	1.184	0.56
	3	0.6216	0.3245	2.138	0.92
6	1	1.2217	0.3887	1.063	0.51
	2	0.9686	0.3505	1.431	0.61
	3	0.5131	0.2756	2.447	1.02
7	1	0.5937	0.0000	1.648	—
	2	1.0944	0.3395	1.207	0.53
	3	0.8304	0.3011	1.695	0.66
	4	0.4332	0.2381	2.731	1.13
8	1	1.1112	0.3162	1.164	0.51
	2	0.9754	0.2979	1.381	0.56
	3	0.7202	0.2621	1.963	0.71
	4	0.3728	0.2087	2.992	1.23
9	1	0.5386	0.0000	1.857	—
	2	1.0244	0.2834	1.277	0.52
	3	0.8710	0.2636	1.574	0.59
	4	0.6320	0.2311	2.226	0.76
	5	0.3257	0.1854	3.237	1.32
10	1	1.0215	0.2650	1.264	0.50
	2	0.9393	0.2549	1.412	0.54
	3	0.7815	0.2351	1.780	0.62
	4	0.5604	0.2059	2.479	0.81
	5	0.2883	0.1665	3.466	1.42

Table 16.7: Butterworth Coefficients

n	i	a_i	b_i	$k_i = f_{Ci}/f_C$	Q_i
1	1	1.0000	0.0000	1.000	—
2	1	1.4142	1.0000	1.000	0.71
3	1	1.0000	0.0000	1.000	—
	2	1.0000	1.0000	1.272	1.00
4	1	1.8478	1.0000	0.719	0.54
	2	0.7654	1.0000	1.390	1.31
5	1	1.0000	0.0000	1.000	—
	2	1.6180	1.0000	0.859	0.62
	3	0.6180	1.0000	1.448	1.62
6	1	1.9319	1.0000	0.676	0.52
	2	1.4142	1.0000	1.000	0.71
	3	0.5176	1.0000	1.479	1.93
7	1	1.0000	0.0000	1.000	—
	2	1.8019	1.0000	0.745	0.55
	3	1.2470	1.0000	1.117	0.80
	4	0.4450	1.0000	1.499	2.25
8	1	1.9616	1.0000	0.661	0.51
	2	1.6629	1.0000	0.829	0.60
	3	1.1111	1.0000	1.206	0.90
	4	0.3902	1.0000	1.512	2.56
9	1	1.0000	0.0000	1.000	—
	2	1.8794	1.0000	0.703	0.53
	3	1.5321	1.0000	0.917	0.65
	4	1.0000	1.0000	1.272	1.00
	5	0.3473	1.0000	1.521	2.88
10	1	1.9754	1.0000	0.655	0.51
	2	1.7820	1.0000	0.756	0.56
	3	1.4142	1.0000	1.000	0.71
	4	0.9080	1.0000	1.322	1.10
	5	0.3129	1.0000	1.527	3.20

Table 16.8: Tschebyscheff Coefficients for 0.5 dB

n	i	a_i	b_i	$k_i = f_{Ci}/f_C$	Q_i
1	1	1.0000	0.0000	1.000	—
2	1	1.3614	1.3827	1.000	0.86
3	1	1.8636	0.0000	0.537	—
	2	0.0640	1.1931	1.335	1.71
4	1	2.6282	3.4341	0.538	0.71
	2	0.3648	1.1509	1.419	2.94
5	1	2.9235	0.0000	0.342	—
	2	1.3025	2.3534	0.881	1.18
	3	0.2290	1.0833	1.480	4.54

Continued

Table 16.8: Tschebyscheff Coefficients for 0.5 dB—cont'd

n	i	a_i	b_i	$k_i = f_{Ci}/f_C$	Q_i
6	1	3.8645	6.9797	0.366	0.68
	2	0.7528	1.8573	1.078	1.81
	3	0.1589	1.0711	1.495	6.51
7	1	4.0211	0.0000	0.249	—
	2	1.8729	4.1795	0.645	1.09
	3	0.4861	1.5676	1.208	2.58
	4	0.1156	1.0443	1.517	8.84
8	1	5.1117	11.9607	0.276	0.68
	2	1.0639	2.9365	0.844	1.61
	3	0.3439	1.4206	1.284	3.47
	4	0.0885	1.0407	1.521	11.53
9	1	5.1318	0.0000	0.195	—
	2	2.4283	6.6307	0.506	1.06
	3	0.6839	2.2908	0.989	2.21
	4	0.2559	1.3133	1.344	4.48
	5	0.0695	1.0272	1.532	14.58
10	1	6.3648	18.3695	0.222	0.67
	2	1.3582	4.3453	0.689	1.53
	3	0.4822	1.9440	1.091	2.89
	4	0.1994	1.2520	1.381	5.61
	5	0.0563	1.0263	1.533	17.99

Table 16.9: Tschebyscheff Coefficients for 1 dB Pa

n	i	a_i	b_i	$k_i = f_{Ci}/f_C$	Q_i
1	1	1.0000	0.0000	1.000	—
2	1	1.3022	1.5515	1.000	0.96
3	1	2.2156	0.0000	0.451	—
	2	0.5442	1.2057	1.353	2.02
4	1	2.5904	4.1301	0.540	0.78
	2	0.3039	1.1697	1.417	3.56
5	1	3.5711	0.0000	0.280	—
	2	1.1280	2.4896	0.894	1.40
	3	0.1872	1.0814	1.486	5.56
6	1	3.8437	8.5529	0.366	0.76
	2	0.6292	1.9124	1.082	2.20
	3	0.1296	1.0766	1.493	8.00
7	1	4.9520	0.0000	0.202	—
	2	1.6338	4.4899	0.655	1.30
	3	0.3987	1.5834	1.213	3.16
	4	0.0937	1.0432	1.520	10.90

Table 16.9: Tschebyscheff Coefficients for 1 dB Pa—cont'd

n	i	a_i	b_i	$k_i = f_{Ci}/f_C$	Q_i
8	1	5.1019	14.7608	0.276	0.75
	2	0.8916	3.0426	0.849	1.96
	3	0.2806	1.4334	1.285	4.27
	4	0.0717	1.0432	1.520	14.24
9	1	6.3415	0.0000	0.158	—
	2	2.1252	7.1711	0.514	1.26
	3	0.5624	2.3278	0.994	2.71
	4	0.2076	1.3166	1.346	5.53
	5	0.0562	1.0258	1.533	18.03
10	1	6.3634	22.7468	0.221	0.75
	2	1.1399	4.5167	0.694	1.86
	3	0.3939	1.9665	1.093	3.56
	4	0.1616	1.2569	1.381	6.94
	5	0.0455	1.0277	1.532	22.26

Table 16.10: Tschebyscheff Coefficients for 2 dB Pa

n	i	a_i	b_i	$k_i = f_{Ci}/f_C$	Q_i
1	1	1.0000	0.0000	1.000	—
2	1	1.1813	1.7775	1.000	1.13
3	1	2.7994	0.0000	0.357	—
	2	0.4300	1.2036	1.378	2.55
4	1	2.4025	4.9862	0.550	0.93
	2	0.2374	1.1896	1.413	4.59
5	1	4.6345	0.0000	0.216	—
	2	0.9090	2.6036	0.908	1.78
	3	0.1434	1.0750	1.493	7.23
6	1	3.5880	10.4648	0.373	0.90
	2	0.4925	1.9622	1.085	2.84
	3	0.0995	1.0826	1.491	10.46
7	1	6.4760	0.0000	0.154	—
	2	1.3258	4.7649	0.665	1.65
	3	0.3067	1.5927	1.218	4.12
	4	0.0714	1.0384	1.523	14.28
8	1	4.7743	18.1510	0.282	0.89
	2	0.6991	3.1353	0.853	2.53
	3	0.2153	1.4449	1.285	5.58
	4	0.0547	1.0461	1.518	18.39

Continued

Table 16.10: Tschebyscheff Coefficients for 2 dB Pa—cont'd

n	i	a_i	b_i	$k_i = f_{Ci}/f_C$	Q_i
9	1	8.3198	0.0000	0.120	–
	2	1.7299	7.6580	0.522	1.60
	3	0.4337	2.3549	0.998	3.54
	4	0.1583	1.3174	1.349	7.25
	5	0.0427	1.0232	1.536	23.68
10	1	5.9618	28.0376	0.226	0.89
	2	0.8947	4.6644	0.697	2.41
	3	0.3023	1.9858	1.094	4.66
	4	0.1233	1.2614	1.380	9.11
	5	0.0347	1.0294	1.531	29.27

Table 16.11: Tschebyscheff Coefficients for 3 dB Pa

n	i	a_i	b_i	$k_i = f_{Ci}/f_C$	Q_i
1	1	1.0000	0.0000	1.000	–
2	1	1.0650	1.9305	1.000	1.30
3	1	3.3496	0.0000	0.299	–
	2	0.3559	1.1923	1.396	3.07
4	1	2.1853	5.5339	0.557	1.08
	2	0.1964	1.2009	1.410	5.58
5	1	5.6334	0.0000	0.178	–
	2	0.7620	2.6530	0.917	2.14
	3	0.1172	1.0686	1.500	8.82
6	1	3.2721	11.6773	0.379	1.04
	2	0.4077	1.9873	1.086	3.46
	3	0.0815	1.0861	1.489	12.78
7	1	7.9064	0.0000	0.126	–
	2	1.1159	4.8963	0.670	1.98
	3	0.2515	1.5944	1.222	5.02
	4	0.0582	1.0348	1.527	17.46
8	1	4.3583	20.2948	0.286	1.03
	2	0.5791	3.1808	0.855	3.08
	3	0.1765	1.4507	1.285	6.83
	4	0.0448	1.0478	1.517	22.87
9	1	10.1759	0.0000	0.098	–
	2	1.4585	7.8971	0.526	1.93
	3	0.3561	2.3651	1.001	4.32
	4	0.1294	1.3165	1.351	8.87
	5	0.0348	1.0210	1.537	29.00
10	1	5.4449	31.3788	0.230	1.03
	2	0.7414	4.7363	0.699	2.94
	3	0.2479	1.9952	1.094	5.70
	4	0.1008	1.2638	1.380	11.15
	5	0.0283	1.0304	1.530	35.85

Table 16.12: All-Pass Coefficients

n	i	a_i	b_i	f_i/f_C	Q_i	T_{gr0}
1	1	0.6436	0.0000	1.554	—	0.2049
2	1	1.6278	0.8832	1.064	0.58	0.5181
3	1	1.1415	0.0000	0.876	—	0.8437
	2	1.5092	1.0877	0.959	0.69	
4	1	2.3370	1.4878	0.820	0.52	1.1738
	2	1.3506	1.1837	0.919	0.81	
5	1	1.2974	0.0000	0.771	—	1.5060
	2	2.2224	1.5685	0.798	0.56	
	3	1.2116	1.2330	0.901	0.92	
6	1	2.6117	1.7763	0.750	0.51	1.8395
	2	2.0706	1.6015	0.790	0.61	
	3	1.0967	1.2596	0.891	1.02	
7	1	1.3735	0.0000	0.728	—	2.1737
	2	2.5320	1.8169	0.742	0.53	
	3	1.9211	1.6116	0.788	0.66	
	4	1.0023	1.2743	0.886	1.13	
8	1	2.7541	1.9420	0.718	0.51	2.5084
	2	2.4174	1.8300	0.739	0.56	
	3	1.7850	1.6101	0.788	0.71	
	4	0.9239	1.2822	0.883	1.23	
9	1	1.4186	0.0000	0.705	—	2.8434
	2	2.6979	1.9659	0.713	0.52	
	3	2.2940	1.8282	0.740	0.59	
	4	1.6644	1.6027	0.790	0.76	
	5	0.8579	1.2862	0.882	1.32	
10	1	2.8406	2.0490	0.699	0.50	3.1786
	2	2.6120	1.9714	0.712	0.54	
	3	2.1733	1.8184	0.742	0.62	
	4	1.5583	1.5923	0.792	0.81	
	5	0.8018	1.2877	0.881	1.42	

k_i is the ratio of the corner frequency of a partial filter, f_{Ci}, to the corner frequency of the overall filter, f_C. This ratio is used to determine the unity-gain bandwidth of the op amp, as well as to simplify the test of a filter design by measuring f_{Ci} and comparing it to f_C,

Q_i is the quality factor of the partial filter,

f_i/f_C ratio is used for test purposes of the all-pass filters, where f_i is the frequency, at which the phase is 180 degrees for a second-order filter, respectively 90 degrees for a first-order all-pass.

T_{gr0} is the normalized group delay of the overall all-pass filter.

Further Reading

[1]   D. Johnson, J. Hilburn, Rapid Practical Designs of Active Filters, John Wiley & Sons, 1975.
[2]   U. Tietze, C. Schenk, Halbleiterschaltungstechnik, Springer–Verlag, 1980.
[3]   H. Berlin, Design of Active Filters with Experiments, Howard W. Sams & Co, 1979.
[4]   M. Van Falkenburg, Analog Filter Design, Oxford University Press, 1982.
[5]   S. Franko, Design with Operational Amplifiers and Analog Integrated Circuits, McGraw–Hill, 1988.

CHAPTER 17

Fast, Simple Filter Design

17.1 Introduction

This chapter is presented, with respect, to the design engineer who is more familiar with the digital area, VHDL, and firmware—for whom analog design on the scale found in the last chapter is a formidable journey back into a difficult course one took back in college and would sooner forget than relive! We all have our specialties, and analog design is not currently in vogue in universities, which emphasize things that are deemed more "modern."

Chapter 16, written by Thomas Kugelstadt, has to be one of the finest references on the topic of filter design that I have ever seen. It condenses a topic that usually deserves its own volume into a single, if long, chapter. However, if you are like me, you wonder if there is further condensation possible—after all not every filter design calls for specialized responses, such as Tsychebyshev or Bessell, requires multiple poles or requires gain.

If one was to start with the following assumptions:

- unity gain (except for band pass)
- Butterworth response
- largest number of poles in a single op amp
- single power supply is all that is available

What would be left? The answers are as follows:

- 90% of filter applications,
- the techniques of this chapter!

I have knowingly kept the material and the techniques simple, for the beginning engineer who needs a practical filter and needs it fast.

17.2 Fast, Practical Filter Design

If you already know the response you want from your filter, you may skip this section and go directly to Section 17.3.

Op Amps for Everyone. http://dx.doi.org/10.1016/B978-0-12-811648-7.00017-0

To quickly design a filter with the techniques in this chapter, some things must be known in advance:

- The frequencies that need to be passed, and those that need to be rejected.
- A transition frequency, the point at which the filter starts to work, or a center frequency around which the filter is symmetrical.
- An initial capacitor value—pick one somewhere from 100 pF for high frequencies to 0.1 μF for low frequencies. If the resulting resistor values are too large or too small, pick another capacitor value.

Filter design—step 1:

For the beginner, the filter responses will be presented pictorially. The area shaded in blue represents the frequencies that will be passed, and the area in white represents the frequencies that will be rejected. Do not be concerned with the exact frequency yet, that will be taken care of in the following sections. Look at the responses below, and pick one where the desired frequencies are in the shaded area, and the rejected frequencies in the unshaded area (Figs. 17.1–17.5).

17.3 Designing the Filter

Now that you have selected your filter response, the next step is to design the filter:

17.3.1 Low-Pass Filter (Fig. 17.6)

Design Procedure:

- Pick C_1
- Calculate $C_2 = C_1 \times 2$
- Calculate R_1 and $R_2 = \dfrac{1}{2\sqrt{2} \times \pi \times C_1 \times \text{Frequency}}$

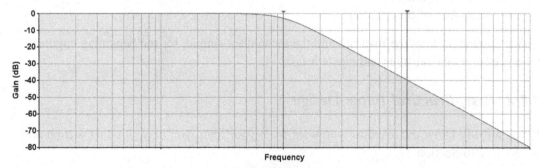

Figure 17.1
Low-pass response—go to Section 17.3.1.

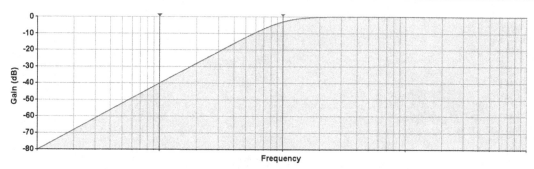

Figure 17.2
High-pass response—go to Section 17.3.2.

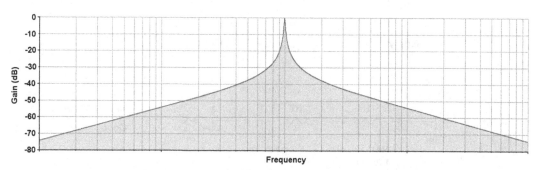

Figure 17.3
Narrow (single-frequency) band pass—go to Section 17.3.4.

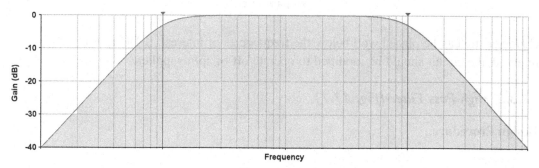

Figure 17.4
Wide band pass—go to Section 17.3.5.

- Calculate $C_{IN} = C_{OUT} = 100-1000$ times C_1 (not critical)
- DONE!

Digging Deeper: The filter selected is a unity gain Sallen–Key Filter, with a Butterworth response characteristic. Note that with the addition of C_{IN} and C_{OUT}, the filter is no longer purely a low-pass filter. It is a wide band-pass filter, but the high-pass response

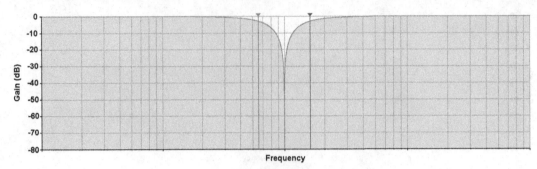

Figure 17.5
Notch filter—single frequency—go to Section 17.3.5.

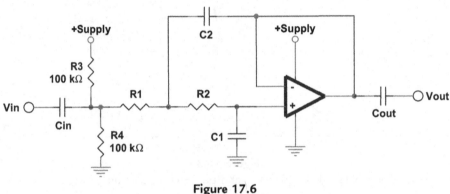

Figure 17.6
Low-pass filter.

characteristic can be placed well below the frequencies of interest. If DC response is required, the circuit should be modified to operate off of split supplies.

17.3.2 High-Pass Filter (Fig. 17.7)

Design Procedure:

- Pick $C_1 = C_2$
- Calculate R_1: $\dfrac{1}{\sqrt{2} \times \pi \times C_1 \times \text{Frequency}}$
- Calculate R_2: $\dfrac{1}{2\sqrt{2} \times \pi \times C_1 \times \text{Frequency}}$
- Calculate $C_{OUT} = 100 - 1000$ times C_1 (not critical)
- DONE!

Digging Deeper: The filter selected is a unity gain Sallen–Key Filter, with a Butterworth response characteristic. Just as was the case with the low-pass filter, there is no such thing as an active high-pass filter, but for a different reason. The gain/bandwidth product of the

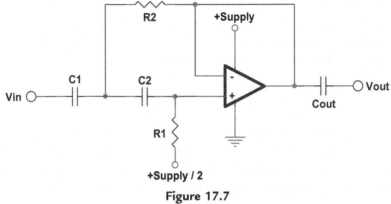

Figure 17.7
High-pass filter.

op amp used will ultimately produce a low-pass response characteristic, making this a wide band-pass filter. It is your responsibility to choose an op amp with a frequency limit well above the bandwidth of interest.

17.3.3 Narrow (Single-Frequency) Band-Pass Filter (Fig. 17.8)

Design Procedure:

- Pick $C_1 = C_2$
- Calculate $R_1 = R_4$: $\frac{1}{2 \times \pi \times C_1 \times \text{Frequency}}$
- Calculate $R_3 = 19 \times R_1$
- Calculate $R_2 = \dfrac{R_1}{19}$
- Calculate $C_{IN} = C_{OUT} = 100 - 1000$ times C_1 (not critical)
- DONE!

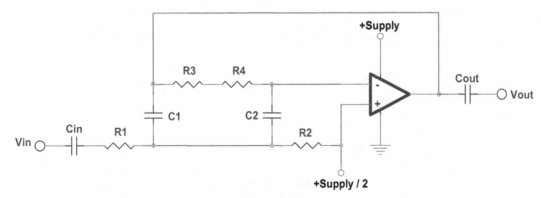

Figure 17.8
Narrow band-pass filter.

Digging Deeper: The filter selected is a modified Deliyannis filter. What is a "Deliyannis Filter?" It turns out to be a special case of the MFB band-pass configuration—one that is very stable and relatively insensitive to component variation. The Q is set at 10, which also locks the gain at 10, as the two are related by the expression:

$$\frac{R_3 + R_4}{2 \cdot R_1} = Q = Gain \qquad\qquad 17.1$$

A higher Q was not selected, because the op amp gain bandwidth product can be easily reached, even with a gain of 20 dB. At least 40 dB of headroom should be allowed above the center frequency peak. The op amp slew rate should also be sufficient to allow the waveform at the center frequency to swing to the amplitude required.

17.3.4 Wide Band-Pass Filter (Fig. 17.9)

Design Procedure:

- Go to Section 17.3.2, and design a high-pass filter for the low end of the band.
- Go to Section 17.3.1, and design a low-pass filter for the high end of the band.
- Calculate $C_{IN} = C_{OUT} = 100-1000$ times C_1 in the low-pass filter section (not critical).
- DONE!

Digging Deeper: This is nothing more than cascaded Sallen—Key high-pass and low-pass filters. The high-pass comes first, so noise from it will be low passed.

Digging Even Deeper—Narrow Versus Wide Band-Pass Filter: At what point is it better to implement a band-pass filter as a narrow/single-frequency filter versus a wide band pass? At high Q values, the single-frequency band-pass is clearly the better choice. However, as Q values decrease, the difference begins to blur. What can be a very sharp peak at resonance erodes to a single-pole rolloff on the low end, and single-pole rolloff on the high end. This results in a lot of unwanted energy in the stop bands.

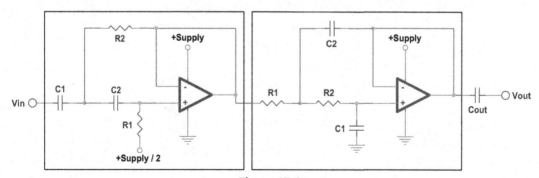

Figure 17.9
Wide band-pass filter.

For Q values of 0.1 (and below), and 0.2, the best implementation is high pass cascaded with low pass. The two implementations have almost an identical pass band response for a Q of 0.5. You are presented with a choice—use a band-pass filter (which can be implemented with a single op amp) to save money, or use a cascaded approach that has better rejection in the stop bands. As the Q becomes higher and higher, however, the response of two separate stages begins to interact, destroying the amplitude of the signal. **A good rule of thumb** is that the start and ending frequencies of a wide band-pass filter should be at least a factor of five different.

17.3.5 Notch (Single-Frequency Rejection) Filter (Fig. 17.10)

Design Procedure:

- Pick Co
- Calculate Ro: $\frac{1}{2 \times \pi \times C_1 \times \text{Frequency}}$
- Calculate $R_Q = 20 \times Ro$
- If you do not want to tune, replace Ro_low and Ro_adj with Ro. If tuning of the center frequency is desired, make a pot part of the value of Ro by going down one standard resistor value, and making sure the pot covers the range of center frequencies. If you do the job right, you can fine-tune the center frequency while preserving the depth of the notch. The nice thing about this topology—you will get a very deep notch somewhere!
- DONE!

Digging Deeper: This is the Fliege filter topology, set to a Q of 10. The Q can be adjusted independently from the center frequency by changing R_Q. Q is related to the center frequency set resistor by the following:

$$R_Q = 2 \times Q \times Ro$$

The Fliege filter topology has a fixed gain of 1. It is best to implement it from split supplies, although it can be operated from a single supply. Inject a reference into R_Q

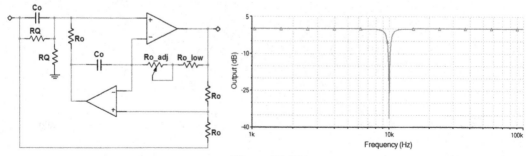

Figure 17.10
Notch filter.

instead of ground to operate off of split supplies. The input and output will have to be isolated by DC blocking capacitors as with other filter types.

Many designers use the "Twin-T" notch topology of Section 17.3.4 for notches. While it is a popular topology, it has many problems. The biggest is that it is not producible. Many runs of simulation with component tolerances of 1% have shown tremendous variation in notch center frequency and notch depth. The only real advantage is that it can be implemented with a single op amp. Some additional stability can be obtained from the two op amp configuration, but if two op amps are used, then why not use a different topology such as the Fliege? To successfully use the Twin-T topology, six precision components are required. The Fliege will produce a deep null at some frequency, and it is easy to tune that frequency by adjusting one of the Ro resistors—the null will remain as deep over a fairly wide range. The following response plot was made from varying the potentiometer on a 10 kHz Fliege filter in 5% increments.

Some key "takeaways" from the Fliege filter response—it does not disturb the frequencies around it to any significant degree. The response in Fig. 17.10 shows that a high Q 10 kHz band-pass filter leaves everything under 9 kHz and above 11 kHz almost unchanged. The response of Fig. 17.11 shows that a pot forming 2% of the Ro value in the position show allows adjustment of the center frequency over about a ±1% range around the center frequency. The depth of the notch remains unchanged over that range of adjustment, making it an ideal way of tuning the notch frequency.

Incidentally, if the reader wants to construct an AM heterodyne filter such as the one shown above, the component values are Ro = 4.42 k, Co = 3600 pF, Ro_low = 4.32 k, Ro_adj = 200 Ω, and R_Q = 88.7 k. The op amps should be at least 100 MHz in bandwidth.

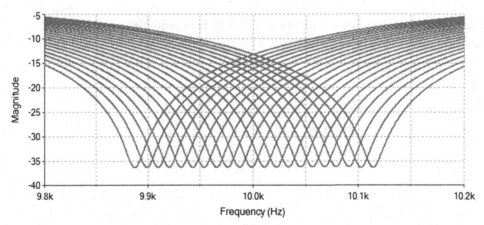

Figure 17.11
Variable frequency notch filter.

The author's version of this design has been absolutely stable for 15 years with no readjustment required.

17.4 Getting the Most Out of a Single Op Amp

As simple and elegant as the procedures are above, I can assure you that there are more elegant things that you can do with simple design techniques. Some of the circuits below are designed for split supplies, they would require modification to work on single supplies.

17.4.1 Three-Pole Low-Pass Filters

Section 17.3.1 above showed how to implement low-pass filters easily and quickly. However, why implement only a two-pole filter, when a single op amp can just as easily implement a three pole-filter? (Fig. 17.12)

The response, in this case, will roll off 60 dB/decade instead of 40 dB/decade as seen in the two-pole filter. This topology also solves a different problem associated with the Sallen–Key architecture, that of feed through. In a two-pole Sallen–Key filter, high frequencies will leak through the filter, especially when the amplifier is turned off. This three-pole topology adds an RC low-pass filter to the input of a two-pole Sallen–Key architecture, thus absolutely guaranteeing at least a 20 dB/decade roll off of high frequencies no matter what happens to the amplifier.

To design this three-pole low-pass filter:

- Pick a value of resistance $R = R_1 = R_2 = R_3$.
- Calculate a base value $fsf = 2 \times \pi \times R \times$ frequency
- Calculate $C_1 = 3.546/fsf$
- Calculate $C_2 = 1.392/fsf$

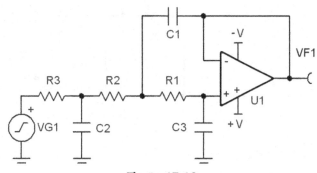

Figure 17.12
Three-pole low-pass filter.

- Calculate $C_3 = 0.2024/\text{fsf}$
- Pick standard value capacitors closest to the calculated ones above.
- DONE!

17.4.2 Three-Pole High-Pass Filters

Just as it is easy to implement three-pole low-pass filters, it is also easy to implement three-pole high-pass filters (Fig. 17.13).

To design this three-pole high-pass filter:

- Pick a value of capacitance $C = C_1 = C_2 = C_3$.
- Calculate a base value $\text{fsf} = 2 \times \pi \times C \times \text{frequency}$
- Calculate $R_1 = 3.546/\text{fsf}$
- Calculate $R_2 = 1.392/\text{fsf}$
- Calculate $R_3 = 0.2024/\text{fsf}$
- Pick standard value resistors closest to the calculated ones above.
- DONE!

17.4.3 Stagger-Tuned and Multiple-Peak Band-Pass Filters

This book has purposely downplayed an interesting topology that is very popular—the Twin-T topology. Part of the reason why this has not been done is that for band-pass filters, it is not a true band-pass topology. It is more of a resonator at the center frequency and also has theoretically infinite gain at its resonance (for ideal components). Therefore, in the real world, it is difficult to control the gain at the center frequency, and the ultimate stop band rejection is 0 dB—therefore unity gain. Not very useful in rejecting out of band signals!

The Twin-T topology for band-pass filters and its response look like Fig. 17.14.

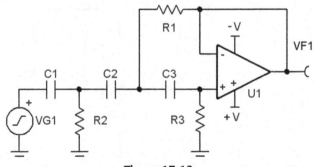

Figure 17.13
Three-pole high-pass filter.

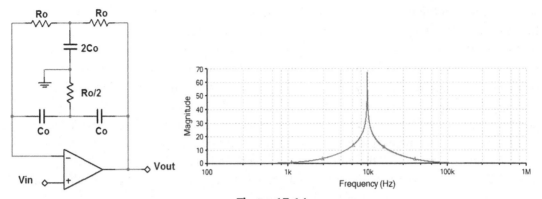

Figure 17.14
Twin-T band-pass filter response.

This topology is very difficult to work with. It requires the user to obtain three resistors, one of which is exactly half the other, and three capacitors, one of which is exactly twice the other. Even if this is possible, the chances are that they would not match exactly, nor track with temperature. Furthermore, the peak is so sharp that real-world components might erode the peak, or miss it entirely. To overcome these shortcomings, the following modifications are made (Fig. 17.15).

This configuration takes advantage of parallel resistance and capacitance characteristics to make your job easier. An additional Ro and Co have been added, but this means that you now only have to find four identical values of resistance and four identical values of capacitance, no more 2 Co and ½ Ro required. Because components manufactured in the same lot oftentimes have virtually identical characteristics, this should be easy.

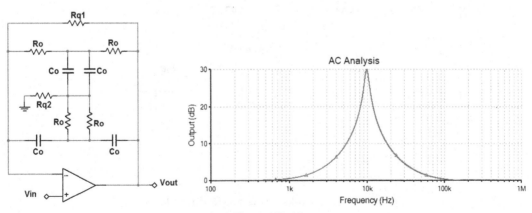

Figure 17.15
Modified Twin-T topology.

A more subtle and unfamiliar change to the familiar Twin-T topology is the addition of R_{Q1} and R_{Q2}. This takes advantage of the only two places the circuit's Q can be adjusted. As long as $R_{Q1} \gg Ro$, it has minimal effect on the center frequency, but it acts to make the two Co's that are in series appear to have more leakage, therefore eroding the amplitude of the peak. Similarly, if $R_{Q2} \ll Ro$, it makes the two parallel Co's appear to have higher equivalent series resistance, also eroding the peak. Acting together, it is possible to have a measure of control over the amplitude and peak, although not as precise as you might like. Comparing the response of the modified and unmodified circuits, you can see that the tendency of the amplitude to be unbounded at resonance has been mitigated, and the response is more reasonable. The Q has also been modified, although it is hard to see on the 4-decade log frequency scale.

Another technique that has been used to adjust the Q of a Twin-T circuit is to unbalance Ro and Co in the legs of the circuit that are in series. This, however, requires you to find multiple precision component values, all of which could potentially track differently over temperature.

With this "brief" introduction to Twin-T band-pass filters complete, now the real fun can begin! There is no reason that two Twin-T networks cannot be placed inside the feedback loop of an op amp (Fig. 17.16).

The two Twin-T sections are effectively in series inside the feedback loop of the op amp. They can be tuned entirely independently, resulting in the following response (Fig. 17.17).

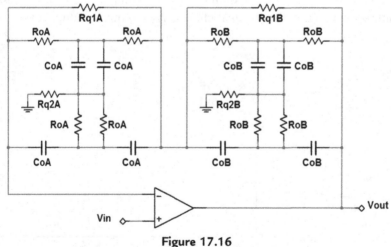

Figure 17.16
Two Twin-T networks inside the feedback loop.

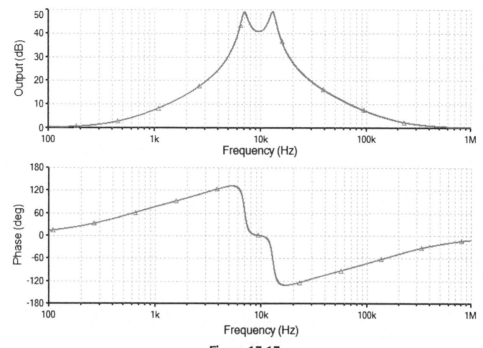

Figure 17.17
Stagger-tuned filter response.

Such a strategy is often employed when it is necessary to have flat phase (and group delay) in a narrow region around the center frequency. The trade-off is that the pass band of the filter is widened to accommodate the flat phase response in the middle. You are also reminded to select a wide band op amp that will be able to handle not only the gain at the center frequency, but also the gain of the upper peak. The open-loop response should be at least 40 dB above the upper peak. In this case, an op amp with 1.5 GHz bandwidth was used to simulate a stagger-tuned 10 kHz filter, in other words a bandwidth more than five orders of magnitude greater than the center frequency!

If the two Twin-T networks are separated enough in frequency, this topology will produce a band-pass filter that has more than one resonant peak as shown in Fig. 17.18.

Notice that the flat phase response has been sacrificed between the peaks, although the phase still crosses through 0 degree. But flat phase between the peaks is no longer the design goal—having two peaks is. The "valley" between the two peaks is not that low, obviously it would get lower if the peaks were further separated in frequency or the "Q" was higher. But for quick detection of multiple tone frequencies, this method is the most economical possible.

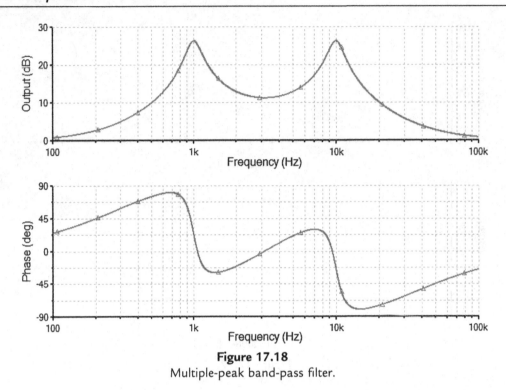

Figure 17.18
Multiple-peak band-pass filter.

You are cautioned again about op amp bandwidth, but in this case the open-loop gain at the upper peak may be less of a constraint because the overall gain is lower.

17.4.4 Single-Amplifier Notch and Multiple Notch Filters

The Twin-T topology can also be used to create a single-amplifier notch filter (Fig. 17.19).

As was the case in the Twin-T band-pass filter, two resistors were added that allow some control over the Q, also affecting notch depth. This is very useful, because the Twin-T notch is difficult to tune for center frequency. If the Q is lowered, the chances are better of actually placing a notch where it is needed to reject an unwanted frequency.

Unfortunately, the Twin-T topology also has the drawback that the notch disturbs the amplitude in the decades above and below the center frequency. Contrast the response above to the response in Fig. 17.10. Where the Fliege topology leaves the surrounding frequencies almost untouched, the Twin-T has very significant effect almost a decade above and below the center frequency, making it a poor choice for a notch filter. So why use it at all? Besides the advantage of being a single-amplifier notch, the other advantage is, like the Twin-T band-pass topology, it is possible to produce two notches from a single

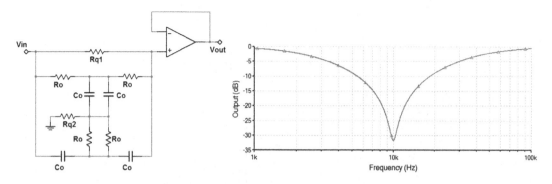

Figure 17.19
Single-amplifier Twin-T notch filter.

amplifier. This can either be a way to produce two individual notches at widely separated frequencies, or a way to produce a band rejection filter if the notch frequencies are closely spaced (Fig. 17.20).

This scheme, however, dramatically affects the audio above 10 Hz and below 300 Hz. It may, however, be suitable for severe cases of hum on speech.

17.4.5 Combination Band-Pass and Notch Filters

There is no reason why a Twin-T notch and a band-pass filter cannot be combined in a single circuit (Fig. 17.21).

Of course more than one peak and/or notch is also possible. In practice, however, you should probably count on no more than two to three sections of Twin-T networks because the number of passive components crowding the op amp will become excessive, leading to parasitic on the board and other problems.

17.5 Design Aids

Just as I did for op amp gain and offset, I have written design aids for filter design.

17.5.1 Low-Pass, High-Pass, and Band-Pass Filter Design Aids

This chapter has presented you with a bewildering variety of filter circuits to ponder. It is impossible (well it would be very difficult) to produce a single circuit that could implement all filter types. I will leave you instead with a single schematic of a universal filter circuit (well low pass, high pass, and band pass) (Fig. 17.22).

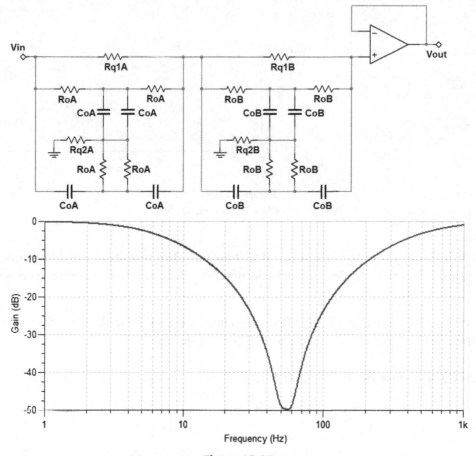

Figure 17.20
Twin-T band reject filter.

The general purpose impedances, or "Z's," can be either resistors or capacitors, depending on the filter topology selected. Not all components will be installed for every filter. Some may be 0 Ω resistors, others may be "open," not installed.

This single circuit can implement low-pass, high-pass, and band-pass stages. Although two-pole Sallen–Key filters can be implemented, why bother when you can have three poles just as easily?

As was the case for gain and offset, I have written a design aid for filters, which is also available in the companion website. Fig. 17.23 shows a screen shot.

A bit of explanation is in order. When the calculator first comes up, all of the impedances and units will be blank, unlike the screen shot above. To use the calculator, first you need to know whether you are designing a low-pass, high-pass, or band-pass filter. These three

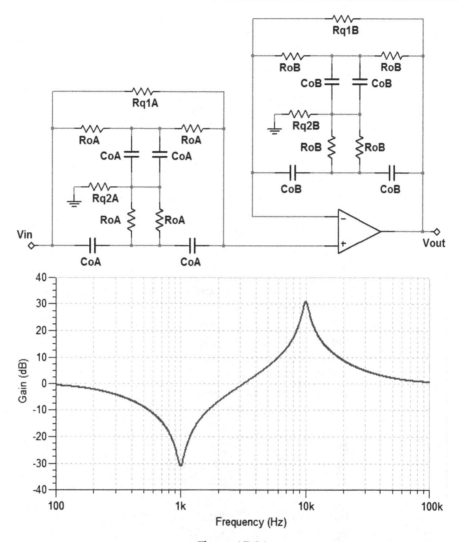

Figure 17.21
Single-amplifier Twin-T notch and band-pass filter.

types are available in a drop-down menu in the "Filter Type" box. Then, there are a series of boxes that depend on the type of filter you are making: For (type of filter), Enter (value) series of lines. For example, all filter types, LP, HP, and BP require a frequency. For low and high pass, it will be the −3 dB cutoff frequency. For band pass, it will be the center frequency. So for *all* filters, Enter Frequency. Because this is a low-pass filter, the only remaining decisions you need to make are the capacitor sequence (E6, E12, or E24), and the seed resistor value, which will scale the capacitors. Click the "Calculate" button and the component values appear.

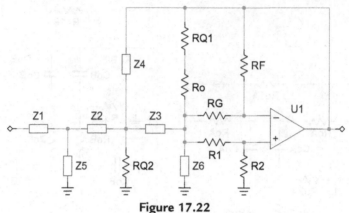

Figure 17.22
Universal filter schematic.

			Filter Type:	LP ▼	
For	**Enter**				
ALL		Frequency (Hz):	15000		
BPF		Q and Gain (V/V):	10		
BPF		Resistor Scale (Ohms):	1000 ▼		
BPF/LPF		Capacitor Sequence:	E12 ▼		
LPF		Seed Resistor (Ohms):	10000		
HPF		Seed Capacitor (pF):	1000		
			Calculate		
Z1	10000	Ohm	**Z4**	3.9	nF
Z2	10000	Ohm	**Z5**	1.5	nF
Z3	10000	Ohm	**Z6**	220	pF
Ro	open				
RQ1	open		**RQ2**	open	
Rg	open		**Rf**	0	Ohm
R1	0	Ohm	**R2**	open	

Figure 17.23
Universal filter calculator.

If you choose a high-pass filter instead, you will notice that capacitors and resistors have swapped left to right. Of course you choose a seed capacitor, instead of a seed resistor, for high-pass filters. The seven resistor values on the bottom, Ro through R_2, are the same for low-pass and high-pass filters, either $0\ \Omega$ or open. These route connections to configure the universal schematic and PC board to be a three-pole Sallen–Key filter topology. This changes dramatically when you choose "band pass" instead. Z1 becomes $0\ \Omega$, Z5 and Z6 become open, and some of the resistor values on the bottom are

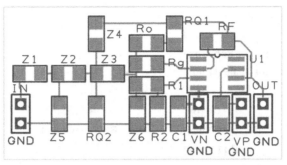

Figure 17.24
Universal filter board.

populated, others are used to configure the circuit for the modified Deliyannis configuration.

In addition, the schematic and circuit notes will appear to the right of the calculator. These change according to topology. It will take some experience on your part to properly scale resistors and capacitors. All capacitors should be 1% NPO/COG dielectric if at all possible, especially if the filter is to encounter different temperatures. Although higher value NPO/COG capacitors are becoming more common, 10,000 pF is generally the limit, and anything above 1000 pF is usually large, expensive, and hard to get. Resist at all costs the temptation to use 5% capacitors, especially in the band-pass filter. Choose 1% tolerance, and make that decision stick with procurement and manufacturing departments.

To further facilitate your designs, I have done a board layout of the universal filter schematic above. It is designed to accommodate SO-8 single op amps and 1206 surface-mount resistors. It is implemented a single PC board layer. The gerbers are also available from the companion website (Fig. 17.24).

You can use this approach to implement a universal arrangement on a PC board. This might be done when multiple filter stages are to be used, and it is unclear whether a given digital signal processor algorithm really needs a high-pass function, or a low-pass function, or whether the application can benefit from two stages of one type, for example.

17.5.2 Notch Filter Design Aids

But wait, there is more! Remember I said it would be difficult to implement a universal filter board that also contained a notch. I have not left you to your own devices. I also have written a notch filter design aid. A screen capture is shown in Fig. 17.25.

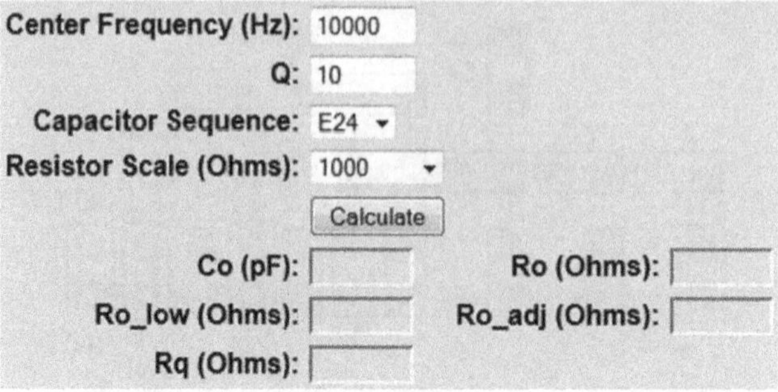

Center Frequency (Hz):	10000
Q:	10
Capacitor Sequence:	E24 ▾
Resistor Scale (Ohms):	1000 ▾

Calculate

Co (pF):		Ro (Ohms):	
Ro_low (Ohms):		Ro_adj (Ohms):	
Rq (Ohms):			

Figure 17.25
Notch filter calculator.

Like the other filter calculator, this one also shows the schematic. This calculator requires only a center frequency—the user can adjust the Q, capacitor sequence, and the resistor scale. As the notes indicate, however, you should not deviate too much from a Q of 10—this is a good compromise value that will give plenty of rejection, yet leave other frequencies untouched. The calculator also includes provisions for tuning the center frequency. If you do not want to do that, just substitute another Ro for Ro_low and Ro_adj.

I have also generated a PC board layout for the notch filter (Fig. 17.26).

This one had to be two layers, mainly so I could route the power. If this is a hardship, just route the power with wires on the backside of a single-layer board.

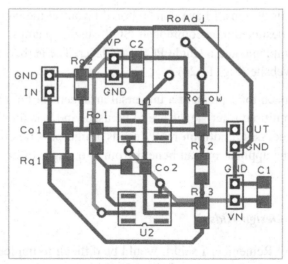

Figure 17.26
Notch filter PCB.

17.5.3 Twin-T Design Aids

I have written separate design aids for Twin-T band-pass and Twin-T notch filters. The screen capture for both of them looks the same (Fig. 17.27).

There is little to decide—enter the center frequency, select the resistor and capacitor sequence, and click "Calculate." There is a resistor scale control. Leave the Q alone, it will allow about 30 dB gain for the band pass and 30 dB of rejection for the notch. Adjustment is possible, but over a relatively narrow range. Remember that Twin-T filters cannot be tuned easily for center frequency, and the response will not be ideal.

I have also generated a PC board layout for Twin-T filters (Fig. 17.28).

This one had to be two layers, mainly so I could route the power. Two of the "Q" adjustment resistors are also on the backside of the board. If the Q resistors are not used, jumpers can be added to ground from the center of the Ro Co pattern, and the Q resistors on the top of the board unused. Remember this will make a very narrow band pass/notch with uncontrolled amplitude and center frequency.

The board is designed to support one section of notch, and one section of band pass. If notch is desired, place a 0 Ω resistor across the R_{Q3} pads. If band pass is desired, place a 0 Ω resistor across the R_{Q1} pads. This board will also support the notch/band-pass configuration of Section 17.4.5.

17.6 Summary

Filter design is a subject plagued by misinformation, endless textbooks with mathematical derivations, tables, and graphs. Many hours have gone into demystifying the topic for you

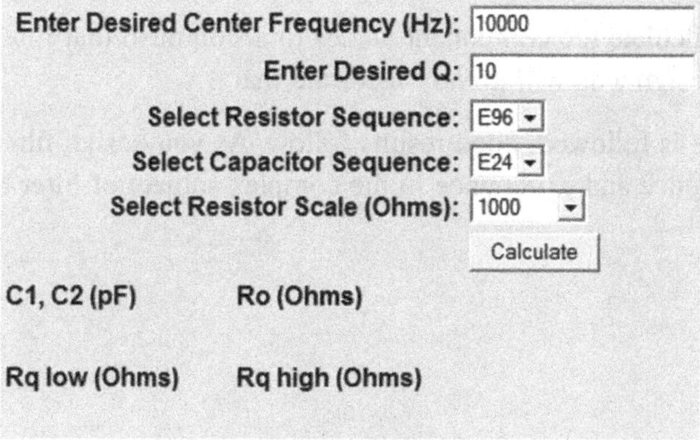

Figure 17.27
Twin-T filter calculator.

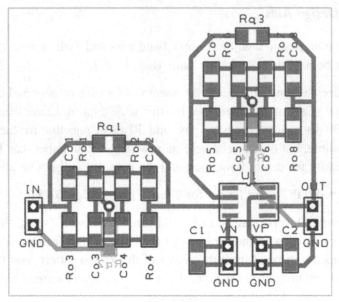

Figure 17.28
Twin-T PC board.

in this book. Gone are the derivations and math exercises, and in their place working filter circuits with real world component values. Many hours have also gone into providing filters that have been tested extensively in the lab and are the simplest possible implementation available.

Filter design in general requires that you know:

- What frequencies you want to pass, and what frequencies you want to reject.
- The ability to choose a filter topology that accomplishes the passing and rejection of those frequencies.
- The ability to calculate the component values to accomplish that filtering function. Hopefully the design aids will greatly facilitate that.

When this procedure is followed, good results follow. As you design filters successfully, you will gain confidence and experience in the complex subject of filter design.

CHAPTER 18

High-Speed Filters

18.1 Introduction

Once the exclusive domain of passive filters and discrete components, op amps have progressed to the point where they can be considered for high-frequency filter design. At high speeds, however, filter design gets particularly challenging. The main limitation of high-speed op amp filter design comes from the op amp open-loop bandwidth. You may remember how the open-loop bandwidth affects op amp gain circuits from Section 7.6. The effect of open-loop gain on filters, however, begins lower in frequency and can produce very nonintuitive—nonworking filters. If you love challenges, high-speed filter design will test the limits of what you can do with analog filters, with the foreknowledge that things will start to get strange!

We will examine the major types of filters, and how op amp bandwidth affects each type (Fig. 18.1).

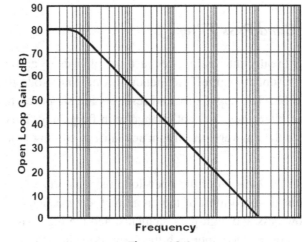

Figure 18.1
Open-loop response.

Op Amps for Everyone. http://dx.doi.org/10.1016/B978-0-12-811648-7.00018-2

18.2 High-Speed Low-Pass Filters

Reviewing Fig. 18.1, which is repeated from earlier in the book.

It is immediately obvious that the general shape of the open-loop response plot is that of a low-pass filter. Therefore, if the low-pass breakpoint is not terribly critical, all you have to do is to select an op amp with a unity gain bandwidth at the desired −3 dB breakpoint! Imagine explaining to a manager how the low-pass filter actually is identical to a unity gain stage. The high-speed low-pass filter can also have gain—it will lower the −3 dB breakpoint by approximately a decade for each 20 dB of gain.

Obviously in the case of low-pass filters, you got off very easily. This will not be the case for the remainder of filter responses!

18.3 High-Speed High-Pass Filters

There is no such thing as an active high-pass filter with good reason. The gain/bandwidth product of the op amp used will ultimately produce a low-pass response characteristic, making this a wide band-pass filter. It is your responsibility to choose an op amp with a frequency limit well above the bandwidth of interest. This is doubly so at high speeds, because you are inevitably closer to the open-loop limitation of the op amp.

You will need to select an op amp that has sufficient open-loop response that it will pass the bandwidth of interest before it starts rolling off due to open-loop gain limitations.

18.4 High-Speed Band-Pass Filters

This is where things get really interesting, because the physics of an op amp can and will actually force the resonant peak off frequency (lower) and erode the peak. You may be asking how can this be? The capacitors and resistors are supposed to define the center frequency of the filter, they will not change with frequency. This is a valid question, and it leads to the answer—the frequency shift is coming from the op amp itself. But again why?

The answer comes from the location of the open-loop response characteristic of the op amp—which is the ultimate speed limit of the op amp at any frequency. A band-pass filter is composed of both low-pass and high-pass elements, and the high pass characteristics will tend to get chopped off as they approach the open-loop characteristic. This will appear as first an amplitude limitation and finally as a frequency shift as the bandwidth limitation of the high-pass elements comes into play and limits the point at which they interact with the low-pass filter elements. The result is a truncated response that appears to be a frequency shift.

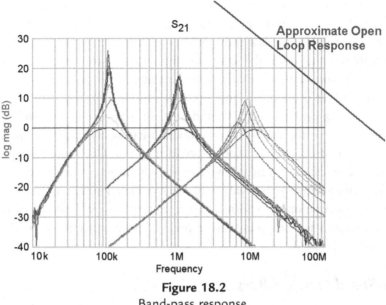

Figure 18.2
Band-pass response.

To illustrate this effect, band-pass filters were constructed using the topology of Section 17.3.3. The results are shown in Fig. 18.2.

Three frequencies were constructed, indicated by the sets of peaks above. For 10 MHz, the third set of peaks, a Q (and gain) of 1, the open-loop response of the op amp was a little over 30 dB above the peak at 10 MHz, and the filter actually worked very well. As the Q (and gain in V/V) was raised in steps of 5, however, things began to change. By the time a Q of 25 was attempted, the gain of the filter was almost back to unity, and the frequency shifted to the left to about 6.5 MHz! Clearly, the proximity of the open-loop response was affecting the op amp. Even the attempts to make a 1 MHz band-pass filter—although not showing the undesirable frequency shift—still show an amplitude compression effect. Only the 100 kHz filter shows anything close to lab results matching theoretical.

Note that the open-loop response of this particular op amp indicates that it is approximately a 1 GHz op amp, in other words close to the state of the art in op amp design! Therefore, there is a practical limit to how fast a band-pass filter can be constructed, about 10 MHz for unity gain and Q of 1. That limit is about 1/100 the rated bandwidth of the op amp. If a higher Q is desired, say 10, then the practical limit is about 1/1000 the rated bandwidth of the op amp. In other words,

$$\text{Center frequency(maximum)} = \text{Aol}/(100 \times Q) \qquad (6.1)$$

This limitation may also hit lower-frequency band-pass filters—be extremely careful! I can assure the reader—this bandwidth limitation has hit more than one time in my career, and the reader is warned that these effects on filter center frequency can occur differently in different batches of the same op amp and may only appear at certain temperatures. Unlike the stability criteria discussed in Chapter 6, there is little or no warning when this effect is about to show up, it is similar to falling off a cliff in the dark! Leave a generous safety margin when designing high-frequency band-pass filters, because the effects of insufficient op amp bandwidth can occur unexpectedly and dramatically. One board may work, another board may not. One layout may work, another layout may not. Even a fairly low-frequency band-pass filter may require a very fast op amp to accommodate higher values of Q. You have two choices if you encounter this limit—either select an op amp with a higher gain/bandwidth product or lower the Q of the band-pass filter. The time to anticipate this is before a product is produced, not after it has failed in the field!

18.5 High-Speed Notch Filters

A very similar bandwidth restriction affects notch filters. Instead of eroding the amplitude of the peak and/or shifting it in frequency as it does band-pass filters, the bandwidth restriction erodes the depth of the notch. A notch filter was constructed using the techniques of Section 17.3.5. A 1 GHz bandwidth op amp was used, for a filter center frequency of 1 MHz. No tuning of center frequency was attempted. Various values of Q were attempted, with the following results (Fig. 18.3).

Obviously, the center frequency is largely unaffected, it is the Q that takes a hit. At a Q of 1, a 30 dB notch is possible at 1 MHz. However, higher Q's only erode the depth of the notch. The circuit as constructed is completely unsuitable for a Q greater than 1.

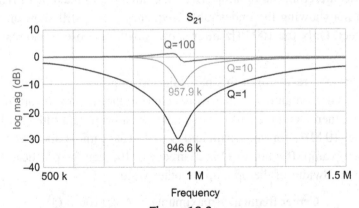

Figure 18.3
Notch filter response for 1 MHz center frequency.

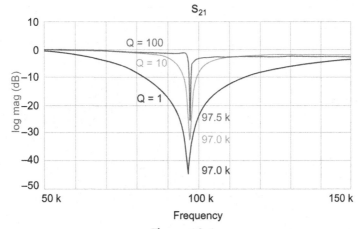

Figure 18.4

Notch filter response for 100 kHz center frequency.

The circuit was retuned for a center frequency of 100 kHz, and the results below were measured (Fig. 18.4).

Notch depth erosion is even evident at 100 kHz. At this point, it is clear that even a 1 GHz op amp can only be used to construct notch filters at 1 MHz and a Q of 1 and 100 kHz with a Q of 10 or so. This amazing degree of limitation was totally unexpected, to say the least!

18.6 10 kHz Notch Filter Results

The results above were so dramatic that I thought I would retune the circuit for a center frequency of 10 kHz to see what the results might be. This is the resulting response for a Q of 100 and a Q of 10 (Fig. 18.5).

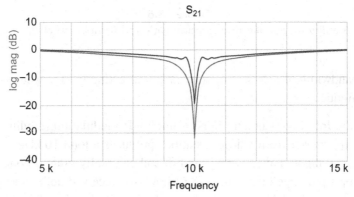

Figure 18.5

Notch filter response for 10 kHz center frequency.

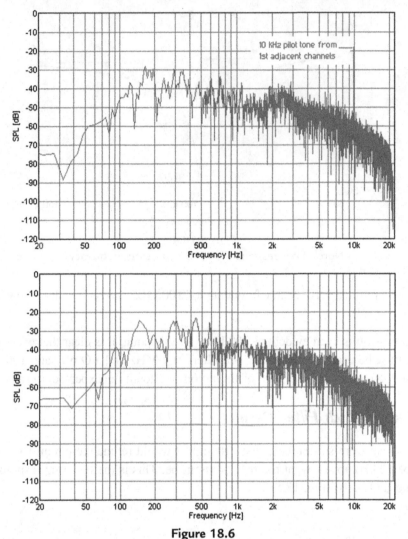

Figure 18.6
Amplitude modulation reception with and without notch filter.

Incredibly, the bandwidth of a 1 GHz op amp is still affecting the Q, even at 10 kHz. That is 5 decades of frequency!

A good application for 10 kHz notch filters is amplitude modulation (medium wave) receivers, where the carrier from adjacent stations produces a loud 10 kHz whine in the audio, particularly at night. This is a real ear-full and can really grate on one's nerves when listening for a prolonged time. Fig. 18.6 shows the received audio spectrum of a station before and after the 10 kHz notch was applied. Note that the 10 kHz whine is the loudest portion of the received audio, although the human ear is less sensitive to it. This

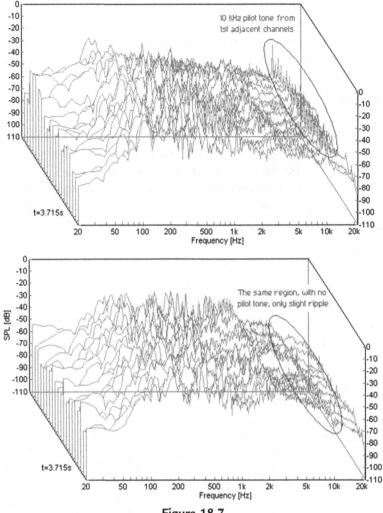

Figure 18.7
Effect of heterodyning and the notch filter.

audio spectrum was taken at night on a local station, which had two strong stations on either side. FCC regulations allow for some variation of the carrier of stations. Therefore, slight errors in carrier frequency of the two adjacent stations will make the 10 kHz tones heterodyne, increasing the unpleasant listening sensation. When the notch filter is applied in the bottom plot of Fig. 18.6, the 10 kHz tone is reduced to the same level as surrounding modulation. Also visible on the audio spectrum are 20 kHz carriers from stations two channels away and a 16 kHz tone from a transatlantic station. These are not a problem, because they are attenuated substantially by the receiver IF. A frequency of 20 kHz is inaudible to the vast majority of people in any event.

Fig. 18.7 shows the same spectrum on a waterfall diagram. In this case, the sample window is widened, and the 10 kHz carrier interference is shown as a string of peaks which vary in amplitude. When the notch is applied, the 10 kHz peaks are eliminated, and there is only a slight ripple in the received audio where 10 kHz has been notched out.

18.7 Conclusions

High-speed filter design is a new frontier—the state of the art of op amps is barely adequate. If proper care is taken, the results can be good. But the designer is cautioned once again—the open-loop bandwidth of an op amp has much more of a limiting effect on an op amp used in a filter configuration than it does on an op amp that is merely used for gain. Especially in the case of band-pass and notch filters, the circuit design must be overly cautious and rigorously tested.

CHAPTER 19

Using Op Amps for RF Design

19.1 Introduction

RF design used to be the exclusive domain of discrete devices. The advent of new generations of high-speed voltage- and current-feedback op amps has made it possible to use op amps for RF design. Op amp–based RF circuitry is easier to design and has less risk associated. "Tweaking" in the lab can be almost eliminated. Although there are many advantages to using op amps, traditional RF designers are reluctant to utilize them. They are confronted with a bewildering array of op amp parameters, many of which do not relate directly to the set of design parameters they are familiar with. This chapter bridges the gap between RF designers and op amp designers, giving the RF designer common ground with which to begin their design.

19.2 Voltage Feedback or Current Feedback?

The RF designer that is considering op amps is presented with a dilemma—are voltage-feedback amplifiers or current-feedback amplifiers better for the design? Frequency of operation is usually the most demanding aspect of RF design, and this makes the op amp bandwidth a critical parameter. The bandwidth specification given in op amp data sheets only refers to the point where the unity gain bandwidth of the device has been reduced by 3 dB by internal compensation and/or parasitics—which is not very useful for determining the actual operating frequency range of the device in an RF application.

Internally compensated voltage-feedback amplifier bandwidth is dominated by an internal "dominant-pole" compensation capacitor. This gives them a constant gain/bandwidth limitation. Current-feedback amplifiers, in contrast, have no dominant-pole capacitor, and therefore can operate much closer to their maximum frequency at higher gain. Stated another way, the gain/bandwidth dependence has been broken but not to the degree most designers would want. In practice, current-feedback op amps offer only a slight advantage over voltage-feedback op amps.

19.3 RF Amplifier Topology

The traditional RF amplifier shown in Fig. 19.1 uses a transistor (or in early days a tube) as the gain element. DC bias ($+V_{bb}$) is injected into the gain element at the load through

Op Amps for Everyone. http://dx.doi.org/10.1016/B978-0-12-811648-7.00019-4

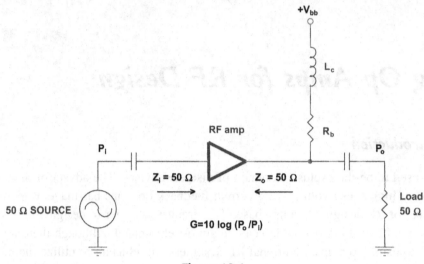

Figure 19.1
A traditional RF stage.

a bias resistor R_b. RF is blocked from being shorted to the supply by an inductor L_c, and DC is blocked from the load by a coupling capacitor.

Both the input impedance and the load are 50 Ω, which ensures matching between stages.

When an op amp is substituted as the active circuit element, several changes are made to accommodate it.

By themselves, op amps are differential input, open-loop devices. They are intended to be operated in a closed-loop topology (different from a receiver's AGC (automatic gain control) loop). The feedback loop for each op amp must be closed locally, within the individual RF stage.

There are two ways close to an op amp locally: inverting and noninverting. These terms refer to whether the output of the op amp circuit is inverted from the input or not. From the standpoint of RF design, this is seldom of any concern. For all practical purposes, either configuration will work and give equivalent results, but RF designers may find the noninverting configuration of Fig. 19.2 easier to work with, because the gain setting elements are not part of impedance matching.

The input impedance of the noninverting input is high, so the input is terminated with a 50 Ω resistor. Gain is set by the ratio of R_F and R_G:

$$G = 20 \cdot \log \frac{1}{2} \left(1 + \frac{R_F}{R_G} \right) dB, \text{ log gain}$$

for a desired gain:

$$1 + \frac{R_F}{R_G} = 2 \left(10^{G/20} \right)$$

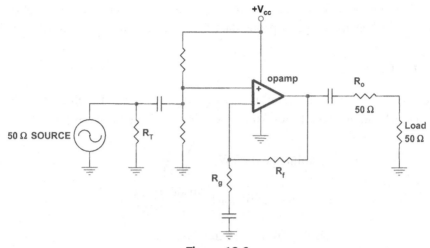

Figure 19.2
Noninverting RF op amp gain stage.

The gain of this stage as shown should never be below ½ (−6 dB), because most op amps are unity gain stable.

The output of the stage is converted to 50 Ω by placing a 50 Ω resistor in series with the output. This combined with a 50 Ω load, means that the gain is divided by 2 (−6 dB) in a voltage divider. A unity gain (0 dB) gain stage would become a gain of ½, or −6 dB.

A virtual ground is generated on the noninverting input after the coupling capacitor, to raise the operating point of the op amp to a virtual ground halfway between the supply voltage and ground.

Coupling capacitors are needed to isolate preceding and succeeding stages, and the gain resistor R_G from the virtual ground DC potential. These two resistors—see in Fig. 19.2 to the right of the input coupling cap—should be large in relationship to RT, but small enough to satisfy the input bias requirements of the op amp. This is seldom a concern to the RF designer, values of 1−10 k are probably correct. They should be equal in value to create a voltage divider of ½ on the supply, creating the DC biasing point (DC operating point) of the stage. The blocking capacitors should be selected to have low impedance at the operating frequency, not so small that they affect the gain of the stage directly, or cause unacceptable variations in gain over the operating range of the stage.

19.4 Op Amp Parameters for RF Designers

When designing for RF, problems can arise because op amp sheets show a totally different set of parameters than are considered for RF designs. This situation has been improving over the years as more and more high-speed amplifiers have been introduced, but there are still differences. This section will go over the more important parameters and specifications and show RF designers how to interpret op amp data sheet specifications to meet their needs.

19.4.1 Stage Gain

Op amp designers think of the gain of an op amp stage in terms of voltage gain. RF designers, in contrast, are used to thinking of RF stage gain in terms of power:

$$\text{Absolute power (W)} = \frac{V_{RMS}^2}{50\,\Omega}$$

$$P_o\,(\text{dBm}) = 10\cdot\log\left(\frac{\text{Absolute power}}{0.001\text{W}}\right)$$

$$\text{dBm} = \text{dBV} + 13 \text{ in a } 50\,\Omega \text{ system}$$

The forward transmission S_{21} is specified over the operating frequency range of interest. S_{21} is never specified on an op amp data sheet because it is a function of the gain, which is set by the input and feedback resistors R_F and R_G. The forward transmission of a noninverting op amp stage shown in Fig. 19.2 is:

$$S_{21} = A_L = \frac{V_L}{V_i} = \frac{1}{2}\left(1 + \frac{R_F}{R_G}\right)$$

Op amp data sheets show open-loop gain and phase. It is the responsibility of the designer to know the closed-loop gain and phase. Fortunately, this is not hard to do. The data sheets many times include excellent graphs of open-loop bandwidth and most of the time include phase. Closing the loop produces a straight line across the graph at the desired gain, curving to meet the limit. The open-loop bandwidth plot should be used as an absolute maximum and should not be approached too closely.

An added benefit of using current-feedback amplifiers is that the values of R_F and R_G are specified in the data sheet. Please note, however, that the gains in the data sheet will not take 50 Ohm matching into account, so the stage gains will be ½ what the data sheet specifies for a given value of R_F and R_G.

19.4.2 Phase Linearity

Oftentimes, a designer is concerned with the phase response of an RF circuit. This is particularly the case with video design, which is a specialized type of RF design. Current-feedback amplifiers tend to have better phase linearity than voltage-feedback amplifiers:

- Voltage-feedback THS 4001: Differential phase = 0.15 degrees
- Current-feedback THS 3001: Differential phase = 0.02 degrees

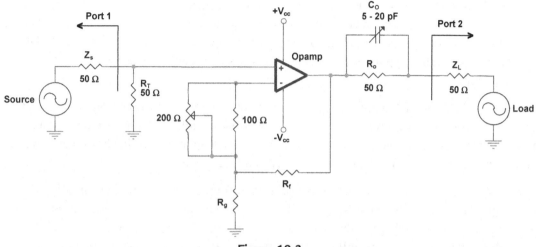

Figure 19.3
Frequency response peaking.

19.4.3 Frequency Response Peaking

Current-feedback amplifiers allow an easy resistive trim for frequency peaking that has no impact on the forward gain. Fig. 19.3 shows this adjustment added to a noninverting circuit. This resistive trim inside the feedback loop has the effect of adjusting the loop gain, and hence the frequency response without adjusting the signal gain, which is still set by R_F and R_G.

Values for R_F and R_G must be reduced to compensate for the addition of the trim potentiometer, although their ratio and hence the gain should remain the same. The adjustment range of the pot, combined with the lower R_F value ensures that the frequency response can be peaked for slight current-feedback amplifier parameter variations.

19.4.4 −1 dB Compression Point

The −1 dB compression point is defined as the output power, at a fixed input frequency, where the amplifier's actual output power is 1 dBm less than expected. Stated another way, it is the output power at which the actual amplifier gain has been reduced by 1 dB from its value at lower output powers. The −1 dB compression point is the way RF designers talk about voltage rails.

Op amp designers and RF designers have very different ways of thinking about voltage rails, which are related to the requirements of the systems that they design:

- An op amp designer, interfacing op amps to data converters, for example, takes great pain not to hit the voltage rail of the op amp, thus losing precious codes.

- An RF designer, on the other hand, is often concerned with squeezing the last half dB out of an RF circuit. In broadcasting, for example, a very slight increase in dBs means a lot more coverage. More coverage means more audience and more advertising dollars. Therefore, slight clipping is acceptable as long as resulting spurs are within FCC regulations.

Standard AC-coupled RF amplifiers show a relatively constant -1 dB compression power over their operating frequency range. For an operational amplifier, the maximum output power depends strongly on the input frequency. The two op amp specifications that serve a similar purpose to -1 dB compression are V_{OM} and slew rate.

At low frequencies, increasing the power of a fixed frequency input will eventually drive the output "into the rails"—the V_{OM} specification. Oftentimes, V_{OM} is broken up into separate low and high clipping levels, which are specified as V_{OL} and V_{OH}, respectively. At high frequencies, op amps will reach a limit on how fast the output can transition (respond to a step input). This is the slew rate limitation of the amplifier. The op amp slew rate specification is divided by two, because of the matching resistor used at the output.

As is the case for op amps used in any other application, it is probably best to avoid operation near the rails, as the inevitable distortion will produce harmonics in the RF signal—harmonics that are probably undesirable for FCC testing. That said, if harmonics are still at an acceptable level at the -1 dB compression point, it can be a very useful way to boost power to a maximum level out of the circuit.

19.4.5 Noise Figure

The RF noise figure is the same thing as op amp noise, when an op amp is the active element. There is some effect from thermal noise in resistors used in RF systems, but the resistor values in RF systems are usually so small that their noise can be ignored.

Noise for an op amp RF circuit is dependent on the following:

- the bandwidth being amplified
- gain

This example assumes the 11.5 $nV/\sqrt{Hz}$ op amp. The application is a 10.7 MHz IF amplifier. The signal level is 0 dBV. The gain is unity.

Fig. 19.4 is extrapolated from real data. The 1/f corner frequency, in this case, is much lower than the bandwidth of interest. Therefore, the 1/f noise can be completely discounted (assuming that filtering removes any noise that would cause the amplifier or data converter to saturate).

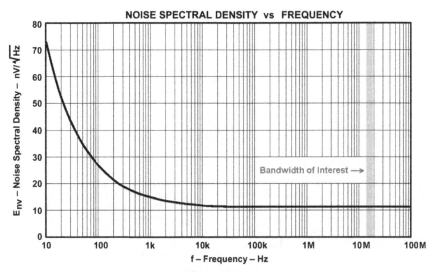

Figure 19.4
Noise bandwidth.

Table 19.1: Noise for Various Bandwidths

BW (kHz)	Ein (µV)	S/N (dB)
280	6.09	−104.3
230	5.52	−105.2
180	4.88	−106.2
150	4.45	−107.0
110	3.81	−108.4
90	3.45	−109.2

For narrow bandwidths, noise may be quite low! Various bandwidths are shown in Table 19.1.

There is an advantage to be had from reducing the bandwidth. Noise is amplified by the gain of the stage. Therefore, if a stage has high gain, care must be taken to find a low noise op amp. If the gain of a stage is lower, then the noise will not be amplified as much, and a less expensive op amp may be suitable.

19.5 Wireless Systems

Fig. 19.5 shows an example of a dual-IF cellular receiver. The receiver contains two mixer stages reminiscent of a classic double conversion superheterodyne receiver.

The RF designer wishing to use op amps for some of the gain stages needs to realize the current state of the art in op amp technology. Op amps are excellent choices for driving

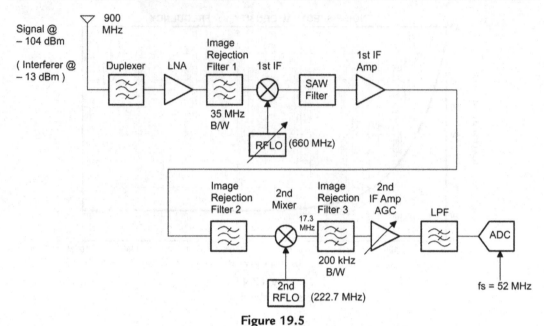

Figure 19.5

Typical GSM cellular base station receiver block diagram.

the analog-to-digital converter (ADC), low-pass filtering, and even second IF gain. But when you attempt to use op amps for the first IF frequency of 660 MHz in Fig. 19.5, you will quickly run out of open-loop gain. This limits the RF designer to using op amps in the final stages of that design, although unity gain buffers can be employed for impedance matching at higher frequencies. For the most part, though, op amps are not suitable for UHF applications.

19.5.1 Broadband Amplifiers

Current-feedback amplifiers are the component of choice for broadband amplifiers. The THS3202 was tested for its wide bandwidth and fast slew rate. A THS3202 evaluation board made a convenient platform on which to construct the circuits presented in here. The key questions explored just how much gain an op amp—based circuit is capable of, and over what frequency range? The circuit in Fig. 19.6 was used to explore these questions. It was first configured as a single stage, consisting of device A of a dual package, a 301 Ω feedback resistor, a 16.5 Ω gain resistor, and a 49.9 Ω back termination resistor. This configuration produces an amplifier gain of 20, a stage gain of 10 when connected to a 50 Ω monitoring device.

Note the simplicity of this circuit compared to traditional RF circuitry. Provide the op amp, termination and decoupling components, and two resistors and the circuit is done!

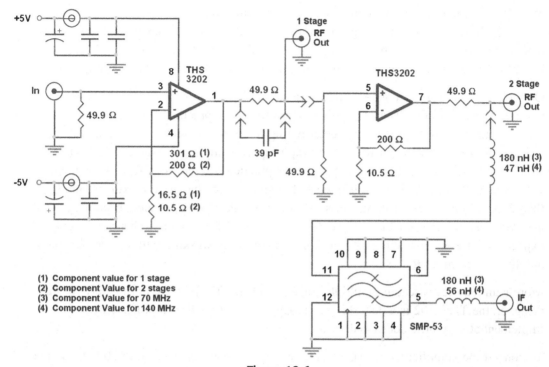

Figure 19.6
Broadband RF IF amplifier.

The 301 Ω (R_F) and 16.5 Ω (R_G) resistors are all that are required to set the stage gain. The stage's gain can be set precisely by the resistors alone, one of the op amp–based design's strong points. This circuit produces the lower amplitude curve in Fig. 19.7.

The op amp stage's voltage gain itself is 20, but this is cut in half by the back termination resistor's action in combination with the load. The RF amplifier's −3 dB point is about

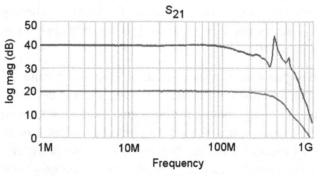

Figure 19.7
Wideband response.

390 MHz. If a flat gain over frequency is required, this circuit is only usable to about 200 MHz. Input and output voltage standing wave ratio (VSWR) values are better than 1.01:1 for most of the bandwidth, only degrading to about 1.1:1 near 200 MHz. S_{12} is −75 dB over most bandwidth, only degrading to −50 dB near the bandwidth limit.

One might wonder if more gain could be coaxed from the stage by lowering the gain resistor (R_G) even more. The answer is yes, but there is a practical limit. Remember that the feedback resistor (R_F) is a large determining factor for current-feedback amplifier stability. Remember also that R_G has to drop proportionally more. One can see that it would not be long until R_G's value becomes impractically small. Lab tests were attempted with various R_F values. The result indicated there is no advantage to making it smaller than 200 Ω. Below that, peaking starts to occur regardless of R_G's value, becoming worse and worse as the resistance is made lower and lower. This is exactly what one would expect, because one of the things a designer cannot do in working with current-feedback amplifiers is to make R_F a short.

More gain requires cascading multiple stages of THS3202 op amps. Fortunately for the designer, the THS3202 is a dual device, making a two-stage RF amplifier easy to implement at very little additional cost.

To convert the amplifier to two stages, the feedback resistor is lowered to 200 Ω, and the gain resistor lowered to 10.5 Ω. A second stage is then connected to the first, using the same values for the feedback and gain resistors. Isolation is accomplished by using interstage termination resistors. The optional 39 pF capacitor provides peaking to compensate for some high-frequency roll-off. Unfortunately, it also creates capacitive load on the first amplifier output. This increases the first stage's tendency to peak as seen in the upper curve of 0. This indicates a tendency toward instability, and it manifests itself by poorer IP3 performance. If maximum IP3 performance is needed, the designer should delete the capacitor and live with less bandwidth from the stage. The other S parameters for this circuit are similar to the single op amp's case.

At this point, it is important to talk about signal levels. All of the above points in the direction of impressive performance, but if these benefits only apply to very small signals they are hardly benefits at all!

The signal level a designer can pass through an op amp is determined by its input and output voltage rails, as described by the device's data sheet. They form a set of high- and low-voltage "hard clipping" points for the signal as it passes through the operational amplifier. Consequently, the −1 dB compression point occurs very soon after any voltage rail limitation has been breached. The wise designer will not attempt to squeeze that last dB from the stage because the hard clipping points may produce substantial harmonics.

For a THS3202 stage, the amplifier's output can swing ± 3.2 V, therefore the output of the stage can swing ± 1.6 V. This corresponds to 14 dBm of output power.

19.5.2 IF Amplifiers

The gain circuit shown in Fig. 19.6 can be cascaded with SAW (surface acoustic wave) filters to form high-performance IF stages. The only design consideration is the insertion loss of the filter, which may not be a constant value from part to part, or from batch of parts to batch of parts. If precise gain is needed from the stage, the designer may need to include a trim resistor in one or both stages. This trim adjustment, however, will not affect the stage's tuning except for a slight effect on its upper frequency limit.

It is easy to cascade the dual wideband RF amplifier with SAW filters. Sawtek conveniently provided an EVM board, which greatly simplified prototyping. All that was required was to provide a short SMA to SMA cable between the two boards. It is important to place the SAW filter element after the gain stage, so noise generated in the op amp circuitry is filtered with the same response as the signal. If the gain stage were placed after the SAW filter, the amplifier stage's broadband noise response would be passed to the next stage instead of a filtered, narrow-band response.

The IF amplifiers of 70 and 140 MHz find use in cellular telephone base stations and satellite communications receivers. A Sawtek 854660 filter was selected for 70 MHz, and a Sawtek 854916 filter was selected for 140 MHz. These filters require input and output inductors and operate with standard 50 Ω input and output. A 70 MHz SAW filter produces the upper response curve, and the 140 MHz filter produces the lower response curve shown in Fig. 19.8:

The most outstanding response curves featured in Fig. 19.8: IF amplifier response is that they are almost identical in shape to the filter curves provided by Sawtek. Narrow-band response is shown in 0, but there is virtually no harmonic content when the broadband response is examined. In other words, the amplifier is providing gain while not adding undesirable harmonic content. The insertion loss of the 70 MHz SAW filter is about 7 dB. The insertion loss of the 140 MHz SAW filter is only 8 dB, but the gain circuit itself is starting to roll off at this frequency, accounting for the rest of the loss. Careful examination of the lower pass-band shows the slight roll-off due to the broadband stage characteristic.

19.6 High-Speed Analog Input Drive Circuits

Communication ADCs, for the most part, have differential inputs and require differential input signals to properly drive the device. Drive circuits are implemented with either RF transformers or high-speed differential amplifiers with large bandwidth, fast settling time, low output impedance, good output drive capabilities, and a slew rate of the order of

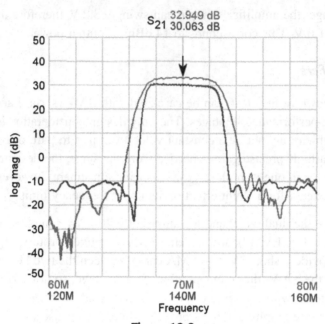

S_{21} 32.949 dB
      30.063 dB

Figure 19.8
IF amplifier response.

1500 V/μS. The differential amplifier is usually configured for a gain of one or two and is used primarily for buffering and converting the single-ended incoming analog signal to differential outputs. Unwanted common-mode signals, such as hum, noise, DC, and harmonic voltages are generally attenuated or canceled out. Gain is restricted to wanted differential signals, which is often 1–2 V.

The analog input drive circuit, as shown in Fig. 19.9, employs a THS4141 device. This device offers fast speed, linear operation over a wide frequency range, and wide power-supply voltage range, but draws slightly more current than a BiCMOS device. The −3 dB bandwidth is 120 MHz measured at the output of the amplifier. The analog input V_{IN} is AC-coupled to the THS4141 and the DC voltage V_{ocm} is the applied input common-mode voltage. The combination R_{47}–C_{57} and R_{26}–C_{34} are selected to meet the desired frequency roll-off. If the input signal frequency is above 5 MHz, higher-order low-pass filtering techniques (third-order or greater) are employed to reduce the op amp's inherent second harmonic distortion component.

19.7 Conclusions

Op amps are suitable for RF design, provided that the cost can be justified by the flexibility they offer the designer. They are more flexible to use than discrete transistors, because the biasing of the op amp is independent of the gain and termination.

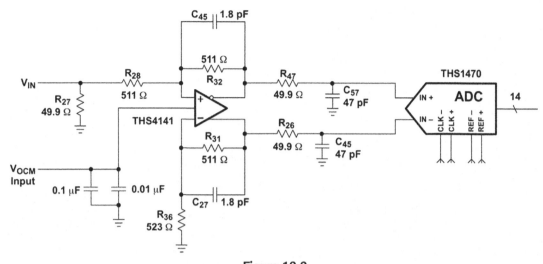

Figure 19.9
Single-ended to differential output drive circuit.

Current-feedback amplifiers are more suitable for high-frequency, high-gain RF design, because they do not have the gain/bandwidth limitation of voltage-feedback op amps.

Scattering parameters for RF amplifiers constructed with op amps are very good. Input and output VSWR are good because the effects of termination and matching resistors can be made independently of stage biasing. Reverse isolation is very good, because the RF stage is made of an op amp consisting of dozens or hundreds of transistors, instead of a single transistor. Forward gain is very good with a current-feedback amplifier.

Special considerations apply to RF design that do not normally apply to op amp design—the phase linearity, the −1 dB compression point (as opposed to voltage rails), the two-tone third-order intermodulation intercept, peaking, and noise bandwidth. In just about every case, the performance of an RF stage implemented with op amps is better than one implemented with a single transistor.

CHAPTER 20

Designing Low-Voltage Op Amp Circuits

20.1 Introduction

Op amp application circuits are a means to an end—the end being a product that is sold. More and more products are being manufactured that operate off of batteries. Consumers want smaller, lighter products that still have the functionality of previous generations of products—the cell phone being an excellent example. The four pound brick-shaped cell phone with an antenna has been supplanted by ever smaller "smart" phones. Cell phones may not be the best example because op amps are integrated inside of ICs that are highly integrated. Make no mistake, though op amp gain and filter circuits still exist inside those ICs! At some point, a designer still had to design gain and filter circuits that were eventually integrated into an IC. Since these IC's operate off of a single battery, they are almost certainly single-supply circuits, harkening back to an earlier chapter.

The trend is not only toward single supplies, but also lower-supply voltages. The first Regency transistor radio operated off of a 22.5 V battery, my childhood transistor radio operated off of a 9 V battery, and my latest portable operates off of a single 1.5 V battery, and yet has longer battery life—and is smaller, lighter, and far more sensitive than my previous radio. A secondary trend is toward lower power consumption, so a product operates longer between battery changes or battery charges, yet with no compromise in performance.

This trend has meant lower voltage rails and power consumption inside of op amps. The familiar ±15 V supplies have been augmented with product offerings that operate off lower and lower supply voltages, oftentimes not even capable of operating off of ±15 V. At first glance, you might think that this is great, no more requirement for separate analog supplies, it is wonderful to be able to run op amps off of the same voltage as logic circuits. But there is a down side to operating off of a lower supply voltage. This book has already introduced you to the op amp output voltage rail specifications—V_{OH} and V_{OL}. These become more and more critical as op amp supply voltages go lower and lower. There are similar limitations on the input voltage range. This chapter will give you the tools you need to make design decisions that will use the voltage ranges available and optimize those designs to work well within a power supply "budget."

Op Amps for Everyone. http://dx.doi.org/10.1016/B978-0-12-811648-7.00020-0

20.2 Critical Specifications

There are several specifications that are critical to understand when working on systems that have a limited power supply range. These will be discussed in order of importance.

20.2.1 Output Voltage Swing

Rail-to-rail output (RRO) voltage swing is desirable for at least two reasons. First, the dynamic range (DR) can achieve the maximum obtainable value if the op amp is RRO. Second, RRO op amps can drive any converter connected to the same power supply if the impedance is compatible. The schematic of an RRO op amp output stage, part of the TLC227X, is shown in Fig. 20.1.

The RRO characteristic is achieved in the construction of the op amp output stage. A totem pole design that has upper and lower output transistors is used, and the output transistors are a complimentary pair. Each transistor in the pair is a "self-locking" type of transistor operating in the common-source mode. Consider the p-channel output transistor; as long as this transistor has a drain-source resistance, it forms a voltage divider with the load resistance. When the load is a very large resistor or if the output current flow is very small, the voltage drop across the output transistor can be neglected. Output current flows through the output transistor, and because current drops a voltage (V_{DS}) across the drain-source resistor, the output voltage swing is reduced. The voltage drop subtracts from the power supply voltage, reducing the output voltage to less than RRO.

20.2.2 Dynamic Range

DR (dynamic range) is affected by the V_{OH} and V_{OL} specifications of the op amp, but is a broader subject. Start with the maximum output voltage swing equals $V_{OUT}(MAX)$.

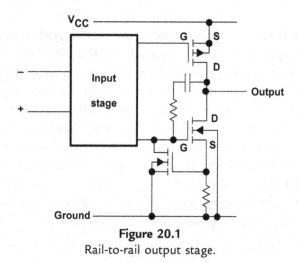

Figure 20.1
Rail-to-rail output stage.

This output voltage swing is defined as the maximum output voltage the op amp can achieve (V_{OH}) minus the minimum output voltage the op amp can achieve (V_{OL}). V_{OH} and V_{OL} are easily obtainable from an op amp IC data sheet. This yields Eq. (20.1):

$$V_{OUT}(MAX) = V_{OH(MIN)} - V_{OL(MAX)} \tag{20.1}$$

Eq. (20.1) can be used to illustrate the role that power supply voltage plays in limiting the DR. $V_{OH(MIN)}$ is the most positive power supply voltage minus the voltage drop across the upper output transistor, thus $V_{OH(MIN)}$ is directly proportional to the most positive power supply voltage. For any op amp, the output voltage swing is directly proportional to the power supply voltage, thus, in a given op amp, the DR is directly proportional to the power supply voltage.

An op amp may have V_{OH} and V_{OL} specifications that are close to its power supply rails, but not necessarily equal. This is because the output transistors are real-world devices that have some voltage drop. So there will never be a true output rail-to-rail op amp, unless it is an op amp that has internal DC−DC converters to boost its internal rails. This is a lot of trouble to go to just to achieve a perfect rail-to-rail specification. Careful system design can achieve required performance without resorting to such extreme measures. As an op amp has smaller V_{OH} and V_{OL} specifications, the semiconductor design challenges get greater and greater, which can lead to increased power consumption and susceptibility to latch-up.

20.2.3 Input Common-Mode Range

Just as there are limitations on the output voltage swing of an op amp, there are limitations on the input voltage range that may be applied to it. This can be troublesome, particularly if the input source is referenced to ground and has small amplitude. Fortunately, there are such things as true rail-to-rail input op amps. There are trade-offs associated with them, however, and it may be sufficient to use an op amp that includes either ground or positive power in its common-mode range, but not both.

Fig. 20.2 shows the simplified input circuitry for an op amp that can go to the ground rail (but not positive power rail). The PNP input transistors are biased by the emitter current source. If the positive input is connected to ground bias current still flows, and the transistor stays active.

An op amp with an NPN input stage works in a similar way around the positive supply rail. It can sense voltages close to V_{CC} and maybe slightly above V_{CC}, but it will not work when it is within 1.5 V of ground. The solution for this problem is to include parallel input circuits as shown in Fig. 20.3.

There are both PNP and NPN differential amplifiers used in the input stages of the rail-to-rail input (RRI) op amp, thus the RRI op amp can operate above and below the power supply

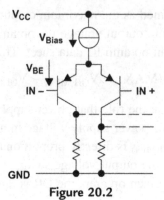

Figure 20.2
Input circuit of a non rail-to-rail input op amp.

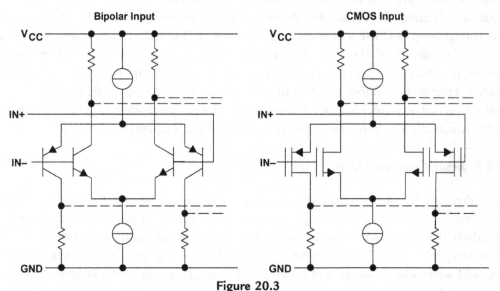

Figure 20.3
Input circuit of a rail-to-rail input op amp.

voltage. As Fig. 20.3 shows, the parallel input stages can be made in bipolar or MOS technology. Inclusion of complementary differential input amplifiers achieves V_{ICR} exceeding the power supply limits, but there is a penalty to pay in input bias current, input offset voltage, and distortion. This has serious implications for low-level DC-coupled systems.

The input stages operate in three different ranges:

- When the input voltage ranges from about -0.2 to 1 V, the PNP differential amplifier is active and the NPN differential amplifier is cutoff.
- When the input voltage ranges from about 1 V to ($V_{CC} - 1$ V), both the NPN and PNP differential amplifiers are active.

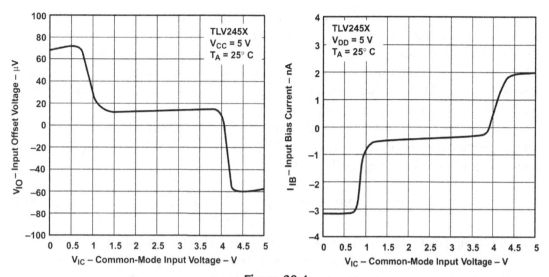

Figure 20.4

Input offset voltage and bias current changes with input common-mode voltage.

- When the input voltage ranges from about ($V_{CC} - 1$ V) to ($V_{CC} + 0.2$ V), the NPN differential amplifier is active and the PNP differential amplifier is cutoff.

Fig. 20.4 shows the input bias current and input offset voltage as a function of the input common-mode voltage.

RRO op amps cannot drive heavy loads and maintain their RRO capability because of the voltage dropped across the output transistors. Load resistance or output current is a test condition when the measurement of an op amp's output voltage swing is made. The size of the load resistor or output current is a measure of the op amp's ability to retain its RRO capability while sourcing or sinking an output current. When selecting an RRO op amp, you must consider the load resistance or output current required because these conditions control the output voltage swing.

When an op amp is made that has RRI and RRO capability, it is called a rail-to-rail input/output op amp. This long name is shortened to RRIO.

20.2.4 Signal-to-Noise Ratio

Noise sets a limit on the information and signals that can be handled by a system. The ability of an amplifier, receiver, or other device to discern a signal is degraded by noise. Noise mixed with the incoming signal, noise generated by the op amp, resistor noise, and power supply noise ultimately determine the size of the signal that can be recovered and measured.

Noise fluctuates randomly over a period of time, so instantaneous signal or noise levels do not describe the situation adequately. Averages over a long period of time (root mean squared or RMS) are used to describe both the signal and the noise. Signal-to-noise ratio (SNR) was initially established as a measure of the quality of the signal that exists in the presence of noise. This SNR was a power ratio, and it was established at the output of a circuit. The SNR that we are interested in is a voltage ratio because the impedance is constant, and it is established at the input to the op amp. This means that all noise voltages, including resistor noise voltage, must be calculated in RMS volts at the op amp input. The SNR is given in Eq. (20.2):

$$SNR = 20 \, \text{Log}_{10} \left(\frac{V_{\text{SIGNAL}}}{V_{\text{NOISE}}} \right) \tag{20.2}$$

A good starting point is to test the analog signal chain with a terminated input, whether that termination is to ground, V_{REF}, or some other common-mode point. If there is a characteristic input impedance, such as 50 Ω, it should be included in the termination. If all goes well, the level of noise induced by the signal chain should be very low. Careful attention to layout and decoupling techniques is important for low-level signals.

Most signals are established by a transducer; a device that senses a change in a variable and converts that change into a voltage change. Transducers also convert some of their physical surroundings into a noise voltage that is combined with the signal. Noise from the physical surroundings of the transducer, unless its nature is well known, is almost impossible to separate from the transducer signal. When transducers are connected to the electronics, cabling picks up noise and crosstalk from other signals, and some transducers like thermocouples can pick up noise from the connecting junctions. Thus, the signal is never clean as it enters the electronics.

Transducers often have a very small output voltage swing, so when the transducer output voltage swing is converted to least significant bits (LSBs) the noise voltage should be very small compared to an LSB. Consider a temperature transducer that has a 10 mV swing over its range. When the transducer output voltage swing is considered to be the full-scale voltage (FSV) of an ADC, the LSB is very small as is shown in Eq. (20.3) for a 12 bit (N) ADC.

$$LSB = \frac{FSV}{2^N} = \frac{10 \, \text{mV}}{2^{12}} = \frac{10 \, \text{mV}}{4096} = 2.44 \, \mu V \tag{20.3}$$

The op amp for this application must be a very low noise op amp because an op amp with a $20 \, \text{nV}/\sqrt{\text{Hz}}$ equivalent input noise voltage and a bandwidth of 4 MHz contributes 40 μV of noise. This high noise contribution is why extensive filtering and "optimally" low bandwidth is found desirable in the input stages of some electronic systems. If there is power supply noise, some of that noise passes through the op amp to its input. The power supply noise is divided by the power supply rejection ratio, but there is always a residual noise component of the power supply on the op amp input as shown in Eq. (20.4) where k_{SVR} is 60 dB

$$V_{PS(INPUT)} = \frac{V_{PS}}{k_{SVR}} = \frac{10 \text{ mV}}{1000} = 10 \text{ μV} \tag{20.4}$$

When interfaced with a data converter, the noise level ultimately determines how many bits of the digital code are meaningful—it does no good to purchase a data converter with a large number of bits when the level of noise in the analog signal chain makes several of the LSBs irrelevant. If this is the case, you can

- replace amplifiers in the signal chain with lower noise devices;
- employ averaging over a large number of samples to reduce the contribution of noise;
- use a data converter with fewer bits to save money, if the noise level is already the lowest that can be obtained, or is acceptable to system requirements.

20.3 Summary

It is extremely hard to achieve large DR when the application is limited to a low power supply voltage. In an attempt to approach the DR obtained by ±15 V power supply designs, the new op amp designs put increased emphasis on the output voltage swing. However, the lower supply voltages have put severe limitations on DR, and attempts by semiconductor manufacturers to lower the V_{OH} and V_{OL} limits have regained only a fraction of that DR, putting the burden on you to be very careful in selecting the operating range of their circuits.

RRI op amps are able to work with transducers connected to the power supply rails. As long as the AC component of the transducer output voltage does not exceed the input common-mode range of the op amp, the design is reliable. RRI op amps are troubled by distortion introduced by the change in bias current, input offset voltage, and gain, but their contribution to the system's signal handling capability is invaluable. RRO op amps yield the highest output voltage swing of any series of op amps. But beware: RRO op amps are specified at a load resistance or current, and the output voltage swing decreases dramatically when the load resistance or current is increased. RRIO op amps contain the input and output features of RRI and RRO op amps. They also contain the drawbacks of both features.

The final thing to be considered is that low power supply voltage invariably means single-supply design, and single-supply design is tougher than split-supply design. Remember to get the two sets of data points, put them in simultaneous equations, solve for the slope and intercept, select the circuit configuration, and calculate the component values. Digital-to-analog converters are little different because you have to account for the polarity of the current, but their design generally follows the same procedure.

CHAPTER 21

Extreme Applications

21.1 Introduction

Not every circuit is destined to be used in a nice environment. Some are destined to go into extreme environments, or applications where failure is simply not an option. While military op amp amps have been available for a long time for deployment in temperatures that range from −55 to 125°C, military environments are no longer the most severe. Space, oil exploration (downhole), and geothermal environments have much greater temperature extremes, as well as potentially damaging mechanical vibration and shock requirements. Space, medical, and automotive applications demand high levels of reliability. Service is impossible for space electronics. Medical devices can put a patient's life in danger if they fail. Automotive applications can cause traffic accidents if they do not work properly. This chapter is intended to be an introduction to the topic—the techniques here will be a starting point for extreme design, not a guarantee of operation of a given system in an extreme environment. That is left to the designer, who must adhere strictly to a set of qualification tests developed by the company for which they work. A system is suitable for extreme environments only after such qualification tests have been passed, not before.

21.2 Temperature

This section will focus mainly on high-temperature design, because that is the environment the author is most familiar with. When dealing with high temperatures, some of op amp parameters will degrade, and the trend will accelerate on some parameters as the temperature increases. Put another way—the degradation of parameters is not necessarily linear, and the temperature at which a given parameter begins to accelerate toward unacceptable may differ between parameters. Curiously, other parameters may remain unchanged. Table 21.1 is based on a real-world op amp, which is available in commercial and high-temperature grades; however, the exact op amp will not be mentioned because its specifications are subject to change by the manufacturer.

Not that some, but not all, parameters have changed at high temperatures.

For a full understanding of these op amp parameters, look at Appendix B. But I will briefly describe what these parameters mean to you for this particular op amp. The news is definitely not all bad!

Op Amps for Everyone. http://dx.doi.org/10.1016/B978-0-12-811648-7.00021-2

Table 21.1: Comparison of Ambient and High-Temperature Parameters

Parameter	Commercial Grade	High Temperature	Units
V_{OS}	±125	±260	µV
$\Delta V_{OS}/°C$	1.5	2	µV/°C
I_B	±200	±250	nA
I_{OS}	±150	±150	nA
e_n	1.1	1.1	$nV/\sqrt{Hz}$
I_n	1.7	1.7	$pA/\sqrt{Hz}$
CMRR	120	113	dB
A_{OL}	110	110	dB
GBW	45	45	MHz
SR	27	27	V/µS
t_s	400	580	nS
THD + N	0.000015	0.000015	%
V_{OH}	(V+) − 0.2	(V+) − 0.2	V
V_{OL}	(V−) + 0.2	(V−) + 0.2	V
I_{SC}	+30/−45	+30/−45	mA
Z_O	5	5	Ω
IQ	6	7.5	mA
PSRR	140	140	dB

21.2.1 Noise

The specifications related to noise—e_n, I_n, and THD + N—do not change at high temperature. A low-noise op amp, in this case, remains a low-noise op amp (which probably drove the selection of this part in the first place). Parameters related to conducted sources of noise such as power supply rejection ratio (PSRR) and common-mode rejection ratio (CMRR) are not affected or barely affected—another lucky break.

21.2.2 Speed

The parameters related to speed—gain bandwidth product (GBW) and slew rate (SR)—do not change at high temperature. While this is not a particularly high-speed op amp, it retains the speed it had at room temperature. In some cases, I have actually seen the bandwidth of a high-speed op amp increase with high temperature! Not a lot, just a few percent. But it is one of those rare cases where the performance of the part may actually be better at high temperature than it is at room temperature.

21.2.3 Output Drive and Stage

There is some more good news for this device—its output stage appears unaffected by high temperature. V_{OH}, V_{OL}, I_{SC}, and Z_O all are the same at high temperature. This is good news if you plan to drive a heavy load, or if you need a rail-to-rail output. But be careful, most rail-to-rail op amps degrade under load conditions. It is always wise to check the graphs in the data sheet, and doubly so for parameters like these that may interact.

21.2.4 DC Parameters

The supply current increases at high temperatures, which will cause battery life to decrease in battery-operated systems. An op amp, which was not targeted for precision DC transducer applications, has higher values of offset voltage and input bias current at high temperature. Beware of the input bias current—it can vary by orders of magnitude at temperature for some devices. You ignore this effect at the risk of a nonfunctional product.

21.2.5 Most Important Parameter of All

The parameter that degrades at temperature is one of the most important: **LIFE**. An integrated circuit operating continuously at high temperature is slowly destroying itself. Integrated circuits operating at high temperatures can have lifetimes in the thousands or even hundreds of hours—orders of magnitude less than the same device operated at ambient temperature. A mission profile that realistically projects the amount of time the circuitry will be operated at elevated temperatures is imperative. A downhole application is a good example. The amount of time electronics is actually subjected to elevated temperature may be relatively short in relation to the amount of time the tool is used downhole. Temperature inside the hole rises (and falls) slowly, not reaching maximum temperature until the electronics is actually at maximum depth. On the other hand, a space application may see very rapid temperature fluctuations that create thermal shock conditions, which can rapidly degrade device packaging.

21.2.6 Final Parameter Comments

Just because these particular parameters varied the way they did for this particular op amp, does not mean these variations will hold true for every op amp. It is your responsibility to do a "what-if" analysis for every parameter you have any doubt about. Designing for worst case is always a good idea, and this goes double for any extreme application, high temperature in particular. You may have to use a commercial grade component for prototyping purposes, because the high-temperature versions are much more expensive. The good news is that the high-temperature version of the part is probably fabricated from a commercial grade die that has been screened for high-temperature performance, so there is a very good chance it will work properly at high temperatures for short periods of time. In cases like that, the difference between a commercial grade device and a high-temperature device is packaging—the commercial grade die is screened for high-temperature performance and then placed in a high-temperature package.

Experts at companies specializing in these applications can be a valuable source of information. Oftentimes, a list of parts that have survived harsh environments in the past is available, and use of those parts is preferred—if at all possible.

21.3 Packaging

Another aspect of extreme applications that needs to be addressed has to do with packaging. This includes the integrated circuit itself, the board on which it is mounted, the solder which attaches it to the board, and other aspects. All of these must be seriously analyzed and understood, or the circuitry may not survive in the extreme environment.

21.3.1 The Integrated Circuit Itself

Many integrated circuit manufacturers are starting to supply products for extreme temperature applications. The portfolios, however, are still tiny, and costs may be very high. The ICs may be packaged in specialized packages that do not match standard packages or may have different pinouts. Fortunately for you, the vast majority of integrated circuits can operate at extreme temperatures, at least for short periods of time.

The silicon die is seldom the limiting factor in high-temperature operation. I have only encountered one IC that does not work at high temperature—all of the others that I have tested do, except those with intentional overtemperature shutdown "protection." Voltage regulators are the most notorious for this "feature." There is usually no need for it, and it severely limits the number of voltage regulators that can be used.

Problems begin with the connection of the bond wire to the die. High temperatures accelerate undesirable metal migration, eventually leading failure at the connection. High currents exaggerate the problem even more. Proper high-temperature bonding techniques are a must at high temperatures. Related to this is the selection of bond wire. Specialized alloys may be necessary for high-temperature operation. Both these alloys and bonding techniques are closely guarded trade secrets, and you will have to work carefully with IC manufacturers to make sure proper attention is paid to both issues.

You may want to look up "Kirkendall voiding" to more fully understand the bond wire/die degradation effect. To gain more reliability from critical parts, some companies employ "die harvesting" techniques, or use specialized die harvesting companies, to rescue good semiconductor die from inadequate bond wire/cheap packages. They then place the die in high-temperature ceramic packages. It is far preferable, of course, to buy the ICs in die form, but not all companies offer this option, including one major supplier of analog ICs—effectively cutting off designers for extreme environments from using their products without die harvesting. If you do harvest die, or employ an outside firm to do the work, 100% testing of harvested die is imperative, as the process does not result in 100% yield.

Obviously, it is extremely important to work with a partner who has experience in high-temperature applications and knows the requirements for the package and bond wires.

21.3.2 The Integrated Circuit Package

One of the fundamental problems in high-temperature design is the thermal expansion characteristics of the various materials used. The silicon die itself has a thermal expansion characteristic, the package enclosing it has a thermal expansion characteristic, the bond wires have thermal expansion characteristics (affecting their length), etc. Adhesives have thermal breakdown characteristics. Plastics may out gas corrosive or alkali components. Repeated temperature cycles allow water to penetrate the packages. Cold temperatures make things brittle, and high vibration and shock may break something mechanically. All of this is a nightmare not only to IC manufacturers but for designers as well.

The best, and least expensive, solution may be to characterize a commercial part in a plastic package for its life span in the extended temperature environment. If a given "mission" is limited to a few hundred, or a few thousand hours and the environment is relatively well understood, a commercial or industrial grade IC in a plastic package may well survive in the vast majority of cases. But an IC destined for space where the "mission" is years long must be packaged in high-temperature packaging if there is to be any hope of survival.

High-temperature packaging is usually ceramic, with gold leads. The ceramic packaging has come to extreme applications from a military background, where ceramic through-hole and flatpack packaging has been used for decades. Both alternatives may be undesirable for applications with limited board space. New surface-mount ceramic packages designed for extreme temperatures are becoming available, but not all integrated circuits are available in these packages.

21.3.3 Connecting the IC

Now that you have your high-temperature IC, how do you place it onto a board? If you are used to soldering it onto an FR-4 PC board with Sn60Pb40 solder, you have just wasted your money. The board will disintegrate to ashes at high temperature, and your IC will fall off because the solder melts! High-temperature operation requires a complete rethinking of PC board materials and solders.

21.3.3.1 PC Board Design

The material of choice for high-temperature PC board is traditionally glass reinforced polyimide resin, but this material is porous and will absorb water. If a board gets wet, it is probably scrap. Gold plating is usually done on the traces, but it requires a nickel barrier between the copper and gold, or copper will dissolve into the gold, causing brittle solder joints. A new option—high-temperature FR-4—has been introduced by manufacturers, primarily for the street lighting application. This material is available in temperature ratings up to 180°C, so it may be suitable for high-temperature applications previously requiring polyimide.

Thermal expansion and contraction also takes place with PC boards, making the selection of hole size, via size and trace widths larger than most PC board designers are accustomed to working with. You may have considerable "pushback" from PC board layout departments, but it is important to hold your ground. Surface-mount components help with the situation by minimizing the number of through-holes, which act as barriers on all layers of the board. The mass of surface-mount components must be minimized, or they may tend to come off the board under high shock and vibration conditions. They also may have to be oriented in a common direction to avoid mechanical stress along an axis that might flex. As boards expand and contract, they may tend to break connections at pads and vias. Teardrops or "necking down" should be used to avoid stress points where traces connect to pads and vias.

21.3.3.2 Solders

Solders for high-temperature applications have been migrating toward or high melting point (HMP) solders. The best of these appears to be Sn05Pb92.5Ag2.5, which has a melting point of 280°C, 536°F. These solders require new tips; it is unwise to contaminate new solder joints with a mixture of HMP and older solders. HMP solder has also been described as acting like "silly putty that melts only once," a very good description coming from someone who has been frustrated by it. I have been forced to a Pb-free solder, Sn96Ag04, on more than one occasion to get a solder joint that behaves more like a Sn60Pb40 solder. It certainly is harder to flow through small holes and requires a new set of skills to work with. Just because you can solder does not mean you can solder well, and it certainly does not mean you can work with HMP solders! It is usually best to leave HMP soldering to technicians, assemblers, and third-party houses that can handle it.

21.3.3.3 Adhesives and Other Methods of Component Retention

A board for an extreme environment is never assembled completely when the components are soldered.

Toroidal inductors and transformers are commonly used, because they are usually custom wound on high temperature cores. Retaining them on a PC board usually involves drilling extra nonplated through-holes on the board, and lashing down with high-temperature lashing, retaining with a screw and top plate, or packaging the toroid in a housing which already has leads. Any of these techniques must ensure that vibration/movement in all possible axes is taken into account. They may be combined with adhesive as well.

Heat sinking is difficult in some applications, such as downhole where space is at a premium. Still, there is a very good heat sink located very close to the electronics—the tool itself. Just remember that it is more of a heat spreader than a heat sink because it is at elevated temperature in the hole. But there is usually a supply of drilling fluid, which will also act to draw heat away. Finding a way to thermally attach a component to the metal of the tool is a challenge, and every company probably has an approved method. Space applications cannot use airflow past a heat sink due to the vacuum of space. Heat will flow out of the heat sink, but at a much reduced rate. Heat sinking is certainly not an option in implantable medical devices, so low-power design is imperative. Chances are, the implantable device is already designed to be at low power due to battery life constraints.

Adhesives and potting materials are oftentimes used in high-shock environments to retain components that do come loose, so they will not be distributed to critical areas as mechanical nuisances or shorts. The board from which they came may or may not be disabled by the loose component, if it was a decoupling capacitor or something. But the loose component must not disable anything else! High temperature adhesives and potting must also be selected for low conductivity if high voltages are present.

21.4 When Failure Is Not an Option

The Mythbuster motto is "failure is always an option." That may be true, but oftentimes our job as designers is to run counter to that truism as much as possible. There are no set design techniques to ensure that failure is never an option, because some set of circumstances will cause any circuit to malfunction. The only way to counter this is to make failure happen only after a specific set of circumstances apply—the default mode if something fails is something safe.

Extreme applications include a class of applications where failure would jeopardize life or property. Medical equipment falls into this category—if failure of an implantable device would jeopardize a patient's life, or malfunction of hospital equipment that could somehow hurt or kill a patient. Property damage could include things like malfunctioning circuitry causing an oil spill in a remote location. Legal implications alone from such events could result in a company going out of business, its owners imprisoned if guilty of negligence. If you are a licensed professional engineer (PE), you could also be held personally liable.

There are many techniques that can be used to reduce the probability of failure. A good place to start is to examine the power supply/battery. A reliable circuit without reliable power is one that is doomed to fail. Reliable power is the cornerstone of reliable electronics and must be done first.

After that, where do you go? "Worst-case" design should already be standard practice—never design a circuit around typical values. Beyond that:

- If an electronic design must work under any set of circumstances, develop a mind-set of "what if." What if power fails—will the system default to a state of least damage?

A variable pulse rate pacemaker should revert to single frequency, not shutting off, if at all possible. A patient would be impaired, but not dead. I remember working on a large wire wrap computer board. To prove its self-diagnostic capabilities, a manager would cut a random wire, and see how long it would take the team to find the failure. Such exercises might seem annoying, but at least one can rule out or compensate for some potential failure modes.

- Beware of transistor-related specifications. A lot of designers like to be clever and use transistors, displaying their prowess at biasing. Some transistor parameters, however, vary by orders of magnitude over temperature extremes. When I utilize such circuits, if my transistor switch or amplifier stage will not work with biasing resistors from 100 Ω to 1 MΩ, I redesign until it does. And do not assume you are immune if you only use op amps. Input bias current is a specification that can vary unexpectedly over temperature. Be wary of it.
- Prototype everything. Simulations are starting points, not ending points.
- Innovation and parts pushing the state of the art are not good ideas. Sticking to older, proven technology is smarter.
- Redundancy is a powerful technique. Two power supplies sharing current, but each one capable of providing full load current independently, will greatly increase reliability. Two completely redundant signal chains will allow a measurement to be made even if one signal chain goes down.
- Rely on some set of circumstances that is counter to randomness. I recall a valve control system for an underwater wellhead that would only actuate valves when a timer operated in a narrow range of frequencies, charging a capacitor at just the right rate. Any other set of circumstances, and the value remained tightly closed, preventing environmental disaster. Pretty darn clever, and simple! Other techniques might make critical actions happen only when a set of prime numbers is sent to a system—sequences of prime numbers being uncommon in nature.
- Exhaustive testing is imperative. When a specific test regimen is in place, following it to the letter will minimize liability to managers and outside agencies. When a test regimen is not already in place, establish one appropriate to the level of risk involved and have it approved by corporate legal departments prior to selling a single product.
- Do not accept arbitrary, over aggressive schedules. When substantial risk of injury and/ or property damage is at stake, it is foolish to allow yourself to be rushed. If something bad happens because you did not take the time to investigate, you are the one who will be liable, especially if you have a PE in front of your name. The very managers who were pushing you to meet the schedule will be running for the hills in the event of problems. If you are not prepared financially to join them, just say *"no!"* Remember the Challenger disaster. Engineers were afraid to speak up. Lives were lost. Unfortunately, this is only one of many cases where lives were lost by engineers that were intimidated by managers into rubber stamping bad decisions.

Some very sobering thoughts, I know. But when you are designing such systems, the buck stops with you. Take ownership of the responsibility you took on.

21.5 When It Has to Work for a Really Long Time

Another extreme application has nothing to do with temperature, shock, or vibration extremes. Extreme duration in time is another very challenging area of design. It might be a space probe like a Mars Rover, an implantable medical device, a monitor on a remote pipeline, a detonator on a mine floating in the ocean. In each case, the mission can be measured in years or even decades. In each case, the system has to work over the entire mission time. And in some cases, like a mine detonator, most of the work is done at the end of the device's life span.

Obviously, this produces extreme requirements on the power system, whether it is battery, inductively coupled super capacitors, nuclear, solar, or some other technique. Chances are that aspect of the design has been done for you, and you do not have to worry about it. But you can greatly enhance your value as a designer by minimizing the load you draw from such critical, limited resources.

Extremely low power consumption is a cooperative effort between power supply/battery engineers, digital designers, software/firmware coders, and, of course, those designing the analog signal chain. Many compromises and trade-offs will be involved, but the primary way to save power is by incorporating power switching into the design. The analog signal chain, for example, is only turned on when needed to acquire a signal from a sensor. The clock frequency of the system is adjusted based on need.

In the aforementioned mine detonator, the clock ran at a very low frequency, then an intermediate frequency when it detected motion, then full frequency to characterize "friend or foe." Clock circuits are also power hogs, and a special low current consumption clock oscillator was constructed. Op amps were very low power, and design compromises were made to accommodate the extra noise, larger offset, and lower bandwidth from such an op amp. The result was a system that only consumed power when it needed to and was extremely low power at all other times. When average power was considered, system lifetime could be measured in decades or even centuries, yet instantaneous power when running complex algorithms was comparable to any commercial product running data acquisition.

One area of concern to you as a designer of extremely low power consumption systems is the switching/monitoring circuitry that has to be "on" for a long time. It is very important that everything be taken into consideration, even the leakage currents through electrostatic discharge protection diodes. Power monitoring/switching circuitry can become quite involved. When in doubt, do not discount small microcontrollers if multiple inputs can turn on a system. Extremely low power consumption microcontrollers exist that can painlessly handle different sets of circumstances from more than one input. Everything, of course, must be rigorously tested and prototyped.

21.6 Conclusions

The author has long and hard experience with systems that have failed in unexpected, catastrophic ways. It is sad that we tend to remember our failures much better than our successes. After decades in this industry, I have had my share of failures on the road to successful designs, have seen others fail, have picked up the pieces after other designers failed. So hopefully some of my insights have helped. But there is no methodology, no formula, no magic bullet that will create an absolutely reliable design. If there were, it would have been developed by now. High-reliability design for extreme applications is an exercise in discerning and managing risk areas—those areas in a design that are most likely to fail, addressing those, and then looking for the next most likely failure modes and addressing those, until the design is reliable enough or has failure modes low enough to allow deployment in the given situation. It is not an area for the inexperienced, nor is it one for those who like to push the envelope in new features and use the newest, fanciest parts. It can, nevertheless, be very rewarding to point to a system that has worked as designed for years or decades and say that you had a hand in creating something that will last. So many pieces of high-tech equipment are obsolete almost as they leave the factory, discarded in a year or two for a newer model. Reliable systems can be a point of pride or a selling point on a resume.

Voltage Regulation

22.1 Introduction

Table 5.1 introduced a comprehensive listing of cases. In between the cases for inverting and noninverting gain was a line on the table when op amp gain m = 0 and offset b is either positive, negative, or zero. These fall into the category of voltage regulators and have their own set of interesting challenges.

I have included some design utilities on the companion website, and these are discussed at the end of the chapter.

22.2 Regulator Cases
22.2.1 Virtual Ground: "b" = 0

Let us start with the simplest case, b = 0. This is the ground reference point of a circuit and requires no active circuitry. Active virtual grounds are sold and find a niche market in line driving applications. Rather than simply leaving you with a simplistic "b = 0 = ground connection" answer, I will refer you to Appendix D, which discusses proper circuit board grounding techniques.

22.2.2 Positive and Negative Voltage Regulators: "b" > 0, "b" < 0

The remaining cases are voltage regulators. When there is a positive offset (b > 0), the circuit is a positive voltage regulator, and when there is a negative offset (b < 0), the circuit is a negative voltage regulator. These circuits are designed for power applications—to provide power to other circuitry. AC gain in regulator circuits (m ≠ 0) is undesirable because it is ripple voltage. You need to be aware of ripple voltages in circuits, and the power supply rejection ratio of the op amp will tell you how much the ripple voltage will be rejected. Be especially careful with high-gain op amp circuits powered by switching regulators, as the ripple voltage may be amplified if the power supply rejection ratio of the op amp is not large enough.

This chapter will concentrate on positive voltage regulators, the next chapter will concentrate on techniques for negative regulation.

Op Amps for Everyone. http://dx.doi.org/10.1016/B978-0-12-811648-7.00022-4

Regulators can be linear or switching (utilizing an internal oscillator to perform voltage level conversion). Entire volumes have been written pertaining to both types of regulators—this chapter makes no effort to go into the details of regulator design other than to introduce the concept of the op amp as a feedback device controlling the output voltage level.

An interesting subset of voltage regulators is the voltage references. These can be as simple as a Zener diode, or they can have the same architecture as a linear voltage regulator, just optimized for voltage output accuracy. There is no reason why a voltage reference cannot be used to power low-power circuitry, as its architecture is identical to that of a linear regulator. However, that is not the intended application of the part. Using it at a substantial portion of its maximum-rated load may compromise its precision specification.

22.3 Make or Buy?

Voltage regulators are ubiquitous devices in many applications. They are available in a variety of output voltage, power levels, and packages. Many of them have multiple outputs. With all the work already done, what is the motivation for you to go to the trouble to design your own?

With highly integrated voltage regulators also come some baggage:

- Downhole and geothermal applications operate at high temperatures. The vast majority of integrated regulator IC's have overtemperature protection—designed to keep the IC from burning up if too much load current is drawn when the device is already at an elevated temperature. Designers for these applications have extensive experience in derating components and are used to components being operated in a manner that limits their usable life. Therefore, overtemperature shutdown is an unwanted feature that renders a voltage regulator IC unusable.
- Cellular telephone base stations are installed at tens of thousands of remote locations, in small buildings, or enclosures. Because a base station is a complex and power-hungry device, the buildings and enclosures require air conditioning, especially in hot climates. A common failure of air conditioning systems is loss of coolant, which will cause the equipment to overheat and shut down. The power supply "bricks" or modules that power the circuitry have an undesirable characteristic—they require power cycling or a manual "reset" to recover from fault conditions. This is unacceptable for systems in remote locations. An air-conditioner repairman will not reset the system. These applications require power supplies that automatically recover from fault conditions. So a designer of these systems might elect to design their own power supply.
- A switching regulator may not be available in the frequency range needed. In the case of the legendary GE/RCA Superradio 3 radios, a DC–DC converter was used to step up 9 V DC to a highly regulated 15 V for the tuning voltage. Not only was the highly

regulated voltage an unusual requirement, but the application forced the switching frequency of the regulator above the AM band to eliminate interference. Hobbyists have found it extremely difficult to reproduce this requirement with commercially available ICs now that the original regulator IC was discontinued. Discrete designs using a regulator IC in combination with a precision 15 V reference design have filled the requirement, although the solutions are bulky compared to the original IC.

There are many other reasons why you might opt to build your own rather than buy an IC to do the job. Cost, lead time, the availability of spare op amps/reference voltages in the system, etc.

22.4 Linear Regulators

Fig. 22.1 shows how a linear voltage regulator functions.

Both the linear and switching regulators operate similarly when the control circuitry (the op amp feedback loop) is concerned. The linear regulator, however, is simpler and will be presented in some detail.

The input voltage V_{IN} is assumed to be a voltage above the voltage supplied to R_{load}. Furthermore, it is assumed to be large enough that the transistor above the op amp can be biased into its linear region. This difference between the input voltage and output voltage can never be zero and is referred to as the "dropout" voltage. Dropout voltages can range from a few tenths of a volt to several volts. If you decide to do your own design, low dropout voltage is probably not going to be easy.

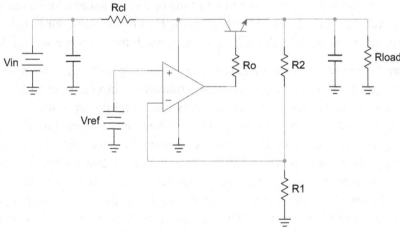

Figure 22.1
Voltage regulator operation.

The input voltage is also assumed to be a bit "messy,"having some ripple or noise on it. Therefore, it is filtered with decoupling capacitors, which are selected to have maximum effect on the ripple. Spend the time necessary to characterize this ripple and choose capacitors that are best at eliminating it—it is time well spent because it makes the job of designing the regulator that much easier. Some regulator designs fail simply because they do not have sufficient bypassing, and break into oscillation.

A current limiting resistor R_{cl} has been added to this circuit—it is good protection against a short circuit on the output in this simple configuration. It will be the element that fuses and protects the rest of the circuit should a short occur. R_{cl} is oftentimes placed elsewhere in the circuit when a regulator IC is used and will be part of a less destructive protection scheme. Suffice it to say—with simplicity comes some trade-offs!

The op amp is the control loop element. It is powered by the input voltage, as is a voltage reference. This reference can be supplied external to the circuit, generated off from a precision Zener diode, or even be a precision voltage reference IC. It is supplied to the noninverting input of the op amp, which should have precision DC specifications, especially if used over a wide temperature range.

I will repeat myself, because it is extremely relevant to this control loop element:

> *"An op amp will do whatever it can to the output voltage in order to make the voltage on its two inputs the same voltage.*
>
> *That statement above can be used to derive everything that has ever been written about op amps, and every application that uses them."*

In the case of a voltage regulator, this statement explains the op amp as a control element in the regulator. R_2 and R_1 form a voltage divider monitoring the output voltage (emitter of the transistor in this example). The inverting input of the op amp is supplied with the voltage-divided output voltage and will adjust the bias on the base of the transistor to balance the two inputs—in other words make the voltage divider match the reference voltage V_{REF}.

Now, for a brief explanation of the transistor. It is called a "pass" transistor in world of power supply regulation. It can be just about any transistor—bipolar or even FET. It should be able to dissipate enough power to supply the output power required by the circuit, plus dissipate its own power caused by the voltage it is expected to drop, plus the load current. Therefore, the pass transistor is the element that takes the beating in a regulator circuit. Many lower power regulators have the pass transistor built-in, others provide a way for using an external pass transistor that can dissipate much more power than an internal transistor could. In the case of this example, the op amp output has been buffered with a resistor Ro to limit the base current in the pass transistor. This is another protection scheme—if the pass transistor were to be short, this would prevent damage to the op amp output.

The response time of the op amp in the control loop is limited only by the bandwidth of the op amp. In some cases, this might result in problems. The op amp response to transients can be speeded up by addition of a capacitor across R_2, or slowed down by a capacitor across R_1. This requires you to do a bit of experimentation. At any rate, the output voltage should be decoupled by capacitors chosen to reduce ripple coming through the regulator. Of course R_{load} is a representation of the circuit load. The circuit load is not necessarily resistive. The output decoupling capacitor already makes it capacitive. If the load has substantial inductive characteristics, it may be wise to put a diode across the pass transistor to protect the circuit in case of back spikes from the load. In normal circuit operation, the diode should be reverse biased, only conducting if V_{OUT} exceeds V_{IN}.

22.5 Switching Power Supplies

The switching regulator shown in Fig. 22.2 is a variation of the circuit of Fig. 22.1.

At first glance, it looks formidable and unfamiliar. But look more closely. The voltage divider network from Fig. 22.1 consisting of R_2 and R_1 is still there. The output connects to an "FB" pin 3 of the IC. If you consult the data sheet for the device, it goes to the inverting input of an "error comparator" in data sheet Figure 1—nothing more than an op amp. The noninverting input of the op amp goes to a 1.5 V reference generated on the IC. All of the sudden, design of this part of the switching power supply circuit is beginning to look a whole lot like that of the linear voltage regulator with good reason. That is because it is exactly the same principle, only instead of controlling gate bias to the pass transistor

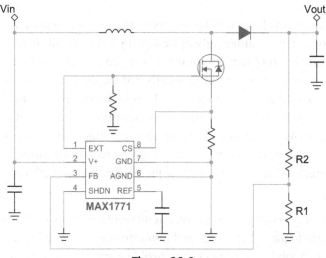

Figure 22.2
Switching regulator.

directly, it is now indirectly controlling the amount of time the transistor is "on," and, in this case, magnetizing the inductor.

This particular IC has yet another easily understood op amp circuit, one that is common to a great many voltage regulator ICs: the current sense circuit. If the reader looks at the circuit in Fig. 22.2, the "CS" input is on pin 8 and on the low side of the transistor switch. So instead of the transistor switching the inductor directly to ground, it is taken through this resistor, which is usually a pretty small value in the 10s or 100s of mΩ. It, therefore, generates a small voltage, which is used by a comparator inside the IC. A 0.1 V reference inside this particular IC determines the trip point—when the CS input exceeds 0.1 V, the IC shuts down, detecting an overcurrent condition.

The venerable old part above has been on the market for over 30 years. It is a "buck-boost" converter, capable of outputting voltage both above and below the input voltage. It is popular with antique radio restorers who use it to power the B+ voltage that would have originally been supplied by a high-voltage battery in "farm radios" that were battery operated. It also finds use by Nixie tube display enthusiasts. It is also one of the very few switching regulators that do not have a thermal shutdown circuit, so it finds use in high-temperature applications. Beware of buck-boost regulators, though—a failure of the IC, which shuts down the internal switching circuitry, will short the input voltage to the output voltage through a single inductor.

Other switching regulators commonly used are "buck" for lowering voltages, and "boost" for raising them.

22.6 A Companion Circuit

Consider the case when a voltage regulator fails— allowing unregulated input voltage to be applied to the load. This is almost always catastrophic, after all, if raw unregulated voltage would not cause a problem why not use it in the first place? Fortunately, it is easy to add an overvoltage protection circuit to greatly reduce the likelihood of this happening. Granted, there are no absolutes, and there are always circumstances where the regulator and the protection circuitry fail at the same time. But playing a game of odds, anything short of a lightning stroke or direct connection to line voltage will probably not produce such a failure. If both the regulator and overvoltage protection circuitry fail at the same time, the whole system should come under suspicion because something really dramatic has happened.

There are two types of overvoltage protection circuits—one that opens the voltage path and does not allow voltage to pass to the load, and a shorting or "crowbar" type of protection that shorts overvoltage to ground, thus blowing any fusible element on the input voltage. This, of

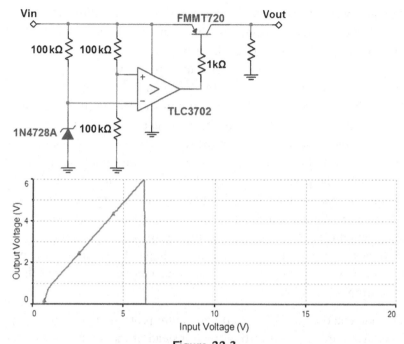

Figure 22.3
Overvoltage protection circuit.

course, is a nonrecoverable condition that may be appropriate for some applications, but why design a nonrecoverable circuit when it is just as easy to design a recoverable one?

Consider the circuit of Fig. 22.3. It is designed to protect the circuitry powered by a 5 V regulator, which is powered by a 20 V input. The load current is 250 mA. The input voltage V_{IN} comes from the voltage regulator output, and the load is applied to V_{OUT} (represented by a resistor). This figure looks very much like Fig. 22.1, but there are some crucial differences. First of all, it is the input voltage that is sampled instead of the output voltage. Secondly, the op amp has been replaced by a comparator.

The pass transistor is a PNP switching type, optimized for low collector–emitter voltage drop and high current. Small package size is achieved even though the transistor is passing high current, its voltage drop is so low that the overall power dissipated in the transistor is low. Looking at the connections to the comparator:

- A Zener diode is used to set a voltage reference at the inverting input.
- The input voltage is monitored by a voltage divider on the noninverting input.
  - The output is connected through a current limiting resistor to the base of the pass transistor.

Looking at the DC sweep, there is, of course, a region of operation where the comparator is not under power. Fortunately, that is not required for the pass transistor to be "on." All that is required is that the base be low. As the input voltage continues to increase, the comparator turns on with its output in the low state (the noninverting input is lower than the inverting input). A 1N4728 is a 3.3 V Zener diode, which is adequate for this circuit to function. Fine adjustments can certainly be made by adjusting resistor values and choosing a different Zener diode. As the input voltage reaches 5 V, the voltage applied to the noninverting input of the comparator is still below the threshold set by the Zener diode on the inverting input, so the comparator output is still low, and the pass transistor conducts. As the input voltage continues to rise, however, the comparator inputs reach equality, and the comparator switches its output high, turning off the pass transistor. V_{OUT} goes to zero, and current goes to zero. There is a small amount of leakage, but it is inconsequential. The comparator output, of course, continues to rise along with the input voltage. This may be a concern if there is a large input voltage, but the base resistance can be increased at the expense of forward voltage drop if need be.

A similar technique could be used for overtemperature protection with the comparator set up to monitor temperature on the inverting input instead of the input voltage. A second pass transistor may not even be needed, if you are clever in using open collector type comparators!

22.7 Another Companion Circuit

This section is offered as a way to test the reader's comprehension of the previous section or at least as a way to give the reader a very useful test circuit.

Take the circuit of Fig. 22.1, and make the following changes:

- Break the connection of the transistor to the input voltage.
- Allow adjustment of the reference voltage.
- Define the load in terms of a resistance either 0.1, 1, 10, or 100 Ω plus the "on resistance" of the pass transistor.

What is left is the active load circuit of Fig. 22.4, which is a very useful way of testing power supply circuits. When power supplies are tested with a power resistor, the load current rises and falls along with the power supply output voltage. When tested with this circuit—an adaptation of a circuit from Maxim Semiconductor—the load current is independent of the power supply voltage.

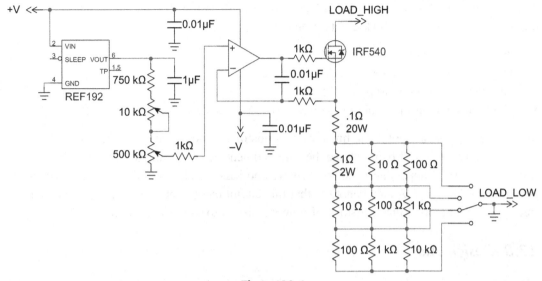

Figure 22.4
Active load.

Consider the case where:

* the voltage applied between LOAD_HIGH and LOAD_LOW is 1 V;
* the switch is in the position indicated, for a total resistance of 10 Ω (plus the on resistance of the IRF540);
* the 10 kΩ pot is adjusted to zero;
* the 500 kΩ pot is adjusted to 500 kΩ.

Therefore:

* The inverting and noninverting inputs of the op amp are both at 1 V.
* The op amp output has reached its positive rail, the transistor is turned on as "hard" as it can be, causing it to effectively be short and drop no voltage.
* Therefore, the 10 Ω resistor drops all of the load, or 100 mA.

Now, increase the voltage to 10 V:

* The op amp inputs are still both at 1 V.
* The op amp output is 4.732 V—probably not a useful number for the reader, but definitely accomplishing something at the transistor gate.
* The transistor is now dropping the applied voltage by 9 V, leaving 1 V at the 10 Ω resistor.
* The total load current is (10−9 V)/10 Ω = 100 mA. Same current as before.

Therefore, the load current is independent of the power supply voltage. One further modification—adjust the 500 kΩ potentiometer to 250 kΩ.

- Both op amp inputs are now at 0.5 V.
- The op amp output is now at 4.2 V.
- The transistor is now dropping 9.5 V.
- The total load current is now (10−9.5 V)/10 Ω = 50 mA.

In practice, the 500 kΩ pot is the fine current adjustment, and the rotary switch is the current range. Be very careful of part wattage, be sure and heat sink the transistor and power resistors properly. This circuit is an inexpensive and handy addition to the test bench of any power supply designer. A side benefit is that this circuit can be set to any load current and frees you from having to have a stock of power resistors to use as dummy loads.

22.8 Design Aid

The voltage regulator circuits shown in Figs. 22.1 and 22.2 require a resistor—divider circuit to set the output voltage to the desired level. I have written a java script calculator Fig. 22.5 to aid in the design of this network. To use it, you have to know the value of the internal voltage reference and the output voltage. There is a drop-down box for resistor sequence and resistor scale. The calculator may produce more than one option for resistors.

Please remember the following:

- The voltage regulator topology may be linear or switching—the switching topology may differ quite a bit from the figures presented here and on the calculator, but the calculator is concerned only with the feedback network needed to set the output voltage—so the exact topology is not important to its operation. Only the correct value of internal reference is important.

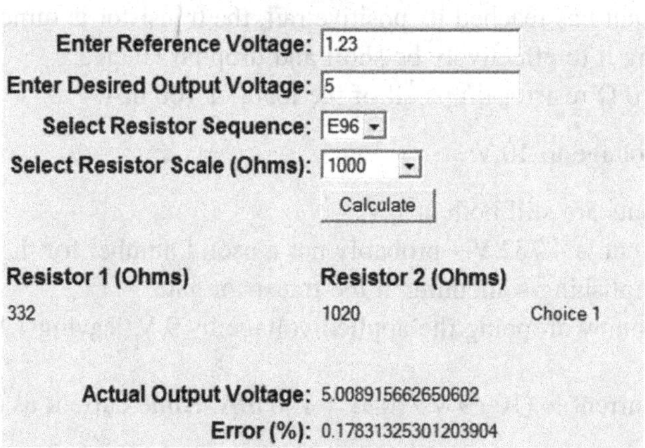

Enter Reference Voltage:	1.23
Enter Desired Output Voltage:	5
Select Resistor Sequence:	E96 ▾
Select Resistor Scale (Ohms):	1000 ▾
	Calculate

Resistor 1 (Ohms)	Resistor 2 (Ohms)	
332	1020	Choice 1

Actual Output Voltage: 5.008915662650602
Error (%): 0.17831325301203904

Figure 22.5
Voltage regulator calculator.

- The regulator may be sensitive to resistor scale. If the data sheet is making recommendations for R_1, stay within a decade of it.
- This calculator is not intended to address the complexities of bypassing on the input, output, or reference. Neither does it address inductor values or transistor selection for switching regulators. This can be tricky, and if you do not know what you are doing, leave those aspects of the design to experts.
- Some regulators may have feedback networks to select undervoltage shutdown. This calculator can be used, in some cases, to design those networks as well. As long as an internal reference voltage is known, the calculator will work.

22.9 Conclusions

Even an op amp designer who does not consider themselves fluent in power supply design can utilize many of their skills in designing parts of the feedback loop. Rather than be intimidated by power supply design, you should embrace it, because you already have most of the basic skills. You can make custom voltage regulator circuits that are specific to the system's needs, and add as much or as little sophistication as necessary to the application—all with just basic op amp design skills!

CHAPTER 23

Negative Voltage Regulation

23.1 Introduction

When looking for a voltage regulator for your design project, the overwhelming majority of parts you will find are positive regulators. This is true whether you are looking for linear or switching regulators. I have talked about the advantages of single-supply design, and you may well have been able to configure your system to work off of positive rails only—so life is good.

What if life is not so good and you need to utilize split supplies—either for a performance advantage or some other design constraint? The number of regulators shrinks to almost none, it is very frustrating indeed!

Fortunately, there are ways around this. I am not aware of any technique for negative linear regulators that does not depend on the presence of a negative input voltage, but in the case of switching regulators it is possible to make use of the energy storage element—usually an inductor—to "fool" the regulator into producing a negative voltage rail.

23.2 Positive Regulators

For the sake of this discussion, we will start with a buck regulator—one that is designed to reduce an input voltage to a lower voltage. Fig. 23.1 shows a typical buck regulator. It is not meant to represent an exact part number; it is generic. There are many variations—consult the data sheet of the device you have in mind.

This is nothing more than a variation on Fig. 22.2, this time with a generic switching buck regulator. Our old friend, the op amp is not seen on this schematic but is well represented inside the IC itself—so this discussion is not far removed from op amps by any means!

Going through the major components:

* Some form of decoupling—C_IN and C_OUT—is required. It can be a single capacitor for noncritical applications, or a combination of electrolytic and ceramic capacitors for more demanding applications. C_IN reduces the sensitivity to conducted emissions; C_OUT reduces output ripple.
* Most IC switching regulators have several forms of protection. Undervoltage lockout is one of them. Short circuit and over temperature are others. In the circuit above,

Op Amps for Everyone. http://dx.doi.org/10.1016/B978-0-12-811648-7.00023-6

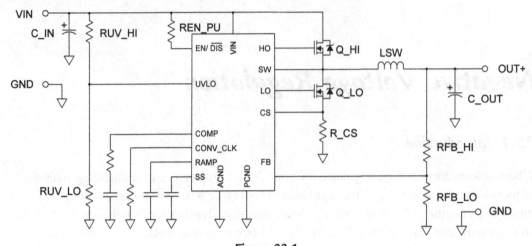

Figure 23.1
Voltage regulator operation.

undervoltage lockout threshold is set by a voltage divider consisting of RUV_HI and RUV_LO. This voltage divider goes to a comparator—a device that should be very familiar to op amp designers!

- The typical switching regulator may have a number of inputs such as soft start (SS), ramp capacitor (RAMP), switching frequency (CONV_CLK), and compensation (COMP). I have represented these with typical circuit elements, some or all may be present or absent, some or all may be configured differently. Suffice it to say, read the data sheet.
- The main outputs of the switching regulator (in this case) are gate drives and switching waveform. The gate drives—high FET gate drive out (HO) and low FET gate drive out (LO)—are very robust digital drive signals. Driving an FET gate at low frequencies is easy, but overcoming gate charge and gate capacitance at high frequencies is NOT! The function of the FETs is to provide energy from V_{IN} to magnetize an inductor LSW through Q_HI, and to collapse the magnetic field of LSW through Q_LO. The profile of this magnetize/demagnetize function is present on a robust analog output—SW. SW is generally a pulse stream controlling the amount of time the inductor is magnetized or demagnetized but may also include sawtooth elements in the waveform.
- This regulator also includes a current sense input. R_CS is usually a small value, and because it is in the direct transistor switching power path, it is important to pay attention to the wattage. Another variation that some switching regulators use is to put the current sense resistor in line with the output voltage—trace routing and parasitics are of prime importance if this technique is used! Regardless of the architecture, the CS input(s) must be amplified—and thus this is the input to an op amp gain circuit—the output of which goes to a comparator. R_CS is selected to provide a setting above the maximum load current to shut down the regulator should the output become shorted.

There may even be a very fast—but not precision—shutdown path to detect hard shorts. Once again, an op amp does the job of determining the short.

- Finally, RFB_HI and RFB_LO—our last op amp circuit—form a voltage divider off of the generated DC level. This goes to an op amp gain stage—the output of which controls the duty cycle of HO and LO, and the waveform SW.

I count at least three or four feedback pathways inside the regulator IC that connect to simple op amp gain circuits driving a comparator. There may be more, depending on the IC! With this introduction, it is time to "hack" the regulator and produce a negative voltage—all courtesy of our little friends—the op amps inside!

There are a few techniques to do this—all have their advantages and disadvantages.

23.3 Parasitic Winding on the Inductor

The waveform that comprises SW is an AC waveform. Depending on the regulator and the load current, it can take many forms, but is generally a pulse stream of varying duty cycle depending on the load current. It may also have some droop on the positive going cycles and sawtooth-like components. The bottom line, however, is that it is some form of analog waveform—one with a lot of power—equal to the peak power demand of the load.

Any time an AC waveform is present; even if it is not a sine wave, it can be used like any other AC waveform. In this first case, it can be used to produce a secondary waveform by converting LSW to a transformer as shown in Fig. 23.2.

LSW has been changed to a transformer as discussed. Pay special attention to the polarity dots in the transformer, it is important to have the polarity in the winding correct. It is also very important to have the windings closely coupled, as the waveform is not optimal for efficient energy transfer.

A low forward drop Schottky diode is also used to establish the correct polarity—because the diode drop is in line with the output voltage, its forward voltage drop will affect the DC level of the negative voltage.

A filter cap is used to filter the output of the diode. This has produced, in effect, a half-wave rectifier on whatever waveform is present in the primary of the transformer.

The advantage to this technique is simplicity, a transformer is used instead of an inductor, a Schottky diode and filter capacitor(s) are added.

Disadvantages to this technique include the following:

- Relative lack of regulation on the negative output voltage.
- The complexity of the transformer as opposed to the inductor.

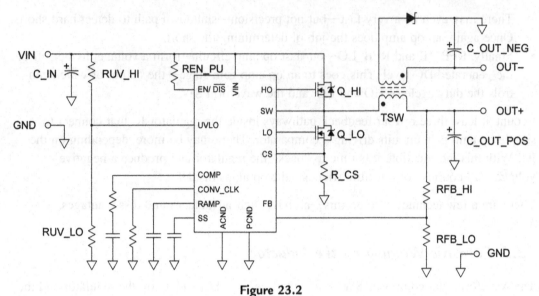

Figure 23.2
Negative output utilizing a parasitic winding.

- This is just a half-wave rectifier—ripple on the negative output will be worse.
- There is a complete dependence on current flowing on the positive leg. If there is no current flowing through the transformer primary, none will be available for the parasitic negative rail either.

23.4 Parasitic Inductor

Making a custom transformer, especially one like that required for the circuit above, is time consuming and expensive. If you want to generate a parasitic rail, you can just as easily use the method in Fig. 23.3.

In this technique, the DC blocking nature of a capacitor is leveraged to eliminate the DC reference for inductor LSW_NEG. It can then be grounded, and the AC waveform across it will be filtered out as a negative voltage, equal in amplitude and opposite in polarity to that across LSW_POS. CSW, of course, does a lot of the hard work and must be very robust to handle the AC current and power presented to it. A Schottky diode is used to eliminate any backspike transients as power is applied and removed.

Advantages of this technique include the following:

- This still has the advantage of simplicity, only a few parts are added.
- No requirement to buy or specify a custom transformer. It uses the same inductor two places.

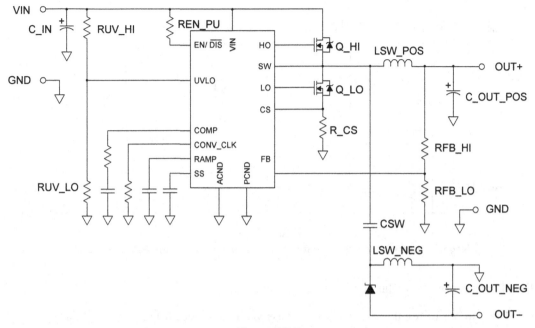

Figure 23.3
Negative output utilizing a parasitic winding.

- Better regulation than the parasitic winding technique. Everything being done in the positive side is mirrored in the negative side. Matching can be fairly good.

Disadvantages of this technique include the following:

- Current limit and short-circuit sense are not good on the negative output. If they work at all, they will be slow and will allow stresses to build up in the FETs.
- The negative rail is still not directly referenced, so regulation on the negative output will not be as good.

23.5 Referencing the Regulator to $-V_{OUT}$ Instead of Ground

This technique involves changing the way you connect the inductor LSW. Instead of it being the positive output voltage, it is connected to ground. Elsewhere in the circuit, everything that used to be ground is now referenced to $-V_{OUT}$ as shown in Fig. 23.4.

Advantages of this technique include the following:

- The negative voltage output is fully regulated, current limited, and short circuit protected by the regulator chip.

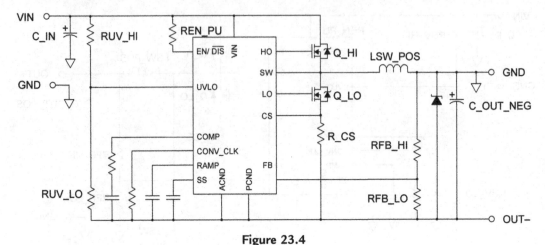

Figure 23.4
Negative output by referencing the regulator to $-V_{OUT}$ instead of ground.

Disadvantages of this technique include the following:

- The number of regulators and support passives is doubled. It takes two regulator ICs—one for the positive output and one for the negative output. This doubles the quiescent power of the power supply, doubles the board space, and doubles the cost.
- If the regulator IC has an enable/disable input, you will not be able to use it on the negative regulator. The logic level is referenced to a negative potential that is not available until the negative regulator operates!
- If you are not extremely careful with layout, you will introduce power supply switching noise into the system ground.
- There are two switching regulator frequencies. They may heterodyne in the audio range.

Things that are different—not advantages or disadvantages—just things, you have to take into account:

- Input capacitors can be shared between the positive and negative regulators. This is about the only thing that can be shared. Although I have shown the input capacitors hooked to ground, some input capacitors directly from V_{IN} to V_{OUT} may be advisable.
- Watch your voltage ratings on capacitors and the regulator itself. Remember that referencing to $-V_{OUT}$, your input range on the regulator is also reduced. If you are producing a -5 V output, and your input range on the regulator was 0–40 V, it is now only 0–35 V because the -5 V reference of the power supply subtracts from the input range. The same applies to the input capacitors if they are hooked to $-V_{OUT}$.
- The undervoltage lockout voltage-divider values will be different, because of the extra input range.
- RFB resistors will be the same.

- R_CS may be different.
- The Schottky diode protects against reverse polarity on power up/power down. If you are operating at low currents, it can be omitted.
- Ripple voltage waveforms will be different, usually the mirror image of each other.

23.6 Other Techniques

If you are using a buck-boost converter, or are considering referencing the negative reference technique of Section 23.5, you may want to adapt the circuit by using a transformer instead of an inductor. The advantage to this technique is that you have two identical output power pathways, both isolated from the input. Other than that, there is no real advantage to doing so.

Some references recommend parasitic FETs to generate secondary outputs. They are generally driven from the HO or LO outputs and may be a complete parallel to the original circuit, or may simply be using the AC waveform to operate the gate of a single FET to create a pseudo—switching supply—actually just a rectified and filter AC waveform and therefore not very well regulated. I do not recommend these techniques, do a web search if you want, but these techniques make the gate drivers in the regulator work much harder to overcome the gate charge and gate capacitance of two FETs instead of one. The regulator, therefore, will be stressed more than it has to be, given that there are other techniques that are better.

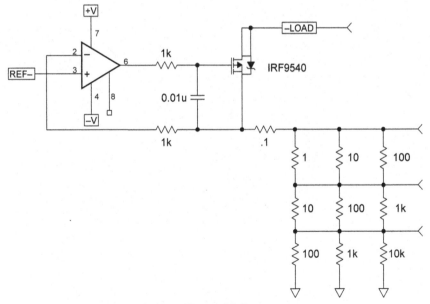

Figure 23.5
Negative active load.

23.7 A Negative Load

The active load circuit of Fig. 22.4 is designed for a positive voltage and is not suitable for use with negative voltages. It is necessary to invert the operation of the circuit. Fortunately, this is easily done. The active negative load circuit of Fig. 23.5 assumes there is a negative reference voltage available—this is easily generated in a unity gain inverting op amp stage. It replaces the N Channel IRF540 with the complementary device, an IRF9540, and therefore can sink a negative voltage.

23.8 Conclusion

Generating a negative rail with a positive regulator is not only feasible but has been done on designs using techniques that have been known for decades. Almost, as soon as switching regulator product appeared on the market, engineers have found new and innovative ways of using them, which include generating a negative rail.

CHAPTER 24

Other Applications

24.1 Op Amp Oscillators

When talking about op amp oscillators, a designer needs to unlearn everything they have previously learned about op amp stability criteria—in other words, they now want to purposely design a circuit which, by its very nature, is unstable.

At this point, this volume could launch back into the theory of feedback design, but to get to the point—the best way of making an op amp unstable is to use positive, not negative, feedback. Negative feedback as used up to this point has operated to limit output voltage excursions in the op amp circuit. Positive feedback reinforces the output voltage excursions, exaggerating and amplifying them, in phase, until the op amp output is saturated, which leads the designer to an important decision—to use op amps or to use a very similar, but different, component, the comparator (Fig. 24.1).

Both circuits above were designed to give an output of approximately 10 kHz. One fact emerged very quickly—the technique used to determine filter cutoff points does not work. If it had, one would have expected both circuits to operate with an RC circuit comprised of 4.42 k and 3600 pF capacitors. The resistor value had to be "tweaked" considerably to slow both of these circuits down, and the resistors are different for the op amp and the comparator. But this is a small matter compared with the suitability of the device.

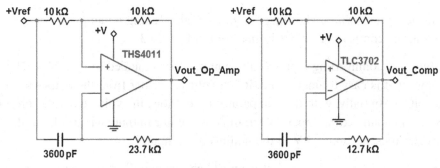

Figure 24.1
Oscillator schematics.

Op Amps for Everyone. http://dx.doi.org/10.1016/B978-0-12-811648-7.00024-8

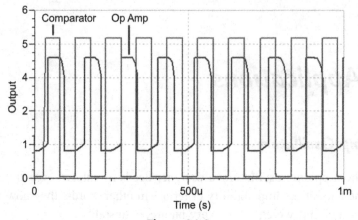

Figure 24.2
Oscillator outputs.

So what is the difference between using an op amp and a comparator in this application? A look at Fig. 24.2 gives a very quick answer.

It is very apparent that the op amp output is very degraded compared with the comparator. After reading this volume, the designer should at this point be familiar with the V_{OL} and V_{OH} specifications of an op amp. In this case, those limitations clamp the voltage excursion of the op amp output to about 0.8—4.5 V. But that is not the end of the story! The leading edge of the waveform is fairly decent, but the trailing edges show a different story—the op amp output stage is tending toward latch-up and takes some time to recover. If the voltage swing of the op amp was not a convincing enough reason to use a comparator, the latch-up tendency surely is! It points to a potential failure of the circuit due to latch-up stresses being repeatedly applied. The comparator, on the other hand, has an output that is designed to switch from rail to rail. This makes it ideal for use in an oscillator circuit. There will be more information about comparators in the next chapter, consider this to be an introduction to that material.

Next, let us delve into what makes this circuit oscillate. Taking op amps out of consideration, the comparator circuit is analyzed in Fig. 24.3:

The inverting and noninverting inputs of the comparator are added to Fig. 24.3. The noninverting input is easy to understand, it is a voltage divider from the output to the reference voltage. Initially, with the comparator output low, the voltage divider presents about 1 V to the noninverting input. When the comparator output switches high, the voltage divider presents about 4 V to the noninverting input.

When the inverting input is analyzed, it is an RC charge and discharge circuit. Initially, the capacitor is charged to the reference voltage, but since the output is low, the capacitor

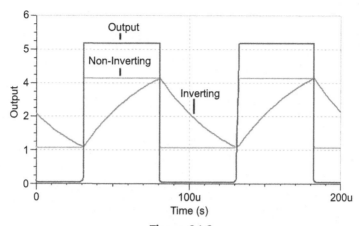

Figure 24.3
Comparator oscillator analysis.

goes into a discharge cycle until it reaches a voltage equal to that on the noninverting input. At that point, the output goes high, and the capacitor is charged until it reaches the higher voltage on the noninverting input, at which time the comparator output switches back low. And that is how the circuit operates—the capacitor is continually charging and discharging to reach the voltage on the noninverting input.

Many designers want a sinusoidal oscillation, not a square wave oscillation. That can be easily accomplished by filtering the circuit above, but not necessarily the output! The designer should instead filter the voltage that appears on the inverting input because a triangle wave's harmonics are lower than a square wave's harmonics, and the voltage swing is within the voltage swing limits of a unity gain op amp stage. This may sound a bit counterintuitive—to use the inverting input of the comparator as its "output," but the designer will be fine as long as the op amp stage does not load the inverting input to any degree. If there is any doubt, use the output of the comparator instead.

24.2 Hybrid Amplifiers and Power Boosters

If a designer is careful, it is possible to create a composite amplifier whose important characteristics are better than either individual op amp.

The circuit in Fig. 24.4 shows a composite amplifier constructed by an OPA277 and an OPA512. The OPA512 is used as a power output buffer in the feedback loop of the OPA277. Characteristics are tabulated in Table 24.1.

The OPA512 has the highest slew rate and therefore is operated within a local closed loop. The slower OPA277 is operated within the outer loop. The 47 pF capacitor provides a small amount of phase shift to help stabilize the system. The resulting performance of the

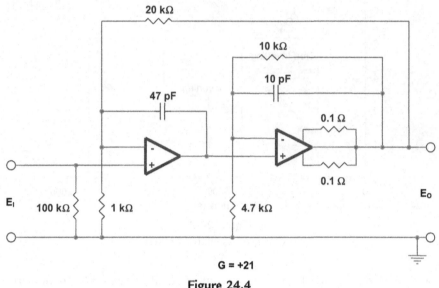

G = +21

Figure 24.4
Composite op amp.

Table 24.1: Composite Amplifier.

Parameter	OPA277	OPA512	Compound
V_{OS}	20 μV	6 mV	20 μV
Drift	0.15 μV/°C	65 μV/°C	0.15 μV/°C
I_B	1 nA	30 nA	1 nA
CMRR	130 dB	100 dB	130 dB
V_{OUT}	± 13 V	± 35 V	± 35 V
I_{OUT}	5 mA	10 A	10 A
SR	0.8 V/ μS	2.5 V/ μS	2.4 V/ μS

compound amplifier shows that the front-end characteristics of the OPA277 are joined with the 35 V at 10 A drive current capabilities out of the OPA512. The slew rate of the OPA277 is 0.8 V/μs. That slew rate is gained up three times in the OPA512 so that there is an effective slew rate for the compound amplifier of 2.4 V/μs.

High-speed hybrid amplifiers can be constructed so they take advantage of the speed and power output of current feedback op amps (Fig.24.5).

In this configuration, the designer should use the values recommended in the data sheet for Rf_2 and Rg_2. This particular schematic shows the current-feedback amplifier used in a gain mode; it is definitely possible to use it in unity gain mode by leaving out Rg_2. Remembering, however, that the least stable op amp configuration is unity gain,

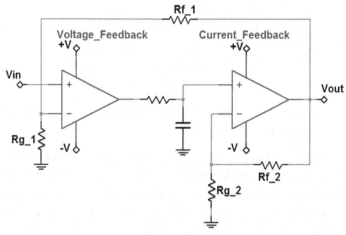

Figure 24.5
Composite op amp.

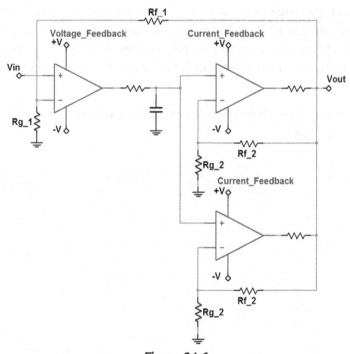

Figure 24.6
Paralleling output op amps.

I usually operate the current-feedback amplifier at a gain of 2—just to slow it down a bit. The RC network in between the two stages is there to slow things down in case the circuit breaks into oscillation. Another place the designer can slow things down is by placing a capacitor in parallel with Rf_1. Remember that the values of Rf_1 and Rg_1 will determine the stage gain, but the values of Rf_2 and Rg_2 will determine the maximum voltage swing possible out of the voltage-feedback amplifier. V_{OL} and V_{OH} of the current-feedback amplifier will determine the voltage swing of the stage. A designer may have to play around with this circuit a bit to get it to operate—there are a lot of things to balance and compromise to get acceptable performance.

What if more power is needed? It is possible to parallel op amps to increase the output power level. Fig. 24.6 shows the technique.

The output of the voltage-feedback amplifier has been connected to an RC low-pass filter as before. But this time, two current-feedback amplifiers are used. Their outputs are connected to small series resistors, between 1 and 5 Ω—the exact value again determined by a bit of experimentation. The series resistors are there to prevent one current-feedback amplifier from driving the output of the other—they compensate for small mismatches in the components. The designer should check the wattage of series resistors, which may have to be power resistors. Like anything else in analog design, there are trade-offs. In this case, the output voltage swing of the stage is reduced because of the series-matching resistors. Because power is V^2/R, this is a fairly serious compromise, but a necessary one. It can be combated by raising the power supply voltage, or by operating the circuit in a bridged configuration. Board parasitics will limit the number of amplifiers that can be paralleled.

There is no reason why these techniques cannot be used at lower speeds with voltage-feedback power amplifiers.

CHAPTER 25

Common Application Mistakes

25.1 Introduction

When one works as an analog applications engineer for many years supporting customer inquiries, some patterns begin to emerge. Inquires run the gamut from newcomers who have no business attempting analog design to those from analog design experts who have encountered something new and unusual that challenges even the best support engineer. There is also a class of inquiries that are bound to elicit a groan—mistakes the author has many times before and unfortunately will see many times again. It is hoped that this chapter will educate people to some of the most common mistakes and hopefully save designers from continuing to make them.

25.2 Op Amp Operated at Less Than Unity (or Specified) Gain

Does the phrase "unity gain stable" ring a bell? In early chapters of this book, the statement was made that an op amp is least stable at its lowest specified gain. Hopefully the engineer has read the material and understands the true nature of stability, and yet, during the book tour for the first edition of this book, a customer approached me with a problem: they had a programmable gain op amp circuit where resistors were switched to program a gain of 1, 1/10, and 1/100 (see Fig. 25.1). The unity gain case worked as expected, but ringing occurred with a gain of 1/10 and sustained oscillation at a gain of 1/100. In no way do I demean the individual who made this mistake; I have been known to grab an amplifier out of a bin to construct a quick circuit, only to have it oscillate uncontrollably. Invariably, a quick glance at the data sheet reveals it is a "gain of ten"

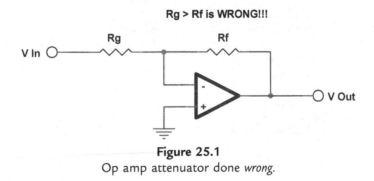

Figure 25.1
Op amp attenuator done *wrong*.

Op Amps for Everyone. http://dx.doi.org/10.1016/B978-0-12-811648-7.00025-X

stable op amp, not unity gain stable. There is no other option at that point but to use a different op amp.

Fortunately, the solution to the problem is exceptionally easy. A voltage divider can be applied to the input of a noninverting op amp buffer as shown in Fig. 25.2.

As far as the op amp is concerned, it is operating at unity gain and is stable. The voltage divider rule is employed to calculate the correct degree of attenuation. The high input impedance of the noninverting op amp input will not affect the voltage divider to any degree unless extremely large value resistors are used.

If the signal must be inverted, then the inverting attenuator of Fig. 25.3 can be used. It is a variation of the approach used in Fig. 25.2, but takes into account the resistors used in for the op amp feedback and input resistance.

A common objection to the solution of Fig. 25.3 is that it introduces resistor noise to the attenuator. This is a fallacy for two reasons—first of all the incorrect inverting attenuator

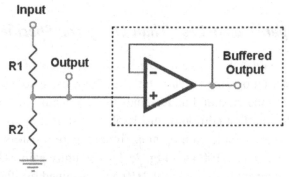

Figure 25.2
Op amp attenuator done correctly.

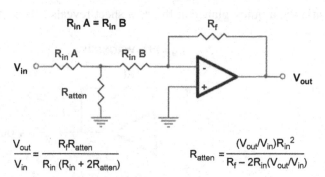

$$\frac{V_{out}}{V_{in}} = \frac{R_f R_{atten}}{R_{in}\,(R_{in} + 2R_{atten})} \qquad R_{atten} = \frac{(V_{out}/V_{in})R_{in}^2}{R_f - 2R_{in}(V_{out}/V_{in})}$$

Figure 25.3
Inverting op amp attenuator.

also contains resistors, that are going to be almost the same value. The second reason why this objection is fallacious is that the carbon composition resistors that caused the problem are a fast-fading memory. Most resistors now are metal film or thick film types that have much better noise specifications.

25.3 Op Amp Used as a Comparator

This misapplication usually occurs in cost-sensitive pieces of equipment when a comparator is needed, and a quad op amp has an unused section. I first encountered it when I discovered that the expensive telephone answering machine I had purchased quit working. "Why"—I asked myself, is there an open-loop op amp circuit on one-fourth of an LM324, and why is it interfaced to a digital logic gate? The answer was somebody, who looked at the schematic symbol of an op amp and a comparator, saw that they look alike and decided that they both work the same way (Fig. 25.4)!

Unfortunately, not even the internal schematic of the parts gives much indication of what is going on (Figs. 25.5 and 25.6).

The input stages look almost identical, except the inputs are labeled opposite (a fine point that will be discussed later). The output stage of the op amp is a bit more complex—which should be a clue that something is different. The output stage of the comparator is obviously different, in that it is a single open collector. But be careful—many newer comparators have bipolar stages that are very similar in appearance to op amp output stages.

So—if very little appears to be different in the schematic symbol or the internal workings—what is the difference?

The difference is in the output stage. An op amp has an output stage that is optimized for linear operation, while the output stage of a comparator is optimized for saturated operation.

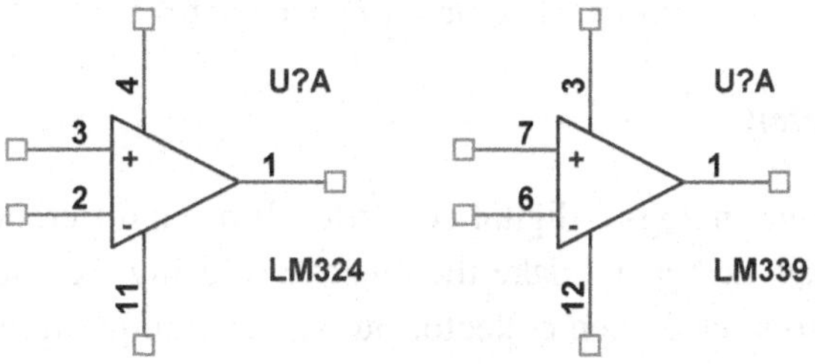

Figure 25.4
Similar schematic symbols, but very different parts!

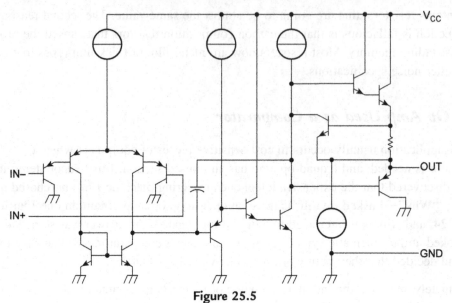

Figure 25.5
Example op amp schematic.

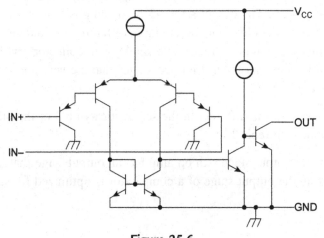

Figure 25.6
Example comparator schematic.

25.3.1 The Comparator

A comparator is a one-bit analog-to-digital converter. It has a differential analog input and a digital output. Very few designers make the mistake of using a comparator as an op amp because most comparators have open collector output. The output transistor of open collector comparators is characterized by low V_{CE} for switching heavy loads. The open collector structure depends on external circuitry to make the connection to power and complete the circuit. Some comparators also bring out the emitter pin as well, relying on the designer to complete the circuit by making both collector and emitter connections. Other comparators substitute an FET, having open-drain outputs instead of open collector. The emphasis is on driving heavy loads.

The comparator is an open-loop device, utilizing no feedback resistors. When applying a comparator, the designer compares the voltage level at two inputs. The comparator produces a digital output that corresponds to the inputs:

- If the voltage on the noninverting (+) input is greater than the voltage on the inverting (−) input, the output of the comparator goes to low impedance "on" for open collector/drain outputs, and "high" for totem-pole outputs.
- If the voltage on the noninverting (+) input is less than the voltage on the inverting (−) input, the output of the comparator goes to high impedance "off" for open collector/drain outputs, and "low" for totem-pole outputs.

25.3.2 The Op Amp

An op amp is an analog component with a differential analog input and an analog output. If an op amp is operated open loop, the output seems to act like a comparator output, but is this a good thing to do?

An op amp, being intended for closed-loop operation, is optimized for closed-loop applications. The results when an op amp is used open loop are unpredictable. No semiconductor manufacturer can or will guarantee the operation of an op amp used in an open-loop application. The analog output transistors used in op amps are designed for the output of analog waveforms, and therefore have large linear regions. The transistors will spend an inordinate amount of time in the linear region before saturation, making the rise and fall times lengthy.

In some cases, the designer may get away with using an op amp as a comparator. When an LM324 is operated in this fashion, it hits a rail and stays there, but nothing "bad" happens. The situation can change dramatically, however, when another device is substituted.

The design of an op amp output stage is bad news for the designer who needs a comparator with fast response time. The transistors used for op amp output stages are not switching transistors. They are linear devices, designed to output accurate representation of analog waveforms. When saturated, they not only may consume more power than expected, but may also latch-up. Recovery time may be very unpredictable. One batch of devices may recover in microseconds, another batch in 10s of milliseconds. Recovery time is not specified because it cannot be tested. Depending on the device, it may not recover at all! Runaway destruction of the output transistors is a distinct possibility in some rail-to-rail devices. Even the best designer might produce a saturated or even open-loop op amp circuit without realizing it.

Oh, and the reason why my answering machine failed: The V_{OL} rail of the open-loop op amp circuit they created was above the logic threshold of the digital gate to which it was interfaced. The two levels were very close—and the slightest drift upward of the op amp output stage caused the low logic threshold to never be reached. V_{OL} is yet another op amp specification that will never be specified under open-loop conditions.

25.4 Improper Termination of Unused Sections

One of the easiest ways to unintentionally misapply an op amp is to misapply unused sections of a multiple section IC. Fig. 25.7 shows the most common ways designers connect unused sections.

Many designers know how to properly terminate unused digital inputs, hooking them to the supply or ground. These designers may not have a clue how to terminate unused op amps. Fig. 25.7 demonstrates techniques Texas Instruments applications actually seen, and I have given them titles:

Brain-dead: This is a common mistake. Designers will assume an op amp is like an audio amp at home, and just leave unused inputs unconnected. This is the absolute *worst* thing that can be done to an op amp! An open-loop op amp will saturate to one voltage rail or the other. Because the inputs are floating and picking up noise, the output of the op amp will switch from rail-to-rail, sometimes at unpredictable high frequencies!
Never: This is another really bad thing that designers occasionally do. Usually one op amp input will be slightly higher than the other due to ground-plane gradients, and the best possible scenario is that the op amp will saturate at one rail or the other. There is

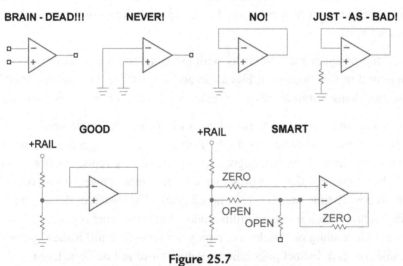

Figure 25.7
Different ways of dealing with unused op amp sections.

no guarantee it will stay there, as a slight change on one pin could cause it to switch to the other rail.

No: This is a little better than the previous case, but not that much. All the designer has accomplished is to ensure that the op amp will hit a rail and stay there. This can wreak havoc—self-heating, increased power consumption.

Just-as-bad: Designers who are designing a board for in-circuit-test commonly do this. It still makes the op amp hit a rail.

Good: This is the minimum recommended circuit configuration. The noninverting input is tied to a potential between the positive and negative rail or to ground in a split-supply system. Virtual ground may already exist in the system, making the resistors unnecessary. The op amp output will also be at virtual ground (or ground in a split-supply system).

Smart: The smart designer will anticipate the possibility of system changes in the future and lay out the board so that the unused op amp section could be used by changing resistors and jumpering. The schematic shows how the unused section could be used for either an inverting or noninverting stage, as required.

25.5 DC Gain

Another way designers create problems is when they forget about DC components on AC signals. Fig. 25.8 illustrates this problem. When an AC signal source has a DC offset, a coupling capacitor isolates the potential in the top circuit. The DC component is rejected, and output voltage is 1 V_{AC}. If the coupling capacitor is omitted, the circuit attempts a gain of -10 on both the AC and DC components, which would be 1 V_{AC}, -50 V_{DC}. Because the power supply of the circuit limits the DC output to $+$ and -15 V_{DC}, the output will be saturated at -15 V_{DC} (minus the voltage rail limitation of the op amp).

25.6 Current Source

The op amp current source circuit shown in Fig. 25.9 *must* always contain the load. Many applications put the load at the end of a cable—and the cable is on a connector. When the cable is unplugged, the op amp has positive feedback! It will hit the negative voltage rail.

The output of the current source is:

$$I_{OUT} = \frac{R_3 \cdot V_{IN}}{R_1 \cdot R_5}$$

$$R_3 = R_4 + R_5$$

$$R_1 = R_2$$

It should be understood that R_1 through R_4 are $\gg R_5$, and $R_5 \gg R_{LOAD}$

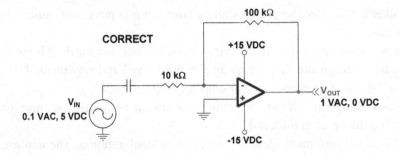

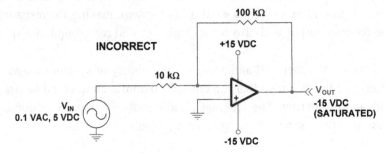

Figure 25.8
Unexpected DC gain.

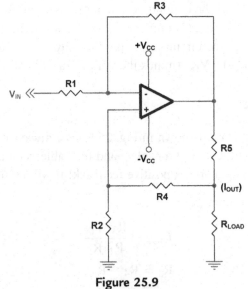

Figure 25.9
Current source.

25.7 Current-Feedback Amplifier: Shorted Feedback Resistor

By far the most common mistake with current-feedback amplifiers (CFAs) is a designer will short the output directly to the inverting input (Fig. 25.10).

The designer is invariably trying to take advantage of the speed and bandwidth of the CFA to make a buffer. Shorting the output pin to the inverting input is always a bad idea, because it will make the CFA unstable. Stability criteria for CFA are different from that of the voltage-feedback amplifier (VFA).

VFA stability criterion:

$$A\beta = \frac{aR_G}{R_F + R_G}$$

CFA stability criterion:

$$A\beta = \frac{Z}{R_F\left(1 + \dfrac{Z_B}{R_F \parallel R_G}\right)}$$

As you can see, VFA stability depends on both R_F and R_G equally. But CFA stability is much more dependent on R_F. In fact, if R_F is zero, the denominator goes zero and the stability criterion fails! Fig. 25.11 shows this graphically in actual data sheet plots:

The effect of changing R_F only slightly has an enormous effect on the CFA response on the left—with an alarming trend as the resistor is lowered. The effect is much smaller and opposite with the VFA plot on the right. Bottom line: stick with the recommended value of feedback resistor for CFA op amps. And that also makes a very easy solution when a noninverting buffer is desired, just put the recommended value of feedback resistor between the output and inverting input, and the stage will work perfectly!

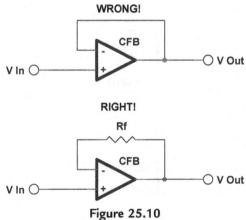

Figure 25.10
Current feedback amplifier incorrectly applied.

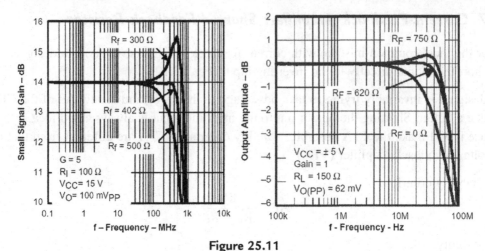

Figure 25.11
Voltage-feedback versus current-feedback amplifier stability versus load resistor.

25.8 Current-Feedback Amplifier: Capacitor in the Feedback Loop

This often occurs when the designer is attempting to do active filter design with CFAs (Fig. 25.12).

There are very few filter topologies that will work with CFAs. Sallen–Key is one, if the proper value of feedback resistor is employed. The bottom line is that CFAs are not the best choice for active filter designs. Choose something else wherever possible.

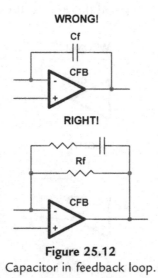

Figure 25.12
Capacitor in feedback loop.

25.9 Fully Differential Amplifier: Incorrect Single-Ended Termination

One of the most common applications for a fully differential op amp is single-ended to fully differential conversion. However, when the input signal must be terminated, the situation gets very complicated!

Looking at the circuit in Fig. 25.13 and the equations that govern it, R_1 and Rt are cross defined. Solving this equation for the correct values requires a goal-seeking algorithm. If the values are calculated incorrectly, the results can be:

- wrong gain
- differential offset
- unmatched differential gain
- incorrect matching impedance

The designer might want to consider the simpler design alternative in Fig. 25.14:

25.10 Fully Differential Amplifier: Incorrect DC Operating Point

Single-supply operation of a fully differential amplifier is very easy to mess up (Fig. 25.15).

What happened? The two outputs have a 3.3 V DC difference between their operating point! Remember that in differential input circuits, there are two potential sources of DC. In this case, the designer correctly put an AC coupling capacitor between V_1 and R_1, but forgot to put one between R_3 and ground. When the second AC coupling capacitor is inserted, the correct DC operating point is established (Fig. 25.16).

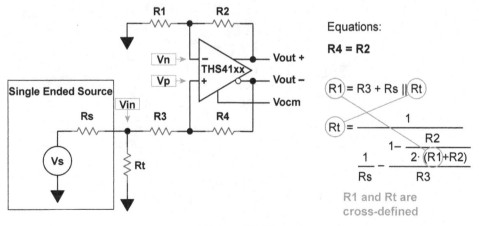

Figure 25.13
Terminating a fully differential amplifier.

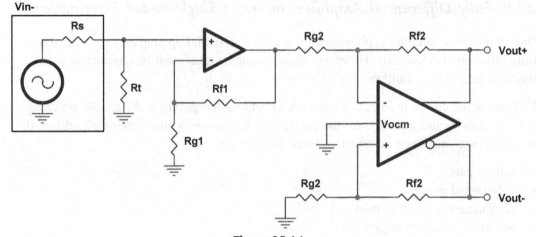

Figure 25.14
Using an input stage.

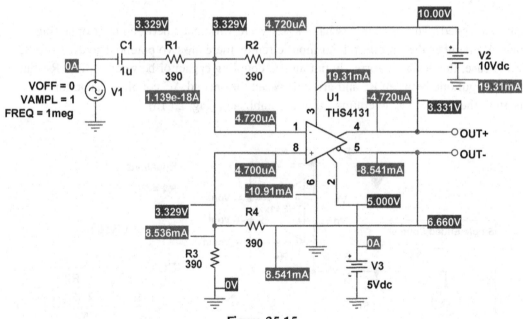

Figure 25.15
Incorrect DC operating point.

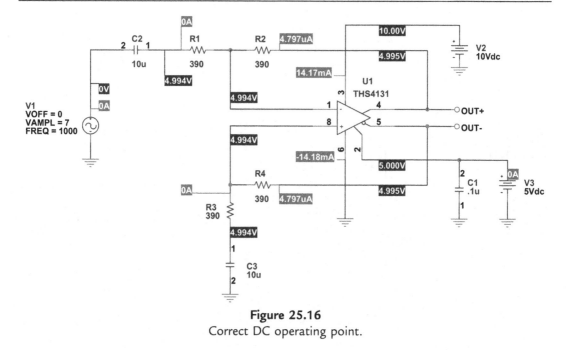

Figure 25.16
Correct DC operating point.

25.11 Fully Differential Amplifier: Incorrect Common-Mode Range

A very subtle, but equally destructive problem often arises from incorrect application of the V_{OCM} input of the fully differential amplifier when the amplifier frequency response has to include DC, making AC coupling capacitors impossible. Consider the circuit shown in Fig. 25.17.

The DC operating point appears to be correctly established, the outputs will swing around the V_{OCM} common-mode point, which is established at 5 V by V_3. But when an AC simulation is done, the results are terrible. What happened?

The problem comes when the input voltage range does not include the negative rail, in this case ground. There are two solutions for the problem. One is to offset the inputs to the same DC level as V_{OCM}. The other is to choose a fully differential amplifier, which includes the negative rail in its common-mode range. Fig. 25.18 illustrates the effect of V_{OCM} on the output signals:

V_{OCM} causes problems when it forces the outputs of the amplifier too close to the power supply rails. It is best to operate both inputs and V_{OCM} as close as possible to the same DC potential!

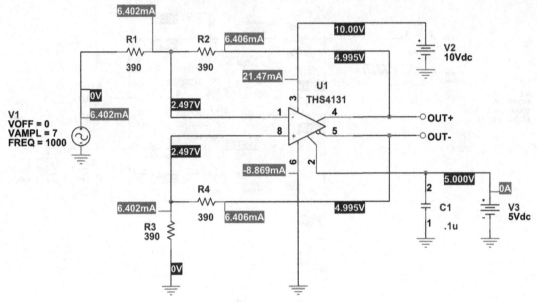

Figure 25.17
Common-mode error.

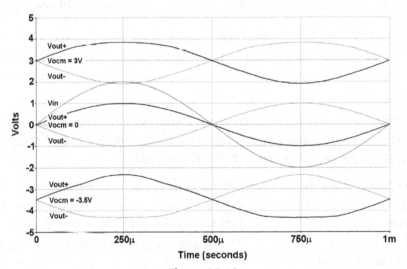

Figure 25.18
The effects of V_{OCM} on outputs.

25.12 The Number One Design Mistake!

I have saved the best, and most common mistake, for last. And it does not even involve an op amp. It involves support components: the decoupling capacitors!

In Chapter 1, I mentioned some part numbers that are etched in the memory of every design engineer, at least those involved in analog design. There is one other:

0.1 µF

Need to decouple? Ok, everybody knows you put a 0.1 µF capacitor on every power supply input and the job is done, right? I can disprove that truism very easily with two words:

Cell Phone

Put your cell phone near your prototype circuit, which is bypassed with 0.1 µF, and make a call while monitoring the output on a high-bandwidth oscilloscope. You will see horrendous 2.4 GHz leakage!

An alternative version of this problem came from some cellular telephone base station installers, who called in a panic: "we have 90 MHz noise running all over our system—and cannot figure out where it is coming from." A suspicion on my part asked them to tell me the exact coordinates where they were installing the system, and they provided the exact latitude and longitude. A quick check of the FCC database revealed the problem. I asked them—whether they were anywhere near the tower for W____ 90.5 FM, a 100,000 W NPR station listed at those coordinates. They told me on the phone that they could see the transmitter 5 ft away, they were colocating with the station!

The point of this is that their board was bypassed with 0.1 µF capacitors. While that worked fine for the digital portions of the board, the analog portions were being clobbered by radiation of the powerful 90.5 MHz FM station. Conventional thinking is that the lower the value of capacitance, the lower the frequencies it will filter. So 0.1 µF should get rid of just about everything because it is a very large value (relatively speaking). This conventional wisdom is wrong! The actual case is the exact opposite.

Where did the value 0.1 µF come from, anyway? A high-technology store near me used to have antiquated computer boards as a wall decoration. Backlit with white light, the translucent green boards made a pretty sight. But they were also populated with 0.1 µF decoupling capacitors. A quick survey of the circuitry revealed that the clock rate of the old computer had been 1 MHz.

So the 0.1 µF capacitor value seems to have come from bypassing transistor–transistor logic in the 1960s! Is not the time to rethink the issue a bit, in light of op amps and other analog components that can operate to frequencies of 3 GHz, especially when virtually every engineer carries a 2 W 2.4 GHz transmitter into the lab (cell phone)?

The reality of the situation is that a good 0.1 µF capacitor with an X7R dielectric exhibits a resonance in the 10 MHz region. This is due to parasitic inductance creating an LC circuit. Below 10 MHz, its impedance is capacitive, decreasing almost linearly on a logarithmic plot until it reaches the resonant frequency. Above the resonant frequency, the impedance is inductive. Since inductors resist the flow of high frequencies and only pass low frequencies, the decoupling capacitor is useless above its resonant frequency.

Looking at representative plots from capacitor manufacturers, at 100 MHz, the venerable 0.1 µF bypass capacitor has become an inductor with an XL of at least 1 Ω. By 2.4 GHz, XL has risen to above 10 Ω.

A good rule of thumb for effective bypassing is to put several capacitors in parallel. The standard 0.1 µF capacitor will do quite nicely for frequencies up to 10 MHz, 1000 pF NPO dielectric will do nicely up to 100 MHz, and 33 pF NPO will eliminate frequencies in the 2.4 GHz region. Bulk decoupling of the power supply as it enters the board will eliminate low-frequency ripple.

Here is a truism to replace the older one: when poor decoupling is suspected, decrease (do not increase) the value of the capacitance.

Review of Circuit Theory

A.1 Introduction

Although this book minimizes math, some algebra is germane to the understanding of analog electronics. Math and physics are presented here in the manner in which they are used later, so no practice exercises are given. For example, after the voltage divider rule is explained, it is used several times in the development of other concepts, and this usage constitutes practice.

Circuits are a mix of passive and active components. The components are arranged in a manner that enables them to perform some desired function. The resulting arrangement of components is called a circuit or sometimes a circuit configuration. The art portion of analog design is developing the circuit configuration. There are many published circuit configurations for almost any circuit task, thus all circuit designers need not be artists.

When the design has progressed to the point that a circuit exists, equations must be written to predict and analyze circuit performance. Textbooks are filled with rigorous methods for equation writing, and this review of circuit theory does not supplant those textbooks. But, a few equations are used so often that they should be memorized, and these equations are considered here.

There are almost as many ways to analyze a circuit as there are electronic engineers, and if the equations are written correctly, all methods yield the same answer. There are some simple ways to analyze the circuit without completing unnecessary calculations, and these methods are illustrated here.

A.2 Ohm's Law

Ohm's law is fundamental to all electronics. It can be applied to a single component, to any group of components, or to a complete circuit. When the current flowing through any portion of a circuit is known, the voltage dropped across that portion of the circuit is obtained by multiplying the current times the resistance (Eq. A.1).

$$V = IR \tag{A.1}$$

The current (I) flows through the resistance (R), and the voltage (V) is dropped across R (Fig. A.1).

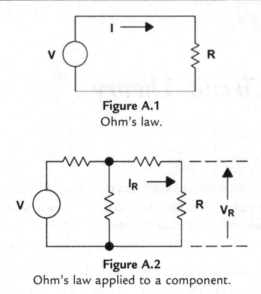

Figure A.1
Ohm's law.

Figure A.2
Ohm's law applied to a component.

In Fig. A.2, Ohm's law is applied to a single component. The current (I_R) flows through the resistor (R) and the voltage (V_R) is dropped across R. Notice, the same formula is used to calculate the voltage drop across R even though it is only a part of the circuit.

A.3 Kirchhoff's Voltage Law

Kirchhoff's voltage law states that the sum of the voltage drops in a series circuit equals the sum of the voltage sources. Otherwise, the source (or sources) voltage must be dropped across the passive components. When taking sums, keep in mind that the sum is an algebraic quantity. Kirchhoff's voltage law is illustrated in Fig. A.3 and Eqs. (A.2) and (A.3).

$$\sum V_{SOURCES} = \sum V_{DROPS} \tag{A.2}$$

$$V = V_{R1} + V_{R2} \tag{A.3}$$

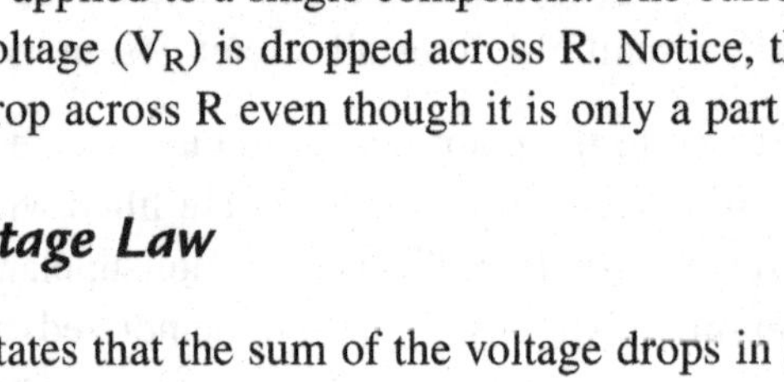

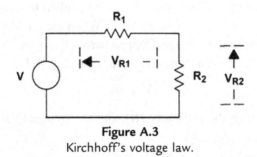

Figure A.3
Kirchhoff's voltage law.

A.4 Kirchhoff's Current Law

Kirchhoff's current law states: the sum of the currents entering a junction equals the sum of the currents leaving a junction. It makes no difference if a current flows from a current source, through a component, or through a wire, because all currents are treated identically. Kirchhoff's current law is illustrated in Fig. A.4 and Eqs. (A.4) and (A.5).

$$\sum I_{\text{IN}} = \sum I_{\text{OUT}} \tag{A.4}$$

$$I_1 + I_2 = I_3 + I_4 \tag{A.5}$$

A.5 Voltage Divider Rule

When the output of a circuit is not loaded, the voltage divider rule can be used to calculate the circuit's output voltage. Assume that the same current flows through all circuit elements (Fig. A.5). Eq. (A.6) is written using Ohm's law. Eq. (A.7) is written as Ohm's law across the output resistor.

$$I = \frac{V}{R_1 + R_2} \tag{A.6}$$

$$V_{\text{OUT}} = IR_2 \tag{A.7}$$

Substituting Eq. (A.6) into Eq. (A.7), and using algebraic manipulation yields Eq. (A.8).

$$V_{\text{OUT}} = V\frac{R_2}{R_1 + R_2} \tag{A.8}$$

A simple way to remember the voltage divider rule is that the output resistor is divided by the total circuit resistance. This fraction is multiplied by the input voltage to obtain the output voltage. Remember that the voltage divider rule always assumes that the output resistor is not loaded; the equation is not valid when the output resistor is loaded by a parallel component. Fortunately, most circuits following a voltage divider are input

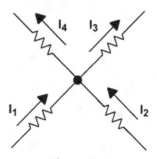

Figure A.4
Kirchhoff's current law.

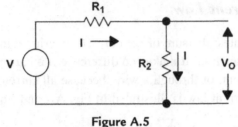

Figure A.5
Voltage divider rule.

circuits, and input circuits are usually high resistance circuits. When a fixed load is in parallel with the output resistor, the equivalent parallel value comprised of the output resistor and loading resistor can be used in the voltage divider calculations with no error. Many people ignore the load resistor if it is 10 times greater than the output resistor value, but this will lead to a 10% error.

A.6 Current Divider Rule

When the output of a circuit is not loaded, the current divider rule can be used to calculate the current flow in the output branch circuit (R_2). The currents I_1 and I_2 in Fig. A.6 are assumed to be flowing in the branch circuits. Eq. (A.9) is written with the aid of Kirchhoff's current law. The circuit voltage is written in Eq. (A.10) with the aid of Ohm's law. Combining Eqs. (A.9) and (A.10) yields Eq. (A.11).

$$I = I_1 + I_2 \tag{A.9}$$

$$V = I_1 R_1 = I_2 R_2 \tag{A.10}$$

$$I = I_1 + I_2 = I_2\frac{R_2}{R_1} + I_2 = I_2\left(\frac{R_1 + R_2}{R_1}\right) \tag{A.11}$$

Rearranging the terms in Eq. (A.11) yields Eq. (A.12).

$$I_2 = I\left(\frac{R_1}{R_1 + R_2}\right) \tag{A.12}$$

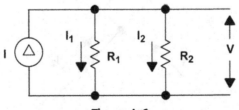

Figure A.6
Current divider rule.

The total circuit current divides into two parts, and the resistance (R_1) divided by the total resistance determines how much current flows through R_2. An easy method of remembering the current divider rule is to remember the voltage divider rule. Then modify the voltage divider rule such that the opposite resistor is divided by the total resistance, and the fraction is multiplied by the input current to get the branch current.

A.7 Thevenin's Theorem

There are times when it is advantageous to isolate a part of the circuit to simplify the analysis of the isolated part of the circuit. Rather than write loop or node equations for the complete circuit, and solving them simultaneously, Thevenin's theorem enables us to isolate the part of the circuit we are interested in. We then replace the remaining circuit with a simple series equivalent circuit, thus Thevenin's theorem simplifies the analysis.

There are two theorems that do similar functions. The Thevenin's theorem just described is the first, and the second is called Norton's theorem. Thevenin's theorem is used when the input source is a voltage source, and Norton's theorem is used when the input source is a current source. Norton's theorem is rarely used, so its explanation is left for the reader to dig out of a textbook if it is ever required.

The rules for Thevenin's theorem start with the component or part of the circuit being replaced. Referring to Fig. A.7, look back into the terminals (left from C and R_3 toward point XX in the figure) of the circuit being replaced. Calculate the no load voltage (V_{TH}) as seen from these terminals (use the voltage divider rule).

Look into the terminals of the circuit being replaced, short independent voltage sources, and calculate the impedance between these terminals. The final step is to substitute the Thevenin's equivalent circuit for the part you wanted to replace as shown in Fig. A.8.

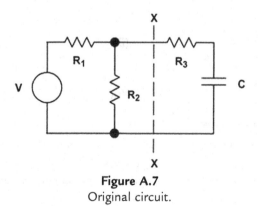

Figure A.7
Original circuit.

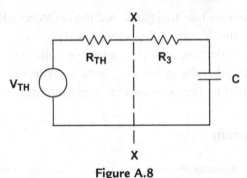

Figure A.8
Thevenin's equivalent circuit for Fig. A.7.

The Thevenin's equivalent circuit is a simple series circuit, thus further calculations are simplified. The simplification of circuit calculations is often sufficient reason to use Thevenin's theorem because it eliminates the need for solving several simultaneous equations. The detailed information about what happens in the circuit that was replaced is not available when using Thevenin's theorem, but that is no consequence because you had no interest in it.

As an example of Thevenin's theorem, let us calculate the output voltage (V_{OUT}) shown in Fig. A.9A. The first step is to stand on the terminals X–Y with your back to the output circuit, and calculate the open circuit voltage seen (V_{TH}). This is a perfect opportunity to use the voltage divider rule to obtain Eq. (A.13).

$$V_{TH} = V\frac{R_2}{R_1 + R_2} \tag{A.13}$$

Still standing on the terminals X–Y, step two is to calculate the impedance seen looking into these terminals (short the voltage sources). The Thevenin impedance is the parallel impedance of R_1 and R_2 as calculated in Eq. (A.14). Now replace the circuit to the left of X–Y with the Thevenin's equivalent circuit V_{TH} and R_{TH}.

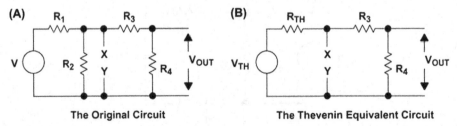

Figure A.9
Example of Thevenin's equivalent circuit. (A) The original circuit (B) the Thevenin's equivalent circuit.

$$R_{TH} = \frac{R_1 R_2}{R_1 + R_2} = R_1 \| R_2 \tag{A.14}$$

Note: Two parallel vertical bars (‖) are used to indicate parallel components as shown in Eq. (A.14).

The final step is to calculate the output voltage. Notice the voltage divider rule is used again. Eq. (A.15) describes the output voltage, and it comes out naturally in the form of a series of voltage dividers, which makes sense. That is another advantage of the voltage divider rule; the answers normally come out in a recognizable form rather than a jumble of coefficients and parameters.

$$V_{OUT} = V_{TH}\frac{R_4}{R_{TH} + R_3 + R_4} = V\left(\frac{R_2}{R_1 + R_2}\right) \frac{R_4}{\frac{R_1 R_2}{R_1 + R_2} + R_3 + R_4} \tag{A.15}$$

The circuit analysis is done the hard way in Fig. A.10, so you can see the advantage of using Thevenin's theorem. Two loop currents, I_1 and I_2, are assigned to the circuit. Then the loop Eqs. (A.16) and (A.17) are written.

$$V = I_1(R_1 + R_2) - I_2 R_2 \tag{A.16}$$
$$I_2(R_2 + R_3 + R_4) = I_1 R_2 \tag{A.17}$$

Eq. (A.17) is rewritten as Eq. (A.18) and substituted into Eq. (A.16) to obtain Eq. (A.19).

$$I_1 = I_2\frac{R_2 + R_3 + R_4}{R_2} \tag{A.18}$$

$$V = I_2\left(\frac{R_2 + R_3 + R_4}{R_2}\right)(R_1 + R_2) - I_2 R_2 \tag{A.19}$$

The terms are rearranged in Eq. (A.20). Ohm's law is used to write Eq. (A.21), and the final substitutions are made in Eq. (A.22).

$$I_2 = \frac{V}{\frac{R_2 + R_3 + R_4}{R_2}(R_1 + R_2) - R_2} \tag{A.20}$$

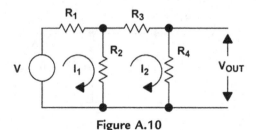

Figure A.10
Analysis done the hard way.

$$V_{OUT} = I_2R_4 \qquad\qquad (A.21)$$

$$V_{OUT} = V\frac{R_4}{\dfrac{(R_2 + R_3 + R_4)(R_1 + R_2)}{R_2} - R_2} \qquad\qquad (A.22)$$

This is a lot of extra work for no gain. Also, the answer is not in a usable form because the voltage dividers are not recognizable, thus more algebra is required to get the answer into usable form.

A.8 Superposition

Superposition is a theorem that can be applied to any linear circuit. Essentially, when there are independent sources, the voltages and currents resulting from each source can be calculated separately, and the results are added algebraically. This simplifies the calculations because it eliminates the need to write a series of loop or node equations. An example is shown in Fig. A.11.

When V_1 is grounded, V_2 forms a voltage divider with R_3 and the parallel combination of R_2 and R_1. The output voltage for this circuit (V_{OUT2}) is calculated with the aid of the voltage divider Eq. (A.23). The circuit is shown in Fig. A.12. The voltage divider rule yields the answer quickly.

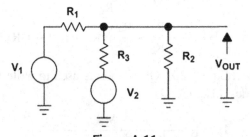

Figure A.11
Superposition example.

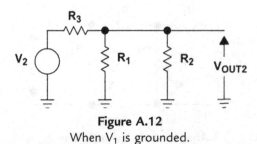

Figure A.12
When V_1 is grounded.

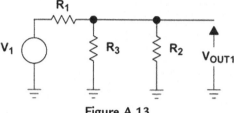

Figure A.13
When V_2 is grounded.

$$V_{OUT2} = V_2 \frac{R_1 \| R_2}{R_3 + R_1 \| R_2} \qquad (A.23)$$

Likewise, when V_2 is grounded (Fig. A.13), V_1 forms a voltage divider with R_1 and the parallel combination of R_3 and R_2, and the voltage divider theorem is applied again to calculate V_{OUT} (Eq. A.24).

$$V_{OUT1} = V_1 \frac{R_2 \| R_3}{R_1 + R_2 \| R_3} \qquad (A.24)$$

After the calculations for each source are made the components are added to obtain the final solution (Eq. A.25).

$$V_{OUT} = V_1 \frac{R_2 \| R_3}{R_1 + R_2 \| R_3} + V_2 \frac{R_1 \| R_2}{R_3 + R_1 \| R_2} \qquad (A.25)$$

The reader should analyze this circuit with loop or node equations to gain an appreciation for superposition. Again, the superposition results come out as a simple arrangement that is easy to understand. One looks at the final equation and it is obvious that if the sources are equal and opposite polarity, and when $R_1 = R_3$, then the output voltage is zero. Conclusions such as this are hard to make after the results of a loop or node analysis unless considerable effort is made to manipulate the final equation into symmetrical form.

Understanding Op Amp Parameters

B.1 Introduction

This appendix explains op amp data sheet parameters. There are usually three main sections of electrical tables in op amp data sheets.

B.1.1 Absolute Maximum Ratings

Absolute maximum ratings are those limits beyond which the life of individual devices may be impaired and are never to be exceeded in service or testing. Limits, by definition, are maximum ratings, so if double-ended limits are specified, the term will be defined as a range (e.g., operating temperature range).

B.1.2 Recommended Operating Conditions

Recommended operating conditions have a similarity to maximum ratings in that operation outside the stated limits could cause unsatisfactory performance. Recommended operating conditions, however, do not carry the implication of device damage if they are exceeded.

B.1.3 Electrical Characteristics

Electrical characteristics are measurable electrical properties of a device inherent in its design. They are used to predict the performance of the device as an element of an electrical circuit. The measurements that appear in the electrical characteristics tables are based on the device being operated within the recommended operating conditions.

Table B.1 lists op amp condition and parameter abbreviations in alphabetical order plus their corresponding description and units.

As the name of this page implies, some of Table B.1 is composed of parameters and some test conditions. Test conditions are conditions placed on the op amp when the parameters are measured. Some abbreviations are used for both a condition and a parameter.

Units listed in the units column of Table B.1 are part of the standard SI units of measure. Multiplier prefixes such as p (pico), M (mega), etc. are often used in data sheets.

373

Table B.1: Op Amp Condition and Parameter Table.

Abbreviations	Parameter	Units
αI_{IO}	Temperature coefficient of input offset current	A/°C
αV_{IO} or α_{VIO}	Temperature coefficient of input offset voltage	V/°C
A_D	Differential gain error	%
A_m	Gain margin	dB
A_{OL}	Open-loop voltage gain	dB
A_V	Large-signal voltage amplification (gain)	dB
A_{VD}	Differential large-signal voltage amplification	dB
B_1	Unity gain bandwidth	Hz
B_{OM}	Maximum-output-swing bandwidth	Hz
BW	Bandwidth	Hz
c_i	Input capacitance	F
C_{ic} or $C_{i(c)}$	Common-mode input capacitance	F
C_{id}	Differential input capacitance	F
C_L	Load capacitance	F
$\Delta V_{DD\pm(or\ CC\pm)}/\Delta V_{IO}$, or k_{SVS}	Supply voltage sensitivity	dB
CMRR or k_{CMR}	Common-mode rejection ratio	dB
f	Frequency	Hz
GBW	Gain bandwidth product	Hz
$I_{CC-(SHDN)}$, $I_{DD-(SHDN)}$	Supply current (shutdown)	A
I_{CC}, I_{DD}	Supply current	A
I_I	Input current range	A
I_{IB}	Input bias current	A
I_{IO}	Input offset current	A
I_n	Input noise current	$A/\sqrt{Hz}$
I_O	Output current	A
I_{OL}	Low-level output current	A
I_{OS}, or I_{SC}	Short-circuit output current	A
CMRR or k_{CMR}	Common-mode rejection ratio	dB
k_{SVR}	Supply rejection ratio	dB
k_{SVS}	Supply voltage sensitivity	dB
P_D	Power dissipation	W
PSRR	Power supply rejection ratio	dB
θ_{JA}	Junction to ambient thermal resistance	°C/W
θ_{JC}	Junction to case thermal resistance	°C/W
r_i	Input resistance	Ω
r_{id}, $r_{i(d)}$	Differential input resistance	Ω
R_L	Load resistance	Ω
R_{null}	Null resistance	Ω
r_o	Output resistance	Ω
R_S	Signal source resistance	Ω
R_t	Open-loop transresistance	Ω
SR	Slew rate	V/S
T_A	Operating temperature	°C
t_{DIS} or $t_{(off)}$	Turn-off time (shutdown)	s
t_{EN} or $t_{(on)}$	Turn-on time (shutdown)	s
t_f	Fall time	s

Table B.1: Op Amp Condition and Parameter Table.—cont'd

Abbreviations	Parameter	Units
THD	Total harmonic distortion	%
THD + N	Total harmonic distortion plus noise	%
T_J	Maximum junction temperature	°C
t_r	Rise time	s
t_s	Settling time	s
T_S or T_{stg}	Storage temperature	°C
V_{CC}, V_{DD}	Supply voltage	V
V_I	Input voltage range	V
V_{ic}	Common-mode input voltage	V
V_{ICR}	Input common-mode voltage range	V
V_{ID}	Differential input voltage	V
V_{DIR}	Differential Input voltage range	V
$V_{IH-SHDN}$ or $V_{(ON)}$	Turn-on voltage (shutdown)	V
$V_{IL-SHDN}$ or $V_{(OFF)}$	Turn-off voltage (shutdown)	V
V_{IN}	Input voltage (DC)	V
V_{IO}, V_{OS}	Input offset voltage	V
V_n	Equivalent input noise voltage	$V/\sqrt{Hz}$
$V_{N(PP)}$	Broad band noise	V P–P
V_{OH}	High-level output voltage	V
V_{OL}	Low-level output voltage	V
$V_{OM\pm}$	Maximum peak-to-peak output voltage swing	V
$V_{O(PP)}$	Peak-to-peak output voltage swing	V
$V_{(STEP)PP}$	Step voltage peak-to-peak	V
X_T	Crosstalk	dB
Z_O	Output impedance	Ω
Z_t	Open-loop transimpedance	Ω
Φ_D	Differential phase error	Degree
Φ_m	Phase margin	Degree
	Bandwidth for 0.1 dB flatness	Hz
	Case temperature for 60 s	°C
	Continuous total dissipation	W
	Differential gain error	%
	Differential phase error	Degree
	Duration of short-circuit current	s
	Input offset voltage long-term drift	V/month
	Lead temperature for 10 or 60 s	°C

B.2 Temperature Coefficient of the Input Offset Current (αI_{IO})

The temperature coefficient of the input offset current, αI_{IO}, is defined as the ratio of the change in input offset current to the change in the die temperature. This is an average value for the specified temperature range.

αI_{IO} specifies the expected input offset current drift over temperature. Its units are $\mu A/°C$. I_{IO} is measured at the temperature extremes of the part, and αI_{IO} is computed as $\Delta I_{IO}/\Delta°C$.

Normal aging in semiconductors causes changes in the characteristics of devices. The input offset voltage long-term drift specifies how I_{IO} is expected to change with time. Its units are amperes/month.

B.3 Temperature Coefficient of Input Offset Voltage (αV_{IO} or α_{VIO})

The temperature coefficient of input offset voltage, αV_{IO} or α_{VIO}, is defined as the ratio of the change in input offset voltage to the change in the die temperature. This is an average value for the specified temperature range.

αV_{IO} specifies the expected input offset drift over temperature. Its units are $V/°C$. V_{IO} is measured at the temperature extremes of the part, and αV_{IO} is computed as $\Delta V_{IO}/\Delta°C$.

Normal aging in semiconductors causes changes in the characteristics of devices. The input offset voltage long-term drift specifies how V_{IO} is expected to change with time. Its units are $\mu V/month$.

B.4 Differential Gain Error (A_D)

The differential gain error parameter, A_D, is defined as the change in AC gain with change in DC level. The AC signal is 40 IRE (0.28 VPK) and the DC level change is ± 100 IRE (± 0.7 V). Typically tested at 3.58 MHz (NTSC) or 4.43 MHz (PAL) carrier frequencies. It is represented in units of percent. With the conversion to digital broadcast video, this parameter is quickly becoming irrelevant.

B.5 Gain Margin Parameter (A_m)

Gain margin, A_m, is defined as the absolute value of the difference in gain between the unity gain point and the gain at the -180 degrees phase shift point. It is measured open loop and expressed in units of decibels, dB.

Gain margin (A_m) and phase margin (Φ_m) are different ways of specifying the stability of the circuit. Since rail-to-rail output op amps have higher output impedance, a significant phase shift is seen when driving capacitive loads. This extra phase shift erodes the phase margin, and for this reason most CMOS op amps with rail-to-rail outputs have limited ability to drive capacitive loads. Fig. B.1 shows the gain margin graphically.

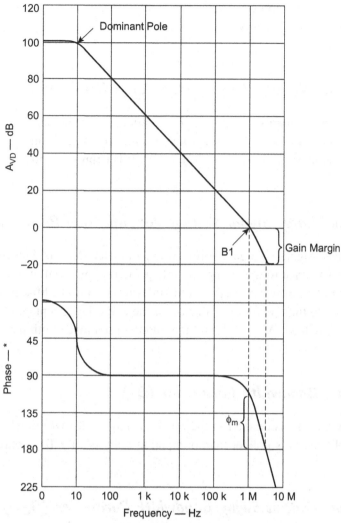

Figure B.1
Gain and phase margin.

B.6 Open-Loop Voltage Gain Parameter (A_{OL})

The open-loop voltage gain parameter, A_{OL}, is defined as the ratio of change in output voltage to the change in voltage across the input terminals. Usually, the DC value and a graph showing the frequency dependence are shown in the data sheet. It is expressed either unitless or in dB.

A_{OL} is similar to the open-loop gain A_{VD} of the amplifier except A_{VD} is usually measured with an output load. A_{OL} is usually measured without any load. Both parameters are measured open loop.

B.7 Large-Signal Voltage Amplification Gain Condition (A$_V$)

The large-signal voltage amplification or gain condition, A_V, is defined as the ratio of change in output voltage to the change in voltage across the input terminals that is set up for a test of parameters such as Z_O or THD + N. It is expressed either unitless or in dB.

B.8 Differential Large-Signal Voltage Amplification Parameter (A$_{VD}$)

The differential large-signal voltage amplification parameter, A_{VD}, is defined as the ratio of change in output voltage to the change in voltage across the input terminals. It is expressed either unitless or in dB. A_{VD} is sometimes referred to as differential voltage gain. A_{VD} is similar to the open-loop gain A_{OL} of the amplifier except A_{OL} is usually measured without any load. A_{VD} is usually measured with a load. Both parameters are measured open loop.

B.9 Unity Gain Bandwidth Parameter (B$_1$)

Unity gain bandwidth, B_1, is defined as the range of frequencies within which the open-loop voltage amplification is greater than or equal to unity (0 dB). B_1 is expressed in units of Hertz.

B.10 Maximum-Output-Swing Bandwidth Parameter (B$_{OM}$)

The maximum-output-swing bandwidth parameter, B_{OM}, is defined as the maximum frequency that the output swing is above a specified value or at the extents of its linear range. B_{OM} is also called full-power bandwidth. B_{OM} is expressed in units of Hertz.

The limiting factor for B_{OM} is slew rate (SR). As the frequency gets higher and higher the output becomes SR limited and cannot respond quickly enough to maintain the specified output voltage swing. The following equation expresses the relationship between B_{OM} and SR.

$$B_{OM} = \frac{SR}{2\pi V_{(PP)}} \tag{B.1}$$

B.11 Bandwidth Parameter (BW)

Bandwidth, BW, is defined as the maximum frequency that an op amp circuit can deliver the specified output. The specified output varies and includes conditions such as small signal (-3 dB), 0.1 dB flatness, and full power. BW is expressed in units of Hertz.

B.12 Input Capacitance Parameter (C_i)

The input capacitance parameter, C_i, is defined as the capacitance between the input terminals of an op amp with either input grounded. It is expressed in units of farads.

C_i is one of a group of parasitic elements affecting input impedance. Fig. B.2 shows a model of the resistance and capacitance between each input terminal and ground and between the two terminals. There is also parasitic inductance, but the effects are negligible at low frequency. Input impedance is a design issue when the source impedance is high. The input loads the source.

Input capacitance, C_i, is measured between the input terminals with either input grounded. C_i is usually a few picofarads. In the figure above, if V_p is grounded, then $C_i = C_d \| C_n$.

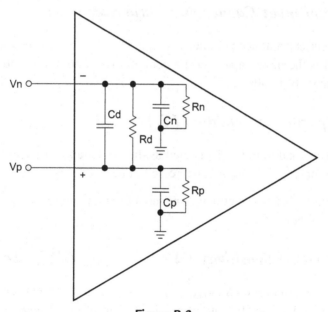

Figure B.2
Input parasitic elements.

Sometimes common-mode input capacitance, C_{ic}, is specified. In the figure above, if V_p is shorted to V_n, then $C_{ic} = C_p \| C_n$. C_{ic} is the input capacitance a common-mode source would see referenced to ground.

B.13 Common-Mode Input Capacitance Parameter (C_{ic} or $C_{i(c)}$)

The common-mode input capacitance parameter, C_{ic} or $C_{i(c)}$, is defined as the input capacitance a common-mode source would see to ground. It is expressed in units of farads.

C_{ic} is one of a group of parasitic elements affecting input impedance. Fig. B.2 shows a model of the resistance and capacitance between each input terminal and ground and between the two terminals. There is also parasitic inductance, but the effects are negligible at low frequency. Input impedance is a design issue when the source impedance is high. The input loads the source.

Input capacitance, C_i, is measured between the input terminals with either input grounded. C_i is usually a few picofarads. In the figure above, if V_p is grounded, then $C_i = C_d \| C_n$.

Sometimes common-mode input capacitance, C_{ic} is specified. In the figure above, if V_p is shorted to V_n, then $C_{ic} = C_p \| C_n$. C_{ic} is the input capacitance a common mode source would see when referenced to ground.

B.14 Differential Input Capacitance Parameter (C_{id})

The differential input capacitance parameter, C_{id}, is the same as the common-mode input capacitance, C_{id}. It is the input capacitance a common-mode source would see to ground. It is expressed in units of farads.

B.15 Load Capacitance Condition (C_L)

The load capacitance condition, C_L, is defined as the capacitance between the output terminal of an op amp and ground. It is expressed in units of farads.

C_L is a capacitive load that is sometimes connected to an op amp when parameters such as SR, t_s, Φ_m, or A_m are being tested.

B.16 Supply Voltage Sensitivity ($\Delta V_{DD\pm(or\ CC\pm)}/\Delta V_{IO}$, or k_{SVS})

The power supply rejection ratio ($\Delta V_{DD\pm(or\ CC\pm)}/\Delta V_{IO}$) is the same as the supply rejection ratio, k_{SVR}. It is defined as the absolute value of the ratio of the change in supply voltages to the resulting change in input offset voltage. Typically both supply voltages are varied symmetrically. It is expressed in dB.

The power voltage affects the bias point of the input differential pair. Because of the inherent mismatches in the input circuitry, changing the bias point changes the offset voltage, which, in turn, changes the output voltage.

For a dual-supply op amp, $k_{SVR} = \Delta V_{CC\pm}/\Delta V_{OS}$ or $\Delta V_{DD\pm}/\Delta V_{OS}$. The term $\Delta V_{CC\pm}$ means that the plus and minus power supplies are changed symmetrically. For a single-supply op amp, $k_{SVR} = \Delta V_{CC}/\Delta V_{OS}$ or $\Delta V_{DD}/\Delta V_{OS}$. Also note that the mechanism that produces k_{SVR} is the same as for common-mode rejection ratio (CMRR). Therefore k_{SVR} as published in the data sheet is a DC parameter like CMRR. When k_{SVR} is graphed versus frequency, it falls off as the frequency increases.

Switching power supplies produce noise frequencies from 50 to 500 kHz and higher. k_{SVR} is almost zero at these frequencies so that noise on the power supply results in noise on the output of the op amp.

Proper bypassing techniques must be used.

B.17 Common-Mode Rejection Ratio Parameter (CMRR or k_{CMR})

The common-mode rejection ratio parameter, CMRR or k_{CMR}, is defined as the ratio of differential voltage amplification to common-mode voltage amplification. This is measured by determining the ratio of a change in input common-mode voltage to the resulting change in input offset voltage. It is expressed in dB.

Ideally CMRR or k_{CMR} would be infinite with common-mode voltages being totally rejected.

The common-mode input voltage affects the bias point of the input differential pair. Because of the inherent mismatches in the input circuitry, changing the bias point changes the offset voltage, which, in turn, changes the output voltage. The real mechanism at work is $\Delta V_{OS}/\Delta V_{COM}$.

A common source of common-mode interference voltage is 50–60 Hz AC noise. Care must be used to ensure that the CMRR of the op amp is not degraded by other circuit components. High values of resistance make the circuit vulnerable to common-mode (and other) noise pick up. It is usually possible to scale resistors down and capacitors up to preserve circuit response.

B.18 Frequency Condition (f)

Frequency condition, f, is the frequency available to a circuit for a specific parameter test. It is expressed in Hertz.

B.19 Op Amp Gain Bandwidth Product Parameter (GBW)

Gain bandwidth product, GBW, is defined as the product of the open-loop voltage gain and the frequency at which it is measured. GBW is expressed in units of Hertz. Fig. B.1 shows the open-loop bandwidth graphically.

GBW is similar to unity-gain bandwidth (B_1). While B_1 specifies the frequency at which the gain of the op amp is 1, GBW specifies the gain-bandwidth product of the op amp at a frequency that may be different than the B_1.

GBW is constant for voltage-feedback amplifiers. It does not have much meaning for current-feedback amplifiers because there is not a linear relationship between gain and bandwidth.

When an op amp is selected for a specific application both the bandwidth and the SR should be taken into account (along with other factors including power consumption, distortion, price, etc.).

B.20 Supply Current (Shutdown) Parameter ($I_{CC(SHDN)}$ or $I_{DD(SHDN)}$)

The supply current (shutdown) parameter, $I_{CC(SHDN)}$ or $I_{DD(SHDN)}$, is defined as the current into the V_{CC+} (V_{DD+}) or V_{CC-} (V_{DD-}) terminal of the amplifier while it is turned off. It is expressed in units of amperes.

B.21 Supply Current Parameter (I_{CC} or I_{DD})

The supply current parameter, I_{CC} or I_{DD}, is defined as the current into the V_{CC+} (V_{DD+}) or V_{CC-} (V_{DD-}) terminal of the op amp while it is operating without load and the input and/or output is at virtual ground. It is expressed in units of amperes.

B.22 Input Current Range Parameter (I_I)

The input current range parameter, I_I, is defined as the amount of current that can be sourced or sinked by the op amp input. It is usually specified as an absolute maximum rating expressed in units of amperes.

B.23 Input Bias Current Parameter (I_{IB})

The input bias current parameter, I_{IB}, is defined as the average of the currents into the two input terminals with the output at a specified level. It is expressed in units of amperes.

The input circuitry of all op amps requires a certain amount of bias current for proper operation. The input bias current, I_{IB}, is computed as the average of the two inputs:

$$I_{IB} = \frac{(I_N + I_P)}{2} \qquad \text{(B.2)}$$

CMOS and JFET inputs offer much lower input current than standard bipolar inputs.

Input bias current is of concern when the source impedance is high. If the op amp has high input bias current, it will load the source and a lower than expected voltage is seen. If the source impedance is high, the best solution is to use an op amp with either CMOS or JFET input. The source impedance can also be lowered by using a buffer stage to drive the op amp that has high input bias current.

B.24 Input Offset Current Parameter (I_{IO})

The input offset current parameter, I_{IO}, is defined as the difference between the currents into the two input terminals of an op amp with the output at the specified level. It is expressed in units of amperes.

B.25 Input Noise Current Parameter (I_n)

The input noise current parameter, I_n, is defined as the internal noise current reflected back to an ideal current source in parallel with the input pins. It is expressed in units of $A/\sqrt{Hz}$.

It is important for a designer to calculate noise that the device will deliver in an application. The simplest way is to calculate this noise is to use the following equation:

$$e_{nt} = \sqrt{V_n^2 + (I_n \times R_s)^2} \qquad \text{(B.3)}$$

where, e_{nt} = total noise voltage; V_n = voltage noise ($nV/\sqrt{Hz}$); I_n = current noise ($pA/\sqrt{Hz}$); R_s = source resistance (Ω).

B.26 Output Current Parameter (I_O)

The output current parameter, I_O, is defined as the amount of current that may be drawn from the op amp output. Usually I_O is expressed in units of amperes.

B.27 Low-level Output Current Condition (I_{OL})

The low-level output current condition, I_{OL}, is defined as the current into an output that is supplied during the test for V_{OL}. It is usually expressed in units of amperes.

B.28 Short-Circuit Output Current Parameters (I_{OS} or I_{SC})

The short-circuit output current parameter, I_{OS} or I_{SC}, is defined as the maximum output current available from the amplifier with the output shorted to ground, to either supply, or to a specified point. Sometimes a low-value series resistor is specified. It is usually expressed in units of amperes.

It is important to observe power dissipation ratings to keep the junction temperature below the absolute maximum rating when the output is heavily loaded or shorted. See the absolute maximum ratings section of the part's data sheet for more information.

B.29 Supply Rejection Ratio Parameter (k_{SVR})

The supply rejection ratio, k_{SVR}, is the same as the power supply rejection ratio, PSRR. It is defined as the absolute value of the ratio of the change in supply voltages to the resulting change in input offset voltage. Typically both supply voltages are varied symmetrically. It is expressed in dB.

The power voltage affects the bias point of the input differential pair. Because of the inherent mismatches in the input circuitry, changing the bias point changes the offset voltage, which, in turn, changes the output voltage.

For a dual-supply op amp, $k_{SVR} = \Delta V_{CC\pm}/\Delta V_{OS}$ or $\Delta V_{DD\pm}/\Delta V_{OS}$. The term $\Delta V_{CC\pm}$ means that the plus and minus power supplies are changed symmetrically. For a single-supply op amp, $k_{SVR} = \Delta V_{CC}/\Delta V_{OS}$ or $\Delta V_{DD}/\Delta V_{OS}$. Also note that the mechanism that produces k_{SVR} is the same as for CMRR. Therefore k_{SVR} as published in the data sheet is a DC parameter like CMRR. When k_{SVR} is graphed versus frequency, it falls off as the frequency increases.

Switching power supplies produce noise frequencies from 50 to 500 kHz and higher. k_{SVR} is almost zero at these frequencies so that noise on the power supply results in noise on the output of the op amp.

Proper bypassing techniques must be used.

B.30 Power Dissipation Parameter (P_D)

The power dissipation, P_D, is defined as the power supplied to the device less any power delivered from the device to a load. Note: At no load: $P_D = V_{CC+} \times I_{CC}$ or $P_D = V_{DD+} \times I_{DD}$. It is expressed in units of Watts.

B.31 Power Supply Rejection Ratio Parameter (PSRR)

The power supply rejection ratio, PSRR, is the same as the supply rejection ratio, k_{SVR}—see Section B.28.

B.32 Junction to Ambient Thermal Resistance Parameter (θ_{JA})

The junction to ambient thermal resistance parameter, θ_{JA}, is defined as the ratio of the difference in temperature from the die junction to the ambient air and the power dissipated by the die. θ_{JA} is expressed in units of °C/W.

θ_{JA} is dependent on the case to the ambient thermal resistance as well as the θ_{JC} parameter. θ_{JA} is a better indicator of thermal resistance when the package is not well thermally sinked to other components in the assembly. θ_{JA} is listed in the data sheet for different packages. It is useful for evaluating which package is least likely to overheat and to determine what the die temperature is when the ambient temperature and power dissipation are known.

B.33 Junction to Case Thermal Resistance Parameter (θ_{JC})

The junction to case thermal resistance parameter, θ_{JC}, is defined as the ratio of the difference in temperature from the die junction to the case and the power dissipated by the die. θ_{JC} is expressed in units of °C/W.

θ_{JC} is not dependent on the case to the ambient thermal resistance as is the θ_{JA} parameter. θ_{JC} is a better indicator of thermal resistance when the package is to be thermally sinked to other components in the assembly.

θ_{JC} is listed in the data sheet for different packages. It is useful for evaluating which package is least likely to overheat and to determine what the die temperature is when the case temperature and power dissipation are known.

B.34 Input Resistance Parameter (r_i)

The input resistance parameter, r_i, is defined as the DC resistance between the input terminals with either input grounded. It is expressed in units of Ω.

r_i is one of a group of parasitic elements affecting input impedance. Fig. B.2 shows a model of the resistance and capacitance between each input terminal and ground and between the two terminals. There is also parasitic inductance, but the effects are negligible at low frequency. Input impedance is a design issue when the source impedance is high. The input loads the source.

Input resistance, r_i, is the resistance between the input terminals with either input grounded. In the figure above, if V_p is grounded, then $r_i = R_d \| R_n$. r_i ranges from 10^7 to 10^{12} Ω, depending on the type of input.

Sometimes common-mode input resistance, r_{ic}, is specified. In the figure above, if V_p is shorted to V_n, then $r_{ic} = R_p \| R_n$. r_{ic} is the input resistance a common-mode source would see referenced to ground.

B.35 Differential Input Resistance Parameter (r_{id} or $r_{i(d)}$)

The differential input resistance, r_{id} or $r_{i(d)}$, is defined as the small-signal resistance between two ungrounded input terminals. It is expressed in units of Ω.

r_{id} is one of a group of parasitic elements affecting input impedance. Fig. B.2 shows a model of the resistance and capacitance between each input terminal and ground and between the two terminals. There is also parasitic inductance, but the effects are negligible at low frequency. Input impedance is a design issue when the source impedance is high. The input loads the source.

Input resistance, r_i, is the resistance between the input terminals with either input grounded. In the figure above, if V_p is grounded, then $r_i = R_d \| R_n$. r_i ranges from 10^7 to 10^{12} Ω, depending on the type of input.

Sometimes common-mode input resistance, r_{ic}, is specified. In the figure above, if V_p is shorted to V_n, then $r_{ic} = R_p \| R_n$. r_{ic} is the input resistance a common mode source would see referenced to ground. In Fig. B.2, $r_{id} = R_d$.

B.36 Load Resistance Condition (R_L)

The load resistance condition, R_L, is defined as the DC resistance that is attached from the output of an op amp to ground during a test for a parameter such as A_{VD}, SR, THD + D, $t_{(on)}$, $t_{(off)}$, GBW, t_s, Φ_m, and A_m. It is expressed in units of Ω.

B.37 Null Resistance Condition (R_L)

The null resistance condition, R_L, is defined as the DC resistance that is attached in series with C_L when testing for parameters such as phase margin and gain margin. It is expressed in units of Ω.

B.38 Output Resistance Parameters (r_o)

The output resistance parameter, r_o, is defined as the DC resistance that is placed in series with the output of an ideal amplifier and the output terminal for simulation of the real device. It is expressed in units of Ω.

B.39 Signal Source Condition (R_S)

The signal source condition, R_S, is defined as the output resistance of a signal source. It is expressed in units of Ω. R_S is used as a test condition when measuring parameters such as V_{IO}, α_{VIO}, I_{IO}, I_{IB}, and CMMR. A typical value for R_S used in these parameter tests is 50 Ω.

B.40 Open-Loop Transresistance Parameters (R_t)

In a transimpedance or current-feedback amplifier, the open-loop transresistance parameter, R_t is defined as the ratio of change in DC output voltage to the change in DC current at the inverting input. It is expressed in units of Ω.

B.41 Op Amp Slew Rate Parameter (SR)

The slew rate parameter, SR, is defined as the rate of change in the output voltage caused by a step change at the input. It is expressed in V/s. The SR parameter of an op amp is the maximum SR it will pass and is generally specified with a gain of 1. Fig. B.3 shows SR graphically.

For an amplifier to pass a signal without distortion due to insufficient SR, the amplifier must have at least the maximum SR of the signal. The maximum SR of a sine wave occurs as it crosses zero. The following equation defines this SR:

$$SR = 2\pi fV \tag{B.4}$$

where, f = frequency of the signal; V = peak voltage of the signal.

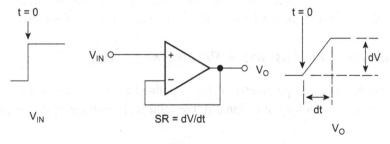

Figure B.3
Slew rate.

The SR is sometimes represented as SR+ and SR−. SR+ is the abbreviation for the SR for a positive transition, and SR− is the abbreviation for the SR for a negative transition. Many applications are best served when SR+ and SR− are the same magnitude.

The primary factor controlling SR in most op amps is an internal compensation capacitor, which is added to make the op amp unity gain stable. When an op amp is selected for a specific application both the bandwidth and the SR should be taken into account.

B.42 Operating Free-Air Temperature Condition (T_A)

The operating free-air temperature condition, T_A, is defined as the free-air temperature over which the op amp is being operated. Some of the other parameters may change with temperature, leading to degraded operation at temperature extremes. T_A is expressed in units of °C.

A range of T_A is listed in a data sheet's absolute maximum ratings table because stress beyond those listed may cause permanent damage to the device. Functional operation to this limit is not implied and may affect reliability.

B.43 Turn-Off Time (Shutdown) Parameter (t_{DIS} or $t_{(off)}$)

The turn-off time (shutdown) parameter, t_{DIS} or $t_{(off)}$, is defined as the time from when the turn-off voltage is applied to the shutdown pin to when the supply current has reached half of its final value. It is expressed in units of seconds.

B.44 Turn-On Time (Shutdown) Parameters (t_{EN})

The turn-on time (shutdown) parameter, t_{EN}, is defined as the time from when the turn-on voltage is applied to the shutdown pin to when the supply current has reached half of its final value. It is expressed in units of seconds.

B.45 Fall Time Parameter (t_f)

The fall time parameter, t_f, is defined as the time required for an output voltage step to change from 90% to 10% of its final value. It is expressed in units of seconds.

B.46 Total Harmonic Distortion Parameter (THD)

The total harmonic distortion parameter, THD, is defined as the ratio of the RMS voltage of the harmonics of the fundamental signal to the total RMS voltage at the output. THD is expressed in dBc or %. THD does not account for the noise as does the total harmonic distortion plus noise parameter.

B.47 Total Harmonic Distortion Plus Noise Parameter (THD + N)

The total harmonic distortion plus noise parameter, THD + N, is defined as the ratio of the RMS noise voltage plus the RMS harmonic voltage of the fundamental signal to the fundamental RMS voltage signal at the output. It is expressed in dBc or %.

THD + N compares the frequency content of the output signal to the frequency content of the input. Ideally, if the input signal is a pure sine wave, the output signal is a pure sine wave. Due to nonlinearity and noise sources within the op amp, the output is never pure.

To simplify further, THD + N is the ratio of all other frequency components to the fundamental.

$$\text{THD} + \text{N} = \left[\frac{(\sum \text{Harmonic voltages} + \text{Noise voltages})}{\text{Fundamental}} \right] \times 100\% \qquad (\text{B.5})$$

B.48 Maximum Junction Temperature Parameter (T_J)

The maximum junction temperature parameter, T_J, is defined as the temperature over which the die may be operated. Some of the other parameters may change with temperature, leading to degraded operation at temperature extremes. T_J is expressed in units of °C.

T_J is listed in the absolute maximum ratings table because stress beyond those listed may cause permanent damage to the device. Functional operation to this limit is not implied and may affect reliability.

B.49 Rise Time Parameter (t_r)

The rise time parameter, t_r, is defined as the time required for an output voltage step to change from 10% to 90% of its final value. It is expressed in units of seconds.

B.50 Settling Time Parameter (t_s)

The settling time parameter, t_s, is defined as the time required for the output voltage to settle within the specified error band of the final value with a step change at the input. It is also known as total response time, t_{tot}. It is expressed in units of seconds.

Settling time is greatly affected by the application, such as a filter circuit where capacitors can store energy. Therefore, it should be measured in-circuit. It is particularly a design issue in data acquisition circuits when signals are changing rapidly. An example is when using an op amp following a multiplexer to buffer the input to an A to D converter. Step changes can occur at the input to the op amp when the multiplexer changes channels. The output of the op amp must settle to within a certain tolerance before the A to D converter samples the signal.

B.51 Storage Temperature Parameter (T_S or T_{stg})

The storage temperature parameter, T_S or T_{stg}, is defined as the temperature over which the op amp may be stored (unpowered) for long periods of time without damage. It is expressed in units of °C.

B.52 Supply Voltage Condition (V_{CC} or V_{DD})

The supply voltage condition, V_{CC} or V_{DD}, is defined as the bias voltage applied to the op amp power supply pin(s). For single-supply applications, it is specified as a positive value; and for split-supply applications, it is specified as a $\pm$ value, referenced to analog ground. It is expressed in units of volts.

V_{CC} or V_{DD} is often defined in the maximum ratings, recommended operating conditions and as a test condition in parameter tables and graphs because the voltage supplied has an important impact on the way a circuit operates. It is also used as one of the axis variables in some of the characteristic graphs.

B.53 Input Voltage Range Condition or Parameter (V_I)

The input voltage range parameter, V_I, is defined as the range of input voltages that may be applied to either the IN+ or IN− inputs. The input voltage range condition, V_I, is defined as the voltage delivered to a circuit input when testing for V_O on a graph such as "large signal inverting pulse response versus time." V_I is expressed in units of volts for either a condition or parameter.

B.54 Common-Mode Input Voltage Condition (V_{ic})

The common-mode input voltage condition, V_{ic}, is defined as the voltage that is common to both input pins. V_{ic} as expressed in units of volts, V. V_{io} set at $V_{DD}/2$ (for single-supply op amps) is often used as a condition when testing for various parameters including V_{IO}, I_{IO}, I_{IB}, V_{OH}, and V_{OL}.

When a two-wire signal is subject to noise and this noise is being received equally on both signal lines, then it can be rejected by a differential amplifier with good common-mode rejection.

B.55 Common-Mode Input Voltage Range Parameter (V_{ICR})

The common-mode input voltage range parameter, V_{ICR}, is defined as the range of common-mode input voltage that, if exceeded, may cause the operational amplifier to cease functioning properly. This sometimes is taken as the voltage range over which the input offset voltage remains within a set limit. V_{ICR} is expressed in units of volts.

The input common voltage, V_{IC}, is defined as the average voltage at the inverting and noninverting input pins. If the common-mode voltage gets too high or too low, the inputs will shut down and proper operation ceases. The common-mode input voltage range, V_{ICR}, specifies the range over which normal operation is guaranteed. The trends toward lower and single-supply voltages make V_{ICR} of increasing concern.

Rail-to-rail input is required when a noninverting unity gain amplifier is used and the input signal ranges between both power rails. An example of this is the input of an analog to digital converter in a low-voltage, single-supply system.

High-side sensing circuits require operation at the positive input rail.

B.56 Differential Input Voltage Parameter (V_{ID})

The differential input voltage parameter, V_{ID}, is defined as the voltage at the noninverting input with respect to the inverting input. V_{ID} is expressed in units of volts.

V_{ID} is usually defined in the absolute maximum ratings table because stress beyond this limit may cause permanent damage to the device.

B.57 Differential Input Voltage Range Parameter (V_{DIR})

The input common-mode voltage range parameter, V_{DIR}, is defined as the range of differential input voltage that, if exceeded, may cause the operational amplifier to cease functioning properly. V_{DIR} is expressed in units of volts.

Some devices have protection built into them, and the current into the input needs to be limited. Normally, differential input mode voltage limit is not a design issue.

B.58 Turn-on Voltage (Shutdown) Parameter ($V_{IH\text{-}SHDN}$ or $V_{(ON)}$)

The turn-on voltage (shutdown) parameter, $V_{IH\text{-}SHDN}$ or $V_{(ON)}$, is defined as the voltage required on the shutdown pin to turn the device on. It is expressed in units of volts.

B.59 Turn-off Voltage (Shutdown) Parameters ($V_{IL\text{-}SHDN}$ or $V_{(OFF)}$)

The turn-off voltage (shutdown) parameter, $V_{IL\text{-}SHDN}$ or $V_{(OFF)}$, is defined as the voltage required on the shutdown pin to turn the device off. It is expressed in units of volts.

B.60 Input Voltage Condition (V_{IN})

The input voltage condition, V_{IN}, is defined as the DC voltage delivered to a circuit input when testing for V_n. V_I is expressed in units of volts.

B.61 Input Offset Voltage Parameter (V_{IO} or V_{OS})

The input offset voltage parameter, V_{IO} or V_{OS}, is defined as the DC voltage that must be applied between the input terminals to cancel DC offsets within the op amp. It is expressed in units of volts.

All op amps require a small voltage between their inverting and noninverting inputs to balance mismatches due to unavoidable process variations. The required voltage is known as the input offset voltage and is abbreviated V_{IO}. V_{IO} is an input referred parameter. This means that it is amplified by the positive closed-loop gain of the circuit.

Input offset voltage is of concern anytime that DC accuracy is required of the circuit. One way to null the offset is to use external null inputs on a single op amp package (Fig. B.4). A potentiometer is connected between the null inputs with the adjustable terminal

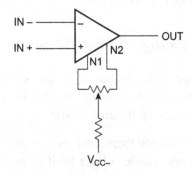

Figure B.4
Offset voltage adjust.

connected to the negative supply through a series resistor. The input offset voltage is nulled by shorting the inputs and adjusting the potentiometer until the output is zero.

B.62 Equivalent Input Noise Voltage Parameter (V$_n$)

The equivalent input noise voltage parameter, V_n, is defined as the internal noise voltage reflected back to an ideal voltage source in parallel with the input pins at a specific frequency. V_n is expressed in units of $V/\sqrt{Hz}$.

When this parameter is measured, the noise measured at the output (with the input connected to virtual ground) is divided by the gain of the amplifier circuit. This is the amplitude of noise at the input that would be amplified by an ideal amplifier to cause an equivalent signal at the output.

V_n is sometimes defined at several frequencies in the operating characteristics table or as a graph.

B.63 High-Level Output Voltage Condition or Parameter (V$_{OH}$)

The high-level output voltage parameter, V_{OH}, is defined as the positive rail of the op amp output for the load current conditions applied to the power pins. When the V_{OH} parameter is tested, it may be defined with I_{OH} of -1, -20, -35, and -50 mA load. When V_{OH} is listed on a data sheet as a test condition, it is used for testing another parameter. Whether V_{OH} is a condition or a parameter it is expressed in units of volts.

B.64 Low-Level Output Voltage Condition or Parameter (V$_{OL}$)

The low-level output voltage parameter, V_{OL}, is defined as the negative rail of the op amp output for the load current conditions applied to the power pins. When the V_{OL} parameter is tested, it may be defined with I_{OL} of -1, -20, -35, and -50 mA load. When V_{OL} is listed on a data sheet as a test condition, it is used for testing another parameter. Whether V_{OL} is a condition or parameter it is expressed in units of volts.

B.65 Maximum Peak-to-Peak Output Voltage Swing Parameter (V$_{OM\pm}$)

The maximum peak-to-peak output voltage swing parameter, $V_{OM\pm}$, is defined as the maximum peak-to-peak output voltage that can be obtained without clipping when the op amp is operated from a bipolar supply. $V_{OM\pm}$ is expressed in units of volts.

B.66 Peak-to-Peak Output Voltage Swing Condition or Parameter $(V_{O(PP)})$

The peak-to-peak output voltage swing condition, $V_{O(PP)}$, is defined as the peak-to-peak voltage set up on the output waveform to test for parameters such as A_{VD} or SR.

The peak-to-peak output voltage swing parameter, $V_{O(PP)}$, is the maximum peak-to-peak output voltage that an op amp can deliver. When it is measured V_{DD}, THD + H, R_L and T_A are the typical test conditions.

$V_{O(PP)}$ is also expressed in units of volts for either a condition or parameter.

B.67 Step Voltage Peak-to-Peak Condition $(V_{(STEP)PP})$

The step voltage peak-to-peak condition, $V_{(STEP)PP}$, is defined as the peak-to-peak voltage step that is used as a test condition for parameters such as t_s. $V_{(STEP)PP}$ is expressed in units of volts.

B.68 Crosstalk Parameter (X_T)

The crosstalk parameter, X_T, is defined as the ratio of the change in output voltage of a driven channel to the resulting change in output voltage from another channel that is not driven. X_T is expressed in units of dB.

X_T is a function of how good the separation is between channels in an IC package or system. It is caused by the signal from one channel being coupled to the other channel inductively, capacitively, through the power supply, etc.

B.69 Output Impedance Parameter (Z_O)

The output impedance parameter, Z_O, is defined as the frequency-dependent small-signal impedance that is placed in series with an ideal amplifier and the output terminal in a closed-loop configuration. Z_O is expressed in units of Ω.

B.70 Open-Loop Transimpedance Parameter (Z_t)

The open-loop transimpedance parameter, Z_t, is defined as the frequency-dependent ratio of change in output voltage to the frequency-dependent change in current at the inverting input in a transimpedance or current-feedback amplifier. Z_t is expressed in units of Ω.

B.71 Differential Phase Error Parameter (Φ_D)

The differential phase error parameter, Φ_D, is defined as the change in AC phase with change in DC level. The AC signal is 40 IRE (0.28 VPK) and the DC level change is ± 100 IRE (± 0.7 V). It is typically tested at 3.58 MHz (NTSC) or 4.43 MHz (PAL) carrier frequencies. Φ_D is expressed in units of degrees. With the advent of digital television transmission, the importance of this parameter may decline.

B.72 Phase Margin Parameter (Φ_m)

The phase margin parameter, Φ_m, is defined as the absolute value of the difference in the phase shift of 180 degrees and the phase shift at unity gain. Φ_m is measured open loop and is expressed in units of degrees, °.

$$\Phi_m = 180° - \Phi@B1 \tag{B.6}$$

Gain margin (A_m) and phase margin (Φ_m) are different ways of specifying the stability of the circuit. Since rail-to-rail output op amps have higher output impedance, a significant phase shift is seen when driving capacitive loads. This extra phase shift erodes the phase margin, and for this reason most CMOS op amps with rail-to-rail outputs have limited ability to drive capacitive loads. Fig. B.1 shows Φ_m graphically.

B.73 Bandwidth for 0.1 dB Flatness

Bandwidth for 0.1 dB flatness is defined as the range of frequencies within which the gain is ± 0.1 dB of the nominal value with full output power. It is expressed in units of Hertz.

B.74 Case Temperature for 60 s

The case temperature for 60 s is defined as the temperature the case may safely be exposed to for 60 s. It is usually specified as an absolute maximum and is meant as a guide for automated soldering processes. It is expressed in °C.

B.75 Continuous Total Dissipation Parameter

The continuous total dissipation parameter is defined as the power that can be dissipated by an op amp package, including loads. It is usually specified as an absolute maximum. This parameter may be broken down by ambient temperature and package style in a table. Continuous total dissipation is expressed in units of Watts.

B.76 Duration of Short-Circuit Current

The duration of short-circuit current parameter is defined as the amount of time that the output can be shorted to network ground. It is usually specified as an absolute maximum. Duration of short-circuit current is usually expressed in seconds.

B.77 Input Offset Voltage Long-Term Drift Parameter

Input offset voltage long-term drift parameter is defined as the ratio of the change in input offset voltage to the change in time. It is the average value for the month and is expressed in units of volts/month.

B.78 Lead Temperature for 10 or 60 s

The lead temperature for 10 or 60 s is defined as the temperature the leads may safely be exposed for 10 or 60 s. It is usually specified as an absolute maximum and is meant as a guide for automated soldering processes. This parameter is expressed in units of °C.

Op Amp Noise Theory

C.1 Introduction

The subject of op amp noise is a complex enough topic that it deserves its own appendix. The purpose of op amp circuitry is the manipulation of the input signal in some fashion. Unfortunately in the real world, the input signal has unwanted noise superimposed on it.

C.2 Characterization

Noise is a purely random signal, the instantaneous value and/or phase of the waveform cannot be predicted at any time. Noise can either be generated internally in the op amp, from its associated passive components, or superimposed on the circuit by external sources.

C.2.1 Root-Mean-Square Versus P—P Noise

Instantaneous noise voltage amplitudes are as likely to be positive as negative. When plotted, they form a random pattern centered on zero. Since noise sources have amplitudes that vary randomly with time, they can only be specified by a probability density function. The most common probability density function is Gaussian. In a Gaussian probability function, there is a mean value of amplitude, which is most likely to occur. The probability that noise amplitude will be higher or lower than the mean falls off in a bell-shaped curve, which is symmetrical around the center (Fig. C.1).

σ is the standard deviation of the Gaussian distribution and the root-mean-square (rms) value of the noise voltage and current. The instantaneous noise amplitude is within $\pm 1\sigma$ 68% of the time. Theoretically, the instantaneous noise amplitude can have values approaching infinity. However, the probability falls off rapidly as amplitude increases. The instantaneous noise amplitude is within $\pm 3\sigma$ of the mean 99.7% of the time.

σ^2 is the average mean-square variation about the average value. This also means that the average mean-square variation about the average value, $\overline{i^2}$ or $\overline{e^2}$, is the same as the variance σ^2.

Thermal noise and shot noise (see below) have Gaussian probability density functions. The other forms of noise do not.

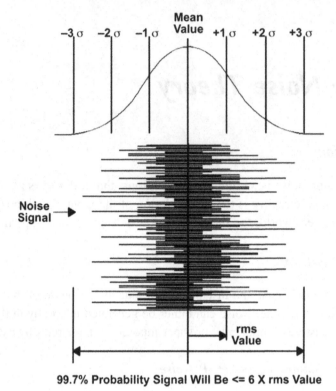

99.7% Probability Signal Will Be <= 6 X rms Value

Figure C.1
Gaussian distribution of noise energy.

C.2.2 Noise Floor

When all input sources are turned off and the output is properly terminated, there is a level of noise called the *noise floor* that determines the smallest signal for which the circuit is useful. The objective for the designer is to place the signals that the circuit processes above the noise floor, but below the level where the signals will clip.

C.2.3 Signal-to-Noise Ratio

The noisiness of a signal is defined as:

$$\frac{S_{(f)}}{N_{(f)}} = \frac{\text{rms signal voltage}}{\text{rms noise voltage}} \tag{C.1}$$

In other words, it is a ratio of signal voltage to noise voltage (hence the name *signal-to-noise ratio*).

C.2.4 Multiple Noise Sources

When there are multiple noise sources in a circuit, the total rms noise signal that results is the square root of the sum of the average mean-square values of the individual sources:

$$E_{Total\ rms} = \sqrt{e_{1\ rms}^2 + e_{2\ rms}^2 + \cdots e_{n\ rms}^2} \qquad (C.2)$$

Put another way, this is the only "break" that the designer gets when dealing with noise. If there are two noise sources of equal amplitude in the circuit, the total noise is not doubled (increased by 6 dB). It only increases by 3 dB. Consider a very simple case, two noise sources with amplitudes of 2 V_{rms}:

$$E_{Total\ rms} = \sqrt{2^2 + 2^2} = \sqrt{8} = 2.83\ V_{rms} \qquad (C.3)$$

Therefore, when there are two equal sources of noise in a circuit, the noise is $20 \times \log \frac{2.83}{2} = 3.01$ dB higher than if there were only one source of noise—instead of double (6 dB) as would be intuitively expected.

This relationship means that the worst noise source in the system will tend to dominate the total noise. Consider a system in which one noise source is 10 V_{rms} and another is 1 V_{rms}:

$$E_{Total\ rms} = \sqrt{10^2 + 1^2} = \sqrt{108} = 10.05\ V_{rms} \qquad (C.4)$$

There is hardly any effect from the 1 V noise source at all!

C.2.5 Noise Units

Noise is normally specified as a spectral density in rms volts or amps per root Hertz, $V/\sqrt{Hz}$ or $A/\sqrt{Hz}$. These are not very "user-friendly" units. A frequency range is needed to relate these units to actual noise levels that will be observed.

For example:

- An op amp with a noise specification of 2.5 $nV/\sqrt{Hz}$ is used over an audio frequency range of 20 Hz–20 kHz, with a gain of 40 dB. The output voltage is 0 dB V (1 V).
- To begin with, calculate the *root Hz* part: $\sqrt{20000 - 20} = 141.35$.
- Multiplying this by the noise spec: $2.5 \times 141.35 = 353.38$ nV, which is the equivalent input noise (E_{IN}). The output noise equals the input noise multiplied by the gain, which is 100 (40 dB).

The signal-to-noise ratio can be now calculated:

$$353.38\ nV \times 100 = 35.3\ \mu V$$
$$\text{Signal-to-noise(dB)} = 20 \times \log(1\ V \div 35.3\ \mu V) = 20 \times \log(28329) = 89\ dB \qquad (C.5)$$

The op amp is an excellent choice for this application. Remember, though, that passive components and external noise sources can degrade performance. There is also a slight increase in noise at low frequencies, due to the 1/f effect (see below).

C.3 Types of Noise

There are five types of noise in op amps and associated circuitry:

1. shot noise,
2. thermal noise,
3. flicker noise,
4. burst noise, and
5. avalanche noise.

Some or all of these noises may be present in a design, presenting a noise spectrum unique to the system. It is not possible in most cases to separate the effects, but knowing general causes may help the designer optimize the design, minimizing noise in a particular bandwidth of interest. Proper design for low noise may involve a "balancing act" between these sources of noise and external noise sources.

C.3.1 Shot Noise

The name *shot noise* is short for Schottky noise. Sometimes it is referred to as *quantum noise*. It is caused by random fluctuations in the motion of charge carriers in a conductor. Put another way, current flow is not a continuous effect. Current flow is electrons-charged particles that move in accordance with an applied potential. When the electrons encounter a barrier, potential energy builds until they have enough energy to cross that barrier. When they have enough potential energy, it is abruptly transformed into kinetic energy as they cross the barrier. A good analogy is stress in an earthquake fault that is suddenly released as an earthquake.

As each electron randomly crosses a potential barrier, such as a pn junction in a semiconductor, energy is stored and released as the electron encounters and then shoots across the barrier. Each electron contributes a little *pop* as its stored energy is released when it crosses the barrier (Fig. C.2).

The aggregate effect of all of the electrons shooting across the barrier is the shot noise. Amplified shot noise has been described as sounding like lead shot hitting a concrete wall.

Some characteristics of shot noise:

- Shot noise is always associated with current flow. It stops when the current flow stops.
- Shot noise is independent of temperature.

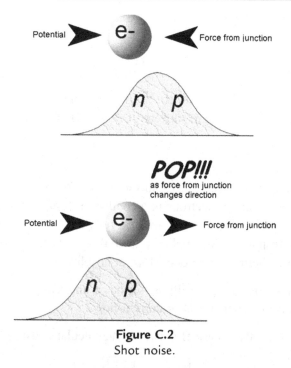

Figure C.2
Shot noise.

- Shot noise is spectrally flat or has a uniform power density, meaning that when plotted versus frequency it has a constant value.
- Shot noise is present in any conductor—not just a semiconductor. Barriers in conductors can be as simple as imperfections or impurities in the metal. The level of shot noise, however, is very small due to the enormous numbers of electrons moving in the conductor, and the relative size of the potential barriers. Shot noise in semiconductors is much more pronounced.

The rms shot noise current is equal to:

$$I_{sh} = \sqrt{(2qI_{dc} + 4qI_o)B} \qquad \text{(C.6)}$$

where q = electron charge (1.6×10^{-19} C), I_{dc} = average forward dc current in A, I_o = reverse saturation current in A, and B = Bandwidth in Hz.

If the pn junction is forward biased, I_o is zero, and the second term disappears. Using Ohm's law and the dynamic resistance of a junction yields Eq. (2.7).

$$r_d = \frac{kT}{qI_{dc}} \qquad \text{(C.7)}$$

The rms shot noise voltage is equal to:

$$E_{sh} = kT\sqrt{\frac{2B}{qI_{dc}}} \tag{C.8}$$

where k is Boltzmann's constant (1.38×10^{-23} J/K); q is electron charge (1.6×10^{-19} C); T is temperature in K; I_{dc} is average dc current in A; and B is bandwidth in Hz.

C.3.2 Thermal Noise

Thermal noise is sometimes referred to as Johnson noise after its discoverer. It is generated by thermal agitation of electrons in a conductor. Simply put, as a conductor is heated, it will become noisy. Electrons are never at rest; they are always in motion. Heat disrupts the electrons' response to an applied potential. It adds a random component to their motion (Fig. C.3). Thermal noise only stops at absolute zero.

Like shot noise, thermal noise is spectrally flat or has a uniform power density (it is *white*), but thermal noise is independent of current flow.

At frequencies below 100 MHz, thermal noise can be calculated using Nyquist's relation:

$$E_{th} = \sqrt{4kTRB} \tag{C.9}$$

or

$$I_{th} = \sqrt{\frac{4kTB}{R}} \tag{C.10}$$

where E_{th} is thermal noise voltage in volts rms; I_{th} is thermal noise current in amps rms; k is Boltzmann's constant (1.38×10^{-23}); T is absolute temperature in Kelvin; R is resistance in ohms; and B is noise bandwidth in Hertz ($f_{max} - f_{min}$).

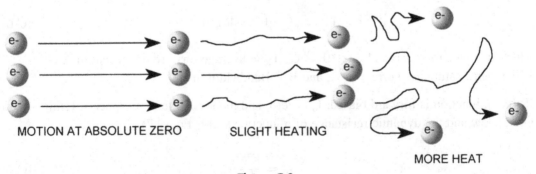

MOTION AT ABSOLUTE ZERO   SLIGHT HEATING

MORE HEAT

Figure C.3
Thermal noise.

The noise from a resistor is proportional to its resistance and temperature. It is important not to operate resistors at elevated temperatures in high gain input stages. Lowering resistance values also reduces thermal noise.

C.3.3 Flicker Noise

Flicker noise is also called 1/f noise. Its origin is one of the oldest unsolved problems in physics. It is pervasive in nature and in many human endeavors. It is present in all active and many passive devices. It may be related to imperfections in crystalline structure of semiconductors, as better processing can reduce it.

Some characteristics of flicker noise:

*   It increases as the frequency decreases, hence the name 1/f.
*   It is associated with a dc current in electronic devices.
*   It has the same power content in each octave (or decade).

$$E_n = K_e \sqrt{\left(\ln \frac{f_{max}}{f_{min}} \right)} \quad I_n = K_i \sqrt{\left(\ln \frac{f_{max}}{f_{min}} \right)} \tag{C.11}$$

where K_e and K_i are proportionality constants (volts or amps) representing E_n and I_n at 1 Hz; and f_{max} and f_{min} are the maximum and minimum frequencies in Hz.

Flicker noise is found in carbon composition resistors, where it is often referred to as excess noise because it appears in addition to the thermal noise that is there. Other types of resistors also exhibit flicker noise to varying degrees, with wire wound showing the least. Since flicker noise is proportional to the dc current in the device, if the current is kept low enough, thermal noise will predominate and the type of resistor used will not change the noise in the circuit.

Reducing power consumption in an op amp circuit by scaling up resistors may reduce the 1/f noise, at the expense of increased thermal noise.

C.3.4 Burst Noise

Burst noise, also called popcorn noise, is related to imperfections in semiconductor material and heavy ion implants. It is characterized by discrete high frequency pulses. The pulse rates may vary, but the amplitudes remain constant at several times the thermal noise amplitude. Burst noise makes a popping sound at rates below 100 Hz when played through a speaker—it sounds like popcorn popping, hence the name. Low burst noise is achieved by using clean device processing, and therefore is beyond the control of the designer.

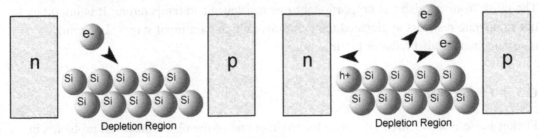

Figure C.4
Avalanche noise.

C.3.5 Avalanche Noise

Avalanche noise is created when a pn junction is operated in the reverse breakdown mode. Under the influence of a strong reverse electric field within the junction's depletion region, electrons have enough kinetic energy that, when they collide with the atoms of the crystal lattice, additional electron-hole pairs are formed (Fig. C.4). These collisions are purely random and produce random current pulses similar to shot noise, but much more intense.

When electrons and holes in the depletion region of a reversed-biased junction acquire enough energy to cause the avalanche effect, a random series of large noise spikes will be generated. The magnitude of the noise is difficult to predict due to its dependence on the materials.

Because the Zener breakdown in a pn junction causes avalanche noise, it is an issue with op amp designs that include Zener diodes. The best way of eliminating avalanche noise is to redesign a circuit to use no Zener diodes.

C.4 Noise Colors

While the noise types are interesting, real op amp noise will appear as the summation of some or all of them. The various noise types themselves will be difficult to separate. Fortunately, there is an alternative way to describe noise, which is called *color*. The colors of noise come from rough analogies to light, and refer to the frequency content. Many colors are used to describe noise, some of them having a relationship to the real world, and some of them more attuned to the field of psycho-acoustics.

White noise is in the middle of a *spectrum* that runs from purple to blue to white to pink and red/brown. These colors correspond to powers of the frequency to which their spectrum is proportional, as shown in Table C.1.

There are an infinite number of variations between the colors. All inverse powers of frequency are possible, as are noises that are narrowband or appear only at one discrete frequency. Those, however, are primarily external sources of noise, so their presence is an

Table C.1: Noise Colors.

Color	Frequency Content
Purple	f^2
Blue	f
White	1
Pink	$\dfrac{1}{f}$
Red/brown	$\dfrac{1}{f^2}$

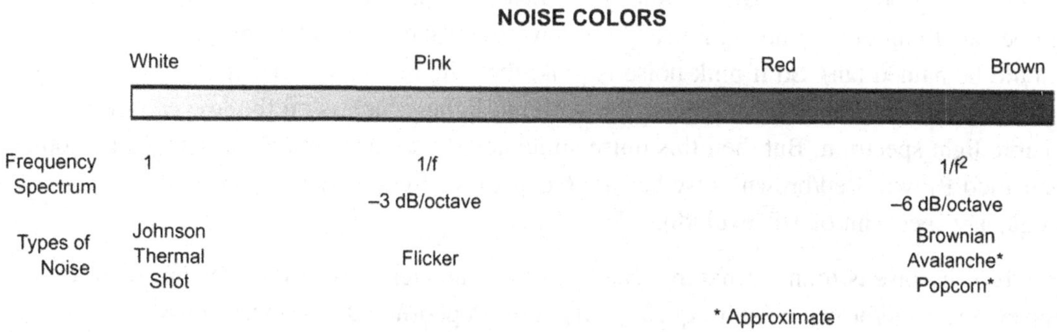

NOISE COLORS

Figure C.5
Noise colors.

important clue that the noise is external, not internal. There are no pure colors; at high frequencies, all of them begin to roll off and become pinkish. The op amp noise sources described above appear in the region between white noise and red/brown noise (Fig. C.5).

C.4.1 White Noise

White noise is noise in which the frequency and power spectrum is constant and independent of frequency. The signal power for a constant bandwidth (centered at frequency f_o), does not change if f_o is varied. Its name comes from a similarity to white light, which has equal quantities of all colors.

When plotted versus frequency, white noise is a horizontal line of constant value.

Shot and thermal (Johnson) noise sources are approximately white, although there is no such thing as pure white noise. By definition, white noise would have infinite energy at infinite frequencies. White noise always becomes pinkish at high frequencies.

Steady rainfall or radio static on an unused channel approximates a white noise characteristic.

C.4.2 Pink Noise

Pink noise is noise with a 1/f frequency and power spectrum excluding dc. It has equal energy per octave (or decade for that matter). This means that the amplitude decreases logarithmically with frequency. Pink noise is pervasive in nature—many supposedly random events show a 1/f characteristic. Flicker noise displays a 1/f characteristic, which also means that it rolls off at 3 dB/octave.

C.4.3 Red/Brown Noise

Red noise is not universally accepted as a noise type. Many sources omit it and go straight to brown, attributing red characteristics to brown. This has more to do with esthetics than it does anything else (if brown noise is the low end of the spectrum, then pink noise should be named tan). So if pink noise is pink, then the low end of the spectrum should be red. Red noise is named for a connection with red light, which is on the low end of the visible light spectrum. But then this noise simulates Brownian motion, so perhaps it should be called Brown. Red/brown noise has a -6 dB/octave frequency response and a frequency spectrum of $1/f^2$ excluding dc.

Red/brown noise is found in nature. The acoustic characteristics of large bodies of water approximate red/brown noise frequency response. Popcorn and avalanche noise approximate a red/brown characteristic, but they are more correctly defined as pink noise where the frequency characteristic has been shifted down as far as possible in frequency.

C.5 Op Amp Noise

This section describes how the noise in op amps and associated circuits is related to the noise types above.

C.5.1 The Noise Corner Frequency and Total Noise

Op amp noise is never specified as shot, thermal, or flicker, or even white and pink. Noise for audio op amps is specified with a graph of equivalent input noise versus frequency. These graphs usually show two distinct regions:

- Lower frequencies where pink noise is the dominant effect.
- Higher frequencies where white noise is the dominant effect.

Actual measurements for a typical op amp show that the noise has both white and pink characteristics (Fig. C.6). Therefore, the noise equations for each type of noise are

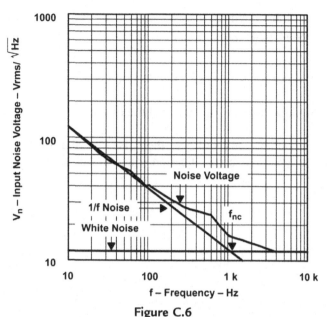

Figure C.6

Typical op amp noise characteristics.

not able to approximate the total noise out of the op amp over the entire range shown on the graph. It is necessary to break the noise into two parts—the pink part and the white part—and then add those parts together to get the total op amp noise using the rms law of yields equation C.7.

C.5.2 The Corner Frequency

The point in the frequency spectrum where 1/f noise and white noise are equal is referred to as the noise corner frequency, f_{nc}. Note on the graph in Fig. C.6 that the actual noise voltage is higher at f_{nc} due to the rms addition of noise sources.

f_{nc} can be determined visually from the graph in Fig. C.6. It appears a little above 1 kHz. This was done by:

- Taking the white noise portion of the curve, and extrapolating it down to 10 Hz as a horizontal line.
- Taking the portion of the pink noise from 10 to 100 Hz, and extrapolating it as a straight line.
- The point where the two intercept is f_{nc}, the point where the white noise and pink noise are equal in amplitude. The total noise is then $\sqrt{2}$ × white noise specification (from Section C.2.4). This would be about 17 nV/$\sqrt{Hz}$ for the op amp shown.

This is good enough for most applications. As can be seen from the actual noise plot in Fig. C.6, small fluctuations make precise calculation impossible. There is a precise method, however:

- Determine the 1/f noise at the lowest possible frequency.
- Square it.
- Subtract the white noise voltage squared (subtracting noise with rms is just as valid as adding).
- Multiply by the frequency. This will give the noise contribution from the 1/f noise.
- Then divide by the white noise specification squared. The answer is f_{nc}.

Circuit Board Layout Techniques

D.1 General Considerations

Prior discussions have focused on how to design op amp circuitry, how to use ICs, and the use of associated passive components. There is one additional circuit component that must be considered for the design to be a success—the printed circuit board (PCB) on which the circuit is to be located.

D.1.1 The Printed Circuit Board Is a Component of the Op Amp Design

Op amp circuitry is analog circuitry and is very different from digital circuitry. It must be partitioned in its own section of the board, using special layout techniques.

PCB effects become most apparent in radio frequency (RF) and high-speed analog circuits, but common mistakes described in this chapter can even affect the performance of audio circuits. The purpose of this chapter is to discuss some of the more common mistakes made by designers and how they degrade performance and provide simple fixes to avoid the problems.

The PCB layout for analog circuitry must be designed such that the effect of the PCB is transparent to the circuit. Any effect caused by the PCB itself should be minimized, so that the operation of the analog circuitry in production will be the same as the performance of the design and prototype.

Long experience in this profession has revealed an extremely disturbing trend—PCB layout being done by specialists in departments dedicated to the task. These specialists are artists, not electrical engineers, and therefore they have neither the experience nor the inclination to take RF and analog performance into account when doing their design. They are used to making all the connections from point to point and calling the job done. PCB layout is thought of as being "beneath" an engineer, or a waste of engineering time.

Considering how critical the PCB is, and how interrelated the performance of the circuitry is with the layout, these attitudes must change! If it is the responsibility of an engineer to see to it an analog circuit works on a PCB, and if they are held accountable should the circuit fail, it is a logical next step to say that the engineer should be the one actually doing the PCB layout. Not only is it necessary, but engineers with the skill set should be commended for taking initiative, not disrespected as "wasting their time," as I have been.

PCB layout software companies are not doing themselves a favor by pricing their products out the reach of engineers. Fortunately, there are free products like Design Spark PCB that have the potential to put engineers back in control of their circuits—at least at the prototype stage. Once the proper layout has been established, the engineer has a layout to pass to the layout department and can say "do it like this and it will work."

Such is my soapbox on the subject of engineers doing PCB layout.

D.1.2 *Prototype, Prototype, Prototype!*

Normal design cycles, particularly of large digital boards, dictate layout of the PCB as soon as possible. The digital circuitry has been simulated, but in most cases, the production PCB itself is the prototype and may even be sold to a customer. Digital designers can correct small mistakes by implementing *cuts and jumpers*, reprogramming gate arrays or flash memories, and go on to the next project. This is not the case with analog circuitry. Some of the common design mistakes discussed in this chapter cannot be corrected by the cut and jumper method.

I have been the unfortunate recipient of a simple analog circuit designed by another engineer who was accustomed to the cut and jumper method of correcting his mistakes. Not only was the op amp hooked up with inverting and noninverting inputs reversed, but an RC time constant had to be added to prevent a race condition. Repercussions from these mistakes, and associated rework problems, caused literally hundreds of hours to be lost from a tight production schedule. Prototyping this circuit would have taken less than a day.

D.1.3 *Noise Sources*

Noise is the primary limitation on analog circuitry performance. Internal op amp noise is covered in , Appendix C. Other types of noise include the following:

- Conducted emissions—noise that the analog circuitry generates through its connections to other circuits. This is usually negligible in analog circuitry, unless it is high power (such as an audio amplifier that draws heavy currents from its power supply).
- Radiated emissions—noise that the analog circuitry generates, or transmits, through the air. This is also usually negligible in analog circuitry, unless it is high frequency such as video.
- Conducted susceptibility—noise from external circuitry that is conducted into the analog circuit through its connections to other circuits. Analog circuitry must be connected to the "outside world" by at least a ground connection, a power connection, an

input, and an output. Noise can be conducted into the circuit through all of these paths, as well as any others that are present.

- Radiated susceptibility—noise that is received through the air (or transmitted into the analog circuitry) from external sources. Analog circuitry, in many cases, resides on a PCB that may have high-speed digital logic including digital signal processor chips. High-speed clocks and switching digital signals create considerable radio frequency interference (RFI). Other sources of radiated noise are endless: the switching power supply in a digital system, cellular telephones, broadcast radio and TV, fluorescent lighting, nearby PCs, lightning in thunderstorms, and so on. Even if the analog circuitry is primarily audio in frequency, RFI may produce noticeable noise in the output. Ever hear GSM tones from a cell phone on a piece of audio gear? That is an excellent example of radiated susceptibility.

Any single or combination of the above sources of noise can render a PCB unusable.

D.2 Printed Circuit Board Mechanical Construction

It is important to choose a PCB with the right mechanical characteristics for the application.

D.2.1 Materials—Choosing the Right One for the Application

PCB materials are available in various grades, as defined by the National Electrical Manufacturers Association (NEMA). It would be very convenient for designers if this organization was closely allied with the electronics industry—controlling parameters such as resistivity and dielectric constant of the material. Unfortunately, that is not the case. NEMA is an electrical safety organization, and the different PCB grades primarily describe the flammability, high-temperature stability, and moisture absorption of the board. Therefore, specifying a given NEMA grade does not guarantee electrical parameters of the material. If this becomes critical for an application, consult the manufacturer of the raw board stock.

Laminated materials are designated with flame resistant (FR) grades. FR-1 is the least flame resistant, and FR-5 is the most as described in Table D.1.

Do not use FR-1. There are many examples of boards with burned spots, where high-wattage components have heated a section of the board for a period of time. This grade of PCB material has more in common with cardboard than anything else.

FR-4 is commonly used in industrial-quality equipment, while FR-2 is used in high-volume consumer applications. These two board materials appear to be industry standards. Deviating from these standards can limit the number of raw board material suppliers and PCB houses that can fabricate the board because their tooling is already set up for these

Table D.1: Printed Circuit Board (PCB) Materials

Grade Designation	Material/Comments
FR-1	Paper/phenolic: room temperature punchable, poor moisture resistance.
FR-2	Paper/phenolic: suitable for single-sided PCB consumer equipment, good moisture resistance.
FR-3	Paper/epoxy: designed for balance of good mechanical and electrical characteristics.
FR-4	Glass cloth/epoxy: excellent mechanical and electrical properties.
FR-5	Glass cloth/epoxy: high strength at elevated temperatures, self-extinguishing.

materials. Nevertheless, there are applications in which one of the other grades may make sense. There are higher temperature—rated FR-4 board materials, which may result in substantial savings if they are used.

For very high-frequency applications, it may even be necessary to consider Teflon or even ceramic board substrate. For high-temperature boards, polyimide is the material of choice. One thing can be counted on, however: the more exotic the board substrate, the more expensive it will be.

In selecting a board material, pay careful attention to the moisture absorption. Just about every desirable performance characteristic of the board will be negatively impacted by moisture. This includes surface resistance of the board, dielectric leakage, high-voltage breakdown and arcing, and mechanical stability. Also, pay attention to the operating temperature. High operating temperatures can occur in unexpected places, such as in proximity to large digital ICs that are switching at high speeds. Be aware that heat rises, so if one of those 500 pin monster ICs is located directly under a sensitive analog circuit, both the PCB and circuit characteristics may vary with the temperature.

D.2.2 Cladding and Plating

Most boards use copper cladding, because it adheres well to the PCB substrate material.

After the board substrate material has been selected, the next decision is how thick to make the copper foil laminated to it. For most applications, 1 oz copper is sufficient. If the circuit consumes a lot of power, 2 oz may be better. Avoid ½ oz copper, because it tends to break between the trace and the pad. Also avoid abrupt changes in traces width, such as a trace intersecting a pad or via. "Tear-dropping" the trace into the pad or via softens the change in width, avoiding a mechanical stress point in thin copper traces. Tear-dropping is a common function in most layout programs. Another common function in most layout programs is trace arcs instead of orthogonal routing. If you use orthogonal routing, use 135 degree angles wherever possible—avoid 90 degree angles.

D.2.3 How Many Layers Are Best?

Depending on the complexity of the overall circuitry being designed, a designer must decide how many layers the PCB should be.

D.2.3.1 Single-Sided

Very simple consumer electronics are sometimes fabricated on single-sided PCBs, keeping the raw board material inexpensive (FR-1 or FR-2) with thin copper cladding. These designs frequently include many jumper wires, simulating the circuit routing on a double-sided board. This technique is only recommended for low-frequency circuitry. For reasons described below, this type of design is extremely susceptible to radiated noise. It is harder to design a board of this type, because many things can go wrong. Many complex designs have been successfully implemented with this technique, but they require a lot of forethought. Be prepared to get creative if the design demands high-volume, low-cost PCBs.

D.2.3.2 Double-Sided

The next level of complexity is double-sided. Although there are some double-sided FR-2 boards, they are more commonly fabricated with FR-4 material. The increased strength of FR-4 material supports better. Doubled-sided boards are easier to route because there are two layers of foil, and it is possible to route signals by crossing traces on different layers. Crossing traces, however, is not recommended for analog circuitry. Wherever possible, the bottom layer should be devoted to a ground plane, and all other signals routed on the top layer. A ground plane provides several benefits as follows:

• Ground is frequently the most common connection in the circuit. Having it continuous on the bottom layer usually makes the most sense for circuit routing.
• It increases mechanical strength of the board.
• It lowers the impedance of all ground connections in the circuit, which reduces undesirable conducted noise.
• It adds a distributed capacitance to every net in the circuit, helping to suppress radiated noise.
• It acts a shield to radiated noise coming from underneath the board.

D.2.3.3 Multilayer

Double-sided boards, in spite of their benefits, are not the best method of construction, especially for sensitive or high-speed designs. The most common board thickness is 0.062 in. This separation is too great for full realization of some of the benefits listed above. Distributed capacitance, for example, is very low due to the separation.

Critical designs call for multilayer boards. Some of the reasons are obvious, some not so obvious:

- Better routing for power as well as ground connections. If the power is also on a plane, it is available to all points in the circuit simply by adding vias.
- Other layers are available for signal routing, making routing easier.
- There will be distributed capacitance between the power and ground planes, reducing high-frequency noise.
- Better EMI/RFI rejection. There is due to the *image plane effect*, which has been known since the time of Marconi. When a conductor is placed close to a parallel conductive surface, most of the high-frequency currents will return directly under the conductor, flowing in the opposite direction. This mirror image of the conductor within the plane creates a transmission line. Since currents are equal and opposite in the transmission line, it is relatively immune to radiated noise. It also couples the signal very efficiently. The image plane effect works equally well with ground and power planes, but they must be continuous. Any gap or discontinuity causes the beneficial effects to quickly vanish. There is more on this in the following paragraphs.
- Reduced overall project cost for small production runs. Although multilayer boards are more expensive to manufacture, EMI/RFI requirements from the FCC or other agencies may require expensive testing of the design. If there are problems, it can force a complete redesign of the PCB, leading to additional rounds of testing. A multilayer PCB can have as much as 20 dB better EMI/RFI performance over a two-layer PCB. If production volumes are going to be small, it makes sense to make a better PCB to begin with than try to cut costs and take the risk of failing $25,000–$50,000 tests.

D.3 Grounding

Good grounding is a system-level design consideration. It should be planned into the product from the first conceptual design reviews.

D.3.1 The Most Important Rule: Keep Grounds and Powers Separate

Separate grounding for analog and digital portions of circuitry is one of the simplest and most effective methods of noise suppression. One or more layers on multilayer PCBs are usually devoted to ground planes. If the designer is not careful, the analog circuitry will be connected directly to these ground planes. The analog circuitry return, after all, is the same net in the netlist as digital return. Autorouters respond accordingly and connect all of the grounds together, creating a disaster. *After the fact* separation of grounds on a mixed digital and analog board is almost impossible.

Just as important as separating ground and power planes is to separate analog and digital power. This can be done from one power rail a number of ways—the most common being small series resistors from the common power rail to analog and digital rails, then

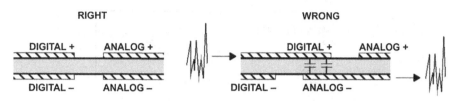

Figure D.1
Digital and analog plane placement.

allowing local bypass capacitors to act as a low-pass filter with the series resistance to filter noise.

D.3.2 Other Ground Rules

- Do not overlap digital and analog ground or power planes (Fig. D.1). Place analog power coincident with analog ground, and digital power coincident with digital ground. If any portion of analog and digital planes overlap, the distributed capacitance between the overlapping portions will couple high-speed digital noise into the analog circuitry. This defeats the purpose of isolated planes. I often put multilayer boards through the "light test" when crosstalk is suspected. No matter how many layers, I should be able to see a gap between digital and analog planes on every layer through the translucent board. If I cannot, something is overlapping that should not be. How much easier it is to do this with the PC board artwork on a viewing program or gerber viewer *before* the board goes out for fabrication!

- Separate grounds does not mean that the grounds are electrically separate on the board. They have to be common at some point, preferably a single, low-impedance point. All ground planes should be connected together at this common connection single point. Oftentimes, this point should be located near the connector. But requirements of mixed signal components, such as A/D converters, may force the common connection point to the proximity of the data converters. Other times, the common connection point should be in the power supply circuitry, if separate windings are used on the transformer of a switching supply, for example. There is usually no way to tell in advance which will be better, so pads should be laid out on the board for common connection points in all of those locations. This will allow quick reconfiguration in case of a problem. I have a further design note. In at least one case, A/D converters were actually destroyed by ground "bounce" if the analog and digital planes were not connected together right at the converters. In this case, the planes actually were split right under the component, with the jumper resistor from analog ground to digital ground being right under the converter. That is how critical common grounding points can be!

- It is important to keep digital signals away from analog portions of the circuit. It makes little sense to isolate planes, keep analog traces short, and place passive components carefully if there are high-speed digital traces running right next to the sensitive analog traces. Digital signals must be routed around analog circuitry, and not overlap

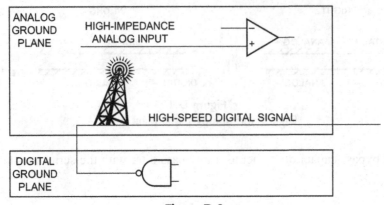

Figure D.2
Broadcasting from printed circuit board traces.

analog ground and power planes. If not, the design will include a new schematic symbol shown in Fig. D.2 — the broadcasting antenna!
- Most digital clocks are high enough in frequency that even small capacitances between traces and planes can couple significant noise. Remember that it is not only the fundamental frequency of the clock that can cause a potential problem, but also the higher frequency harmonics.
- It is a good idea to locate analog circuitry as close as possible to the I/O connections of the board. Digital designers, used to high current ICs, will be tempted to make a 50 mil trace run several inches to the analog circuitry thinking that reducing the resistance in the trace will help get rid of noise. What they have actually done is create a long, skinny capacitor that couples noise from digital ground and power planes into the op amp, making the problem worse! If this is an absolute requirement of the system, place a corresponding "keep out" area on every other layer of the board so that there is no possibility of digital planes or signals crossing the analog power traces and/or planes.

D.3.3 A Good Example

Fig. D.3 shows one possible board layout. In this system, all electronics, including the power supply, reside on one PCB. Three separate and isolated ground/power planes are employed: one for power, one for digital, and one for analog. Power and ground connections from digital and analog sections of the board are combined only in the supply section and are combined in close proximity. High-frequency conducted noise on the power lines is limited by inductors (chokes). In this case, the designer has even located low-frequency analog circuitry close to low-speed digital, keeping high-frequency digital and analog physically apart on the board. This is a good, careful design that has a high likelihood of success—providing that good layout and decoupling rules are also followed.

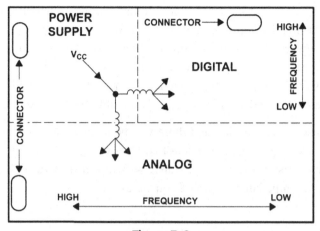

Figure D.3
A careful board layout.

D.4 The Frequency Characteristics of Passive Components

Choosing the right passive components for an analog design is important. In most cases, a *right* passive component will fit on the same pads as a *wrong* passive component, but not always. Start the design process by carefully considering the high-frequency characteristics of passive components, and putting the correct part outline on the board from the start.

Be aware of the frequency limitations of any passive components you use in analog circuitry. Passive components have limited frequency ranges, and operation of the part outside of that range can have some much unexpected results. One might think that this discussion only applies to high-speed analog circuits. But high frequencies that are radiated or conducted into a low-speed circuit will affect passive components as well. For example, a simple op amp low-pass filter may well turn into a high-pass filter at RF frequencies.

D.4.1 Resistors

High-frequency performance of resistors is approximated by the schematic shown in Fig. D.4.

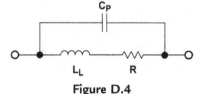

Figure D.4
Resistor high-frequency model.

Resistors are typically one of three types: wire-wound, carbon composition, and film. It does not take a lot of imagination to understand how wire-wound resistors can become inductive because they are coils of resistive wire. Most designers are not aware of the internal construction of film resistors, which are also coils of thin metallic film. Therefore, film resistors are also inductive at high frequencies. The inductance of film resistors is lower, however, and values under 2 kΩ are usually suitable for high-frequency work.

The end caps of resistors are parallel, and there will be an associated capacitance. Usually, the resistance will make the parasitic capacitor so "leaky" that the capacitance does not matter. For very high resistances, the capacitance will appear in parallel with the resistance, lowering its impedance at high frequencies.

D.4.2 Capacitors

High-frequency performance of capacitors is approximated by the schematic shown in Fig. D.5.

Capacitors are used in analog circuitry for power supply decoupling, as filter components, and as stage coupling components. For an ideal capacitor, reactance decreases by the formula:

$$X_C = \frac{1}{(2\pi fC)} \tag{D.1}$$

where, X_C, capacitive reactance in Ohms; F, frequency in Hertz; C, capacitance in microfarads.

Therefore, a 10 μF electrolytic capacitor has a reactance of 1.6 Ω at 10 kHz, and 160 μΩ at 100 MHz. Right?

In reality, one will never see the 160 μΩ with the electrolytic capacitor. Film and electrolytic capacitors have layers of material wound around each other, which creates a parasitic inductance. Self-inductance effects of ceramic capacitors are much smaller, giving them a higher operating frequency. There is also some leakage current from plate to plate, which appears as a resistance in parallel with the capacitor, as well as resistance within the plates themselves, which add a parasitic series resistance.

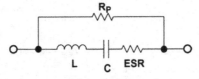

Figure D.5
Capacitor high-frequency model.

Capacitors used for critical analog and RF circuitry should preferentially be the highly stable and low-temperature coefficient COG/NPO type or even silver mica. These are vastly superior for systems that will be subjected to wide variation in temperature/ temperature extremes. Unfortunately, these capacitors are limited to lower capacitance values, with capacitors in these dielectrics becoming large, expensive, and hard to get in larger values.

The next level in quality is usually X7R dielectric. These can vary widely in capacitance over temperature, particularly above 125°C. Larger values of ceramic capacitors available in X7R have become available, due to demands from switching power supply designers. X7R capacitors have much lower values of equivalent series resistance (ESR) than electrolytic varieties, meaning a much lower value can be used for equivalent ripple reduction. But the most common failure mode under conditions of vibration and shock is breakage, which can lead to shorts, so exercise caution. The ESR of ceramic capacitors can be so low that it causes some voltage regulator circuits to become unstable, so small series resistors are sometimes used to add ESR back in.

D.4.3 Inductors

High-frequency performance of inductors is approximated by the schematic shown in Fig. D.6.

Inductive reactance is described by the formula:

$$X_L = 2\pi fL \tag{D.2}$$

where, X_L, inductive reactance in Ohms; f, frequency in Hertz; L, inductance in Henrys.

Therefore, a 10 mH inductor has a reactance of 628 Ω at 10 kHz, which increases to 6.28 MΩ at 100 MHz. Right?

In reality, one will never see the 6.28 MΩ with this inductor. Parasitic resistances are easy to understand—the inductor is constructed of wire, which has a given resistance per unit length. Parasitic capacitance is harder to visualize, unless one considers the fact that each turn of wire in the inductor is located next to adjacent turns, forming a capacitor. This

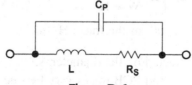

Figure D.6
Inductor high-frequency model.

parasitic capacitance limits the upper frequency of this inductor to under 1 MHz. Even small wire-wound inductors start to become ineffective in the 10–100 MHz range.

D.4.4 Unexpected Printed Circuit Board Passive Components

In addition to the obvious passive components above, the PCB itself has characteristics that form components every bit as real as those discussed previously—just not as obvious.

D.4.4.1 Printed Circuit Board Trace Characteristics

The layout pattern on a PCB can make it susceptible to radiated noise. A good layout is one that minimizes the susceptibility of analog circuitry to as many radiated noise sources as possible. Unfortunately, there is always a level of RF energy that will be able to upset the normal operation of the circuit. If good design techniques are followed, that level will be one that the circuit never encounters in normal operation.

D.4.4.2 Trace Antennas

A board is susceptible because the pattern of traces and component leads form antennas. Antenna theory is a complex subject, well beyond the scope of this book. Nevertheless, a few basics are presented here.

D.4.4.2.1 Whip Antennas

One basic type of antenna is the whip or straight conductor. This antenna works because a straight conductor has parasitic inductance, and therefore can concentrate flux from external sources. The impedance of any straight conductor has a resistive and an inductive component:

$$Z = R + j\omega L \tag{D.3}$$

For DC and low frequencies, resistance is the major factor. As the frequency increases, however, the inductance becomes more important. Somewhere in the range of 1–10 kHz, the inductive reactance exceeds the resistance, so the conductor is no longer a low resistance connection, but rather an inductor.

The formula for the inductance of a PCB trace is:

$$L(\mu H) = 0.0002X\left[\ln\left(\frac{2X}{W+H}\right) + 0.2235\left(\frac{W+H}{X}\right) + 0.5\right] \tag{D.4}$$

where, X length of the trace; W, width of the trace; H, thickness of the trace.

The inductance is relatively unaffected by the diameter, since it varies as the logarithm of the circumference. Common wires and PCB traces vary between 6 and 12 nH/cm.

For example, a 10 cm PCB trace has a resistance of 57 mΩ and an inductance of 8 nH/cm. At 100 kHz, the inductive reactance reaches 50 mΩ. At frequencies above 100 kHz, the trace is inductive, not resistive.

A rule of thumb for whip antennas is that they begin to couple significant energy at about 1/20 of the wavelength of the received signal, peaking at ¼ the wavelength. Therefore, the 10 cm conductor of the previous paragraph will begin to be a fairly good antenna at frequencies above 150 MHz. Remember that although the clock generator on a digital PCB may not be operating at a frequency as high as 150 MHz, it approximates a square wave. Square waves will have harmonics throughout the frequency range where PCB conductors become efficient antennas. If through-hole components are mounted in a way that leaves significant lead length, those component leads also become antennas, particularly if they are bent—which introduces the next topic—loop antennas.

D.4.4.2.2 Loop Antennas

Another major type of antenna is the loop. The inductance of a straight conductor is dramatically increased by bending it into partial or complete loops. Increased inductance lowers the frequency at which the conductor couples radiated signals into the circuit.

Without realizing it, most digital designers are well versed in loop antenna theory. They know not to make loops in critical signal pathways. Some designers, however, would never think of making a loop with a high-speed clock or reset signal will turn right around and create a loop by the technique they use for layout of the analog section of the board. Loop antennas constructed as loops of PCB traces are easy to visualize. What is not as obvious is that slot antennas are just as efficient. Consider the three cases shown in Fig. D.7.

Version **A** is a poor design. It does not use an analog ground plane at all. A loop is formed by the ground and signal traces. An electric field E and perpendicular magnetic field H are

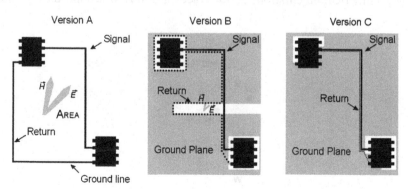

Figure D.7
Loop and slot antenna board trace layouts.

created and form the basis of a loop antenna. A rule of thumb for loop antennas is that the length of each leg is equal to half the most efficiently received wavelength. Remember, however, that even at 1/20 of the wavelength, the loop will still be a fairly efficient antenna.

Version **B** is a better design, but there is intrusion into the ground plane, presumably to make room for a signal trace.

Version **C** is the best design. Signal and return are most coincident with each other, eliminating loop antenna effects completely.

D.4.4.3 Trace Reflections

Reflections and matching are closely related to loop antenna theory, but different enough to warrant their own discussion.

It is given that not all PCB traces can be straight, and so they will have to turn corners. Fig. D.8 shows progressively better techniques of rounding corners. Most CAD systems have all of these routing methods available.

- Sharp 90 degrees corners are used to facilitate high-density digital routing. When a PCB trace turns a corner at a 90 degree angle, a reflection can occur. This is primarily due to the change of width of the trace. At the apex of the turn, the trace width is increased to 1.414 times its width. This upsets the transmission line characteristics, especially the distributed capacitance and self-inductance of the trace, resulting in the reflection.
- 45 degrees orthogonal routing is better, but still does not maintain constant width as the trace rounds the corner.
- Rounded corners maintain the width of the trace as it changes direction, minimizing reflections due to trace width variation. This does not stop loop antenna effects, however. A suggestion for the advanced PCB layout engineer: leave rounding to the last step before tear-dropping and flood-filling. Otherwise, the CAD program will slow down doing numerical calculations as the traces are moved around during routing.

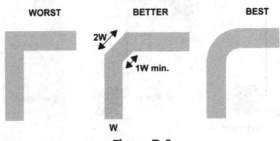

Figure D.8
Printed circuit board trace corners.

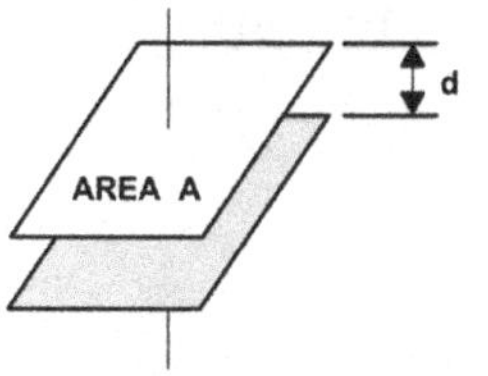

$$C = 0.0085 \times \epsilon_R \times \frac{A}{d}$$

where:

C = capacitance (pF)

ϵ_R = dielectric constant

A = area of plate (mm^2)

d = separation of plates (mm)

Figure D.9
Printed circuit board trace-to-plane capacitance formula.

D.4.4.4 Trace-to-Plane Capacitors

PCB traces, being composed of foil, form capacitance with other traces that they cross on other layers. For two traces crossing each other on adjacent planes, this is seldom a problem. Coincident traces (those that occupy the same routing on different layers) form a long, skinny capacitor. The formula for capacitance is shown in Fig. D.9.

D.4.4.5 Trace-to-Trace Capacitors and Inductors

PCB traces are not infinitely thin. They have some finite thickness, as defined by the *ounce* parameter of the copper clad foil. The higher the number of ounces, the thicker the copper. If two traces run side-by-side, then there will be capacitive and inductive coupling between them (Fig. D.10). The formulas for these parasitic effects can be found in transmission line and/or microstrip references, but are too complex for inclusion here.

Signal lines should not be routed parallel to each other, unless transmission line or microstrip effects are desired. Otherwise, a gap of at least three times the signal trace width should be maintained.

Capacitance between traces in an analog design can become a problem if fixed resistors in the design are large (several MΩ). Capacitance between the inverting and noninverting inputs of an op amp could easily cause oscillation.

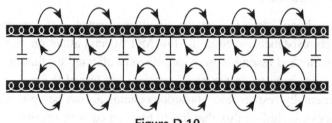

Figure D.10
Coupling between parallel signal traces.

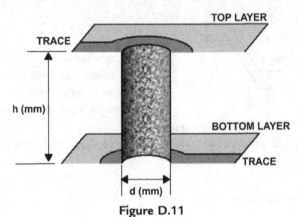

Figure D.11
Via inductance measurements.

D.4.4.6 Inductive Vias

Whenever routing constraints force a via (connection between layers of a PCB, Fig. D.11), a parasitic inductor is also formed. At a given diameter (d) the approximate inductance (L) of via at a height of (h) may be calculated as follows:

$$L = \approx \frac{h}{5} \times \left(1 + \ln\left(\frac{4h}{d}\right)\right) nH \qquad (D.5)$$

One of the best methods of combating via capacitance is to route as many signals as possible without vias. This leads to one of my PCB layout truism—time spent on placement is never wasted. I actually spend a lot of my placement time on the schematic—arranging subcircuits neatly and logically together, using schematic symbols that approximate the true appearance of the part and do not try to functionally group pins together on large ICs. If the schematic can be drawn that minimizes crossovers, vias and routing on other layers will also be minimized on the PCB layout, because you have already developed strategies to avoid it. I can truthfully say that a lot of my placement and routing time is actually done on the schematic, making placement and routing a cinch on the PCB.

D.4.4.7 Flux Residue Resistance

Yes, even an unclean board can affect analog circuit performance. Be aware that if the circuit has very high resistances, even in the low MΩ, special attention may need to be paid to cleaning. A finished assembly may be adversely affected by flux or cleansing residue. The electronics industry in the past few years has joined the rest of the world in becoming environmentally responsible. Hazardous chemicals are being removed from the manufacturing process, including flux that has to be cleaned with organic solvents. Water-soluble fluxes are becoming more common, but water itself can become contaminated easily with impurities. These impurities will lower the insulation characteristics of the PCB substrate. It is vitally important to clean with freshly distilled water every time a high-impedance circuit is cleaned. There are applications that may call for the older organic fluxes and solvents, such as very low-power battery—powered equipment with resistors in the 10 s of megaohm range. Nothing can beat a good vapor defluxing machine for ensuring that the board is clean.

D.5 Decoupling

Noise, of course, can propagate into analog circuitry through the power pins of the circuit as a whole and op amp itself. Bypass capacitors are used to reduce the coupled noise by providing low-impedance power sources local to the analog circuitry.

D.5.1 Digital Circuitry—A Major Problem for Analog Circuitry

If analog circuitry is located on the same board with digital circuitry, it is important to understand a little about the electrical characteristics of digital gates. A typical digital output consists of two transistors connected in series between power and ground (Fig. D.12). One transistor is turned on and the other turned off to produce logic high and vice versa for logic low. Because one transistor is turned off for either logic state, the power consumption for either logic state is low, while the gate is static at that level.

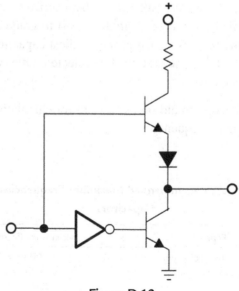

Figure D.12
Typical logic gate output structure.

The situation changes dramatically whenever the output switches from one logic state to the other. There may be a brief period of time when both transistors conduct simultaneously. During this period of time, current drawn from the power supply increases dramatically, since there is now a low-impedance path through the two transistors from power to ground. Power consumption rises dramatically and then falls, creating a droop on the power supply voltage, and a corresponding current spike. The current spike will radiate RF energy. There may be dozens, hundreds, or even thousands of such outputs on a PCB, so the aggregate effect may be quite dramatic. It is impossible to predict the frequencies of these spikes. Digital switching noise will be broadband, with harmonics throughout the spectrum. A general rejection technique is required, rather than one that rejects a specific frequency.

D.5.2 Choosing the Right Capacitor

Table D.2 is a rough guideline describing the maximum useful frequencies of common capacitor types.

Obviously from Table D.2, tantalum electrolytic capacitors are useless for frequencies above 1 MHz. Effective high-frequency decoupling at higher frequencies demands a ceramic capacitor. Self-resonances of the capacitor must be known and avoided or the capacitor may not help, or even make the problem worse. Fig. D.13 illustrates the typical self-resonance of two capacitors commonly used for bypassing—10 μF tantalum electrolytic and 0.01 μF ceramic.

Consider these resonances to be typical values, the characteristics of actual capacitors can vary from manufacturer to manufacturer and grade of part to grade of part. The important thing is to make sure that the self-resonance of the smallest capacitor occurs at a frequency above the range of the noise that must be rejected. Otherwise, the capacitor will enter a region where it is inductive.

Kemet has a particularly nice design aid on its website for calculation of its product's specifications like self-resonant frequency.

Table D.2: Recommended Maximum Frequencies for Capacitors

Type	Maximum Frequency
Aluminum electrolytic	100 kHz
Tantalum electrolytic	1 MHz
Mica	500 MHz
Ceramic	1 GHz

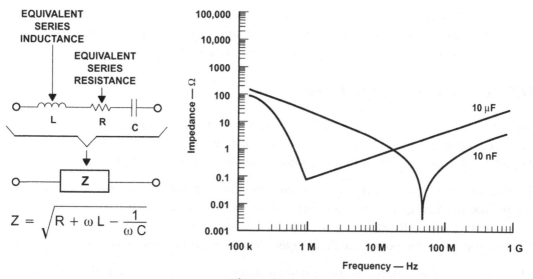

Figure D.13
Capacitor self-resonance.

D.5.3 Decoupling at the IC Level

The method most often used to decouple the high-frequency noise is to include a capacitor, or multiple capacitors connected from the op amp power pin to the op amp ground pin. It is important to keep the traces on this decoupling capacitor short. If not, the traces on the PCB will have significant self-inductance, defeating the purpose of the capacitor.

A decoupling capacitor must be included on every op amp package, whether it contains 1, 2, or 4 devices per package. The value of capacitor must be picked carefully to reject the type of noise present in the circuit.

In particularly troublesome cases, it may be necessary to add a series resistor in a 10–100 Ω range into the power supply line connecting to the op amp. This resistor is in addition to the decoupling capacitors, which are the first line of defense. The resistor should be located before, not after the capacitors. The resistor forms a low-pass filter with the decoupling capacitors. There is a penalty to pay for this technique depending on the power consumption of the op amp, it will reduce the rail-to-rail voltage range. The resistor forms a voltage divider with the op amp as a resistive active component in the lower leg of the divider. Depending on the application, this may or may not be acceptable.

D.5.4 Decoupling at the Board Level

There is usually enough low-frequency ripple on the power supply at the board input to warrant a bulk decoupling capacitor at the power input. This capacitor is used primarily to reject low-frequency signals, so an aluminum or tantalum capacitor is acceptable. An additional ceramic cap at the power input will decouple any stray high-frequency switching noise that may be coupled off of the other boards.

D.6 Input and Output Isolation

Many noise problems are the result of noise being conducted into the circuit through its input and output pins. Due to the high-frequency limitations of passive components, the response of the circuit to high-frequency noise may be quite unpredictable.

In situations in which conducted noise is substantially different in frequency from the normal operating range of the circuit, the solution may be as simple as a passive RC low-pass filter that rejects RF frequencies while having negligible effect at audio frequencies. A good example is RF noise being conducted into an audio op amp circuit.

The effect of radiated energy coupling into an analog circuit can be so bad that the only solution to the problem may be to completely shield the circuit from radiated energy. This shield is called a *Faraday Cage* and must be carefully designed so that frequencies that are causing the problem are not allowed to enter the circuit. This means that the shield must have no holes or slots larger that 1/20 the wavelength of the offending frequency.

Fig. D.14 is a good example of a PCB shield. It has to have holes to allow access to adjustment points, but those holes are too small to allow interference at the frequencies of interest (AM and FM radio). It is a good idea to design a PCB from the beginning to have enough room to add a metal shield if it becomes necessary. If a shield is used, frequently the problem will be severe enough that ferrite beads will also be required on all connections to the circuit.

Figure D.14
Printed circuit board shield.

D.7 Packages

Op amps are commonly supplied one, two, or four per package. Single op amps often contain additional inputs for features such as offset nulling. Op amps supplied two and four per package only offer inverting and noninverting inputs, and the output. If the additional features are important, the only package choice is single. Be aware, though, that the offset-nulling pins on a single op amp package can act as secondary inputs and must be treated carefully. Consult the data sheet on the particular device being used (Fig. D.15).

The single op amp package places the output on the opposite side from the inputs. This can be a disadvantage at high speeds, because it forces longer PCB traces. Some high-speed amplifiers are now adding a second connection to the output on pin 1, so feedback path length can be shortened.

It is popular to use dual op amps for stereophonic circuits, and quad op amps for filter stages with many sections. There is a penalty for doing so, however. Although modern processing techniques provide high levels of isolation between amplifiers on the same piece of silicon, there will be some crosstalk. If isolation between amplifiers is important, then single packages should be considered. Crosstalk problems are not limited to the IC—the dual and quad packages place a high density of passive components in close proximity to each other. This proximity will lead to some crosstalk.

Dual and quad op amp packages offer some additional benefits beyond density. The amplifier stages tend to be mirror images of each other. If similar stages are to be laid out on the PCB, the layout only needs to be done once, then it can be mirror-imaged to form the other stage. Fig. D.16 illustrates this effect for four inverting op amp stages implemented in a quad package.

These illustrations, however, do not show all connections required for operation, in particular, the half-supply generator for single-supply operation. Modifying the diagram of Fig. D.16 to use the fourth op amp as a half-supply generator is shown in Fig. D.17.

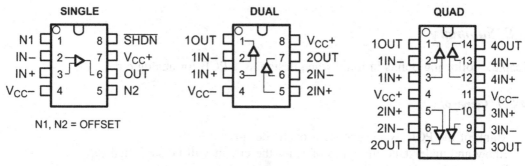

Figure D.15
Common op amp pinouts.

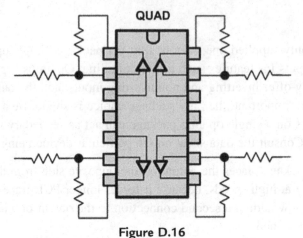

Figure D.16
Mirror-image layout for quad op amp package.

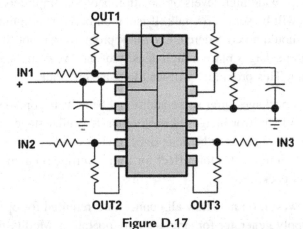

Figure D.17
Quad op amp package layout with half-supply generator.

D.8 Summary

Keep the following points in mind when designing a PCB for analog circuitry.

D.8.1 General

- Think of the PCB as a component of the design.
- Know and understand the types of noise the circuit will be subjected to.
- Prototype the circuit.

D.8.2 Board Structure

- Use a high-quality board material such as FR-4.
- Multilayer boards are as much as 20 dB better than double-sided boards.
- Use separate, nonoverlapping ground and power planes.
- Place power and ground planes to the interior of the board instead of exterior.

D.8.3 Components

- Be aware of frequency limitations of traces and other passive components.
- Use surface mount for high-speed analog circuitry.
- Keep traces as short as possible.
- Use narrow traces if long traces are required.
- Terminate unused op amp sections properly.

D.8.4 Routing

- Never route digital traces through analog sections of the board or vice versa.
- Make sure that traces to the inverting input of the op amp are short.
- Make sure that traces to the inverting and noninverting inputs of the op amp do not parallel each other for any significant length.
- It is better to avoid vias, but the self-inductance of vias is small enough that a few should cause few problems.
- Do not use right angle traces, use curves if at all possible.

D.8.5 Bypass

- Use the correct type of capacitor to reject the conducted frequency range.
- Use tantalum capacitors at power input connectors for filtering power supply ripple.
- Use ceramic capacitors at power input connectors for high-frequency conducted noise.
- Use ceramic capacitors at the power connections of every op amp IC package. More than one capacitor may be necessary to cover different frequency ranges.
- Change the capacitor to a smaller value, not larger, if oscillation occurs.
- Add series resistors for stubborn cases.
- Bypass analog power only to analog return, never to digital return.

Single-Supply Circuit Collection

E.1 Introduction

This appendix is a collection of single-supply op amp circuits. These are presented here because they are somewhat unusual and did not fit well into the material presented in the main portion of the book.

E.2 Attenuation

An inverting attenuation circuit[1] can be thought of as a T network in the R_G resistor. It is shown in Fig. E.1.

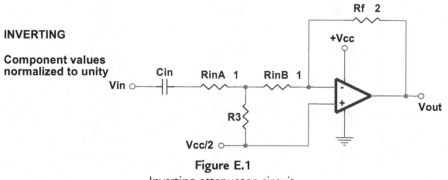

Figure E.1
Inverting attenuator circuit.

R_G is replaced by a T network consisting of $R_{IN}A$, $R_{IN}B$, and R_3. A set of normalized values of the resistor R_3 for various levels of attenuation is shown in Table E.1. For non-tabulated attenuation values, the resistance is:

$$R_3 = \frac{V_O/V_{IN}}{2 - 2(V_O/V_{IN})} \tag{E.1}$$

---

[1] This circuit is taken from the design notes of William Ezell.

Table E.1: Normalization Factors

DB Pad	V_{OUT}/V_{IN}	R_3
0	1.0000	∞
0.5	0.9441	8.4383
1	0.8913	4.0977
2	0.7943	0.9311
2	0.7079	1.2120
3.01	0.7071	1.2071
3.52	0.6667	1.000
4	0.6310	0.8549
5	0.5623	0.6424
6	0.5012	0.5024
6.02	0.5000	0.5000
7	0.4467	0.4036
8	0.3981	0.3307
9	0.3548	0.2750
9.54	0.3333	0.2500
10	0.3162	0.2312
12	0.2512	0.1677
12.04	0.2500	0.1667
13.98	0.2000	0.1250
15	0.1778	0.1081
15.56	0.1667	0.1000
16.90	0.1429	0.08333
18	0.1259	0.07201
18.06	0.1250	0.07143
A1.08	0.1111	0.06250
20	0.1000	0.05556
25	0.0562	0.02979
30	0.0316	0.01633
40	0.0100	0.005051
50	0.0032	0.001586
60	0.0010	0.0005005

To work with normalized values, do the following:

- Select a base-value of resistance, usually between 1 and 100 kΩ for R_F and R_{IN}.
- Divide R_{IN} into two for $R_{IN}A$ and $R_{IN}B$.
- Multiply the base value for R_F and R_{IN} by 1 or 2, as shown in Fig. E.1.
- Look up the normalization factor for R_3 in the table below, and multiply it by the base-value of resistance.

For example, if R_F is 20 kΩ, $R_{IN}A$ and $R_{IN}B$ are each 10 kΩ, and a 3-dB attenuator would use a 12.1-kΩ resistor.

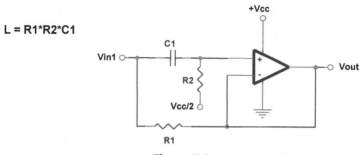

Figure E.2
Simulated inductor.

E.3 Simulated Inductor

The circuit in Fig. E.2 reverses the operation of a capacitor, thus making a simulated inductor. An inductor resists any change in its current, so when a dc voltage is applied to an inductance, the current rises *slowly*, and the voltage falls as the external resistance becomes more significant.

An inductor passes low frequencies more readily than high frequencies, the opposite of a capacitor. An ideal inductor has zero resistance. It passes dc without limitation, but it has infinite impedance at infinite frequency.

If a dc voltage is suddenly applied to the inverting input through resistor R_1, the op amp ignores the sudden load because the change is also coupled directly to the noninverting input via C_1. The op amp represents high impedance, just as an inductor does.

As C_1 charges through R_2, the voltage across R_2 falls, so the op amp draws current from the input through R_1. This continues as the capacitor charges, and eventually the op amp has an input and output close to virtual ground ($V_{cc}/2$).

When C_1 is fully charged, resistor R_1 limits the current flow, and this appears as a series resistance within the simulated inductor. This series resistance limits the Q of the inductor. Real inductors generally have much less resistance than the simulated variety.

There are some limitations of a simulated inductor:

- One end of the inductor is connected to virtual ground.
- The simulated inductor cannot be made with high Q, due to the series resistor R_1.
- It does not have the same energy storage as a real inductor. The collapse of the magnetic field in a real inductor causes large voltage spikes of opposite polarity. The simulated inductor is limited to the voltage swing of the op amp, so the flyback pulse is limited to the voltage swing.

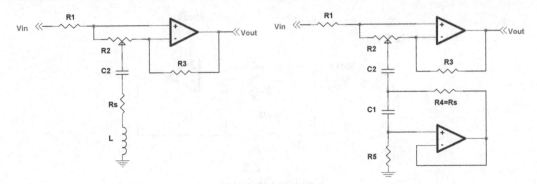

Figure E.3
Graphic equalizer.

These limitations limit the use of simulated inductors, but there is one application that is perfect for simulated inductors—graphic equalizers.

To make a graphic equalizer, start with the basic op amp circuit shown in Fig. E.2. The inductor L is shown with a parasitic resistance R_S. It resonates with C_2, and depending on the setting of potentiometer R_2, the stage either produces a gain or a loss at the resonant frequency. The parasitic resistance of the inductor R_S also sets the Q of the resonant circuit. Therefore, it will determine the number of stages of equalization required to cover the audio band. In the right hand side of Fig. E.3, the inductor L has been replaced by a simulated inductor circuit. To form the graphic equalizer, multiple stages of equalization are added in parallel by placing more potentiometers in parallel with R_2.

E.4 Precision Rectifier

$$E_O \text{ peak} = \frac{-R_O}{R_I} E_I \text{ peak} = -5E_I \text{ peak} \tag{E.2}$$

Figure E.4 shows a half wave rectifier that may contain amplification if desired.

E.5 AC to DC Converter

$$E_O \text{ average} = 0.9E_I \text{ rms}$$

$$E_I = 6 \text{ mV to } 6 \text{ V}; \quad \text{rms at } 10-1000 \text{ Hz}$$

Precision conversion for measurement or control. Full-wave rectifier with a smoothing filter (Fig. E.5).

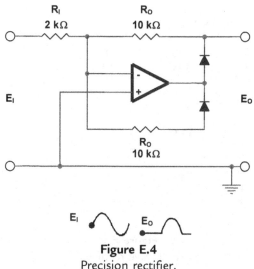

Figure E.4
Precision rectifier.

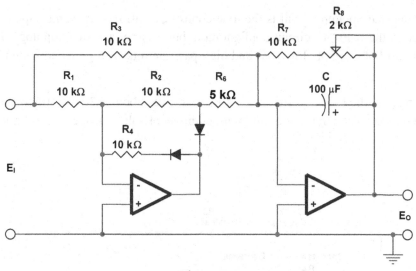

Figure E.5
AC to DC converter.

E.6 Full-Wave Rectifier

Precision absolute value circuit (Fig. E.6).

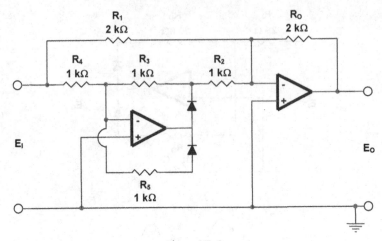

Figure E.6
Full-wave rectifier.

E.7 Tone Control

One rather unusual op amp circuit is the tone control circuit. It bears some superficial resemblance to the twin "T" circuit configuration, but it is not a twin T topology. It is actually a hybrid of one pole low-pass and high-pass circuits with gain and attenuation (Fig. E.7).

The mid range for the tone adjustments is 1 kHz. It gives about ±20 dB of boost and cut for bass and treble. The circuit is a minimum component solution, seeking to limit cost.

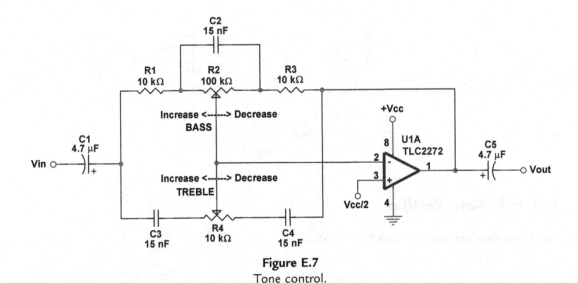

Figure E.7
Tone control.

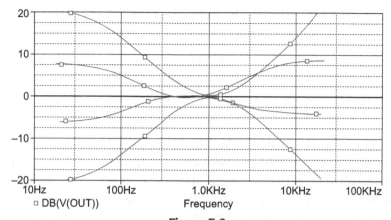

Figure E.8
Tone control response.

This circuit, unlike other similar circuits, uses linear pots instead of logarithmic. Two different potentiometer values are unavoidable, but the capacitors are the same value except for the coupling capacitor. The ideal value of capacitor is 0.016 µF, which is an E-24 value—so the more common E-12 value of 0.015 µF is used instead. Even that value is a bit odd, but it is easier to find an oddball capacitor value than it is an oddball potentiometer value (Fig. E.8).

The plots above show the response of the circuit with the pots at the extremes and at ¼ and ¾ positions. The mid position, although not shown, is flat to within a few milli-dB. The compromises involved in cost reducing the circuit and using linear potentiometers lead to some slight nonlinearities. The ¼ and ¾ positions are not exactly 10 and −10 dB, meaning that the pots are most sensitive towards the end of their travel. This may be preferable to the listener, giving a fine adjustment near the middle of the potentiometers, and more rapid adjustment near the extreme positions. The center frequency shifts slightly, but this should be inaudible. The frequencies nearer the midrange are adjusted more rapidly than the frequency extremes, which also may be more desirable to the listener. A tone control is not a precision audio circuit, and therefore the listener may prefer these compromises.

E.8 Curve Fitting Filters

Analog designers are often asked to design low-pass and high-pass filter stages for maximum rejection of frequencies that are out of band. This is not always the case, however. Sometimes, the designer is asked to design a circuit that will conform to a specified frequency response curve. This can be a challenging task, particularly if the designer knows that a single pole filter rolls off 20 dB per decade, double pole 40 dB per decade. How does the designer implement a different roll-off?

To begin with, it is not possible to get more out of a filter than it is designed to produce. A single pole will give no more than 20 dB per decade and cannot be increased or decreased. More roll-off demands a double pole filter with 40 dB per decade. Obtaining different roll-off characteristics is done by allowing filters at closely spaced frequencies to overlap.

One popular curve fitting application is the RIAA equalization, which compensates for equalization applied to vinyl record albums during manufacture. Many newer pieces of audio gear have omitted the RIAA equalization circuit completely, assuming that the majority of users will not desire the function. In spite of the enormous popularity of audio CDs, there is still a dedicated group of audiophiles that have a large library of record albums—titles that are not available on CDs, or are out of print (Fig. E.9).

RIAA has three breakpoints:

• 17 dB from 20 to 50 Hz,
• 0 dB from 500 to 2120 Hz, and
• −13.7 dB at 10 kHz.

RIAA equalization curves often include another breakpoint at 10 Hz to limit low frequency "rumble" effects that could resonate with the turntable's tonearm. The standard input impedance in the circuits shown here is 47 kΩ. This impedance makes a convenient place to inject dc offset into single-supply circuits, so it is isolated from the phono cartridge by an input capacitance. The phono cartridge output is assumed to be 12 mV.

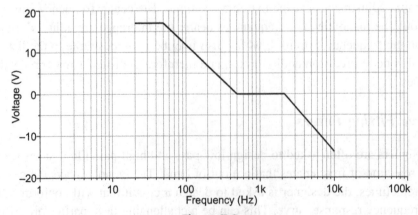

Figure E.9
RIAA equalization curve.

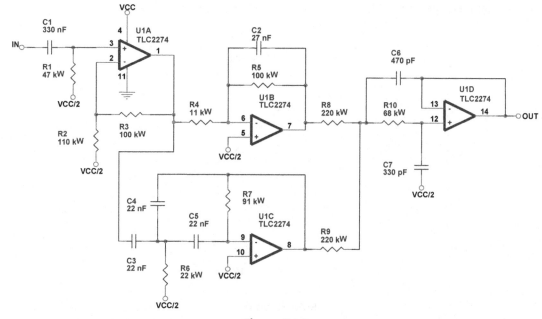

Figure E.10
RIAA equalization preamplifier.

Application circuits were evaluated from many sources in print and on the web. Many of these either did not work at all, did not easily translate to single-supply operation, or deviated markedly from the RIAA specification. There are many circuits that have been proposed for this function, in fact competitions have been held to propose the best. To this fray, this volume respectfully submits Fig. E.10.

This circuit topology is very flexible—most of the RIAA breakpoints are independently adjustable:

* R_1 and C_1 set the low frequency response.
* U1A, R_2, R_3 control the overall gain of the circuit.
* R_4 and R_5 control the low frequency gain.
* R_5 and C_2 control the 50 Hz low frequency breakpoint.
* C_3, C_4, C_5, R_6, R_7, and U1C form a 500 Hz high-pass filter that reverses the effect of the 50 Hz low-pass filter and flattens the response through 1 kHz until the 2120 Hz LPF filter begins to affect the response.
* R_8, R_9, R_{10}, C_6, C_7, and U1D form the 2120 Hz low-pass filter, the input resistor has been split into a summing resistor.

The overall response of the filter is shown below (Fig E.11):

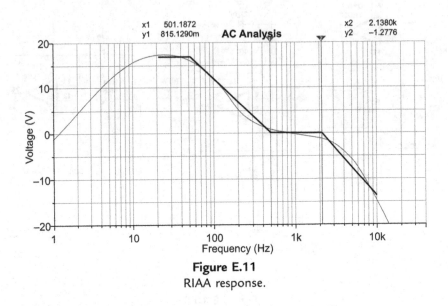

Figure E.11
RIAA response.

The 500 Hz response is above the ideal curve by 0.8 dB, and the 2120 Hz response is below the ideal curve by 1.3 dB. This circuit is about the best that can be done without many more op amps and complex design techniques. It should produce very aesthetically pleasing sound reproduction.

Index

'*Note*: Page numbers followed by "f" indicate figures, "t" indicate tables.'

Printed in the United States
By Bookmasters